# Science The first

## Prof. Dr. Sami AL-Mudhaffar

# Table of contents Page NO.

# Chapter One

# Concept and methodology of science

**Contents**

# Chapter One Concept and methodology of science

A. Scientific concepts

B. The importance of science

C. Theories of the creation and formation of the universe.

D. Intellectual life of human beings

E Model of the concept of a philosophical mystery of life

F. Theory of knowledge and secret of life

G. The science and thought in the understanding of life

H. Scientific disciplines

I. Frontier sciences

J. Living organism and the secret of life

K. Recent scientific trends in Iraq.

L. Techniques used in science.

M. Futurasim of Pure Science

## A. Scientific concepts

The most important of these concepts are the imagination a world without science, that includes, governmental ministries , agencies policy in the areas of education, health, agriculture, industry, the environment, research centers, universities and technical institutes, and so on. Humans are curious creatures and, while at first they only tried to figure out how the most basic things needed for their survival worked.

There is an urgent need to modernize the educational programs of the university in terms of encouraging and supporting the development of knowledge and skills that occurred in the field of science,. To deal with the developments contents for example in chemistry it provides improved or new interpretations of the phenomena and events, such as chemical interactions of environmental pollution, noise pollution in most parts of the world.

The underdevelopment countries in different sectors indicate weakness on the scientific level because of the international system that depends on the investment of science and not to misuse it. It must be reached to create a new international system focuses on the current and future of global problems and takes into consideration underdevelopment suffered by about three quarters of humanity.

Thus it requires the intellectual-Basis and then study the challenges of culture and assimilate the concepts of science and its role in economic development. The study and analysis of underdevelopment is necessary in the Arab region, to find out the relationship between science and technology and its evolution through history.

Curiosity and thirst for knowledge lead to the apparition of multiple branches of science, but all these can be grouped in two big groups: natural sciences and social sciences. While the first group of sciences focused on discovering as much as possible about our planet, the latter focused on studying its inhabitants. Today, each of these groups contains a dozen other sub-genres, and with every major breakthrough, new sub-genres appear. The overall outlook studies relied on detailed statistical information within the mathematicians programs on computer models.

We have the technology to simulate important events in the past, such as the Big Bang, and we have the technology to fly into space orbiting into space to prove it. We have the means, the technology and the will to take things further, and that's exactly what we're doing.

The technology generally preceded science and then started to rely more and more on the achievements of science, and that the discovered science is personal characteristic which has several areas to illustrate the invention process, and in the light of these aspects it can be understood the role of technological progress in the process of growth and development and to understand the process of research and development to its requirements, and that contribute to the building concept of Technology and the role of multinational corporations in the absence of appropriate technology .

## B.The importance of science

Basically, the science is understanding how things work and why, and gave us to survive, and improve our lifestyle in the process, that science is the most important element of our existence.

The recognition of the importance of science is historically known through different civilizations (Pharaonic, Sumerian and others), passing through the efforts of Muhammad Ali in Egypt and study of Shibley Achammal (1853-1917) of the theory of evolution and cell biology which he analyzed the reasons for the weakness of the Ottoman Empire.

That science is a cultural activity achieves the objectives of economic and political community, and according to that science and technology are major forces support the historical, social, national and international change, vary from one community to another, and the internal and external challenges vary from culture to culture.

The Middle East, which intersect three continents of Asia, Africa and Europe it represents the advantages and disadvantages of his power. The beginning of science started with human himself, where the scientific knowledge passed on from generation to generation through the professions, and through the end of the nineteenth century it was dominated by the religious character of scientific knowledge and focused in the church.

It has been focusing on enriching the life with scientific research and according role of science as social power and after that science curricula evolved and reform movement was appeared for the need for the development of curricula for scientific creativity, and in the year (1983) a report was issued entitled "A Nation at Risk "in the United States, and it becomes a warning signal to the community.

Science education should contain the principles and conditions, including early learning of child and in primary, preparatory stage, to develop the ability to use the scientific method to solve problems and to gain scientific knowledge and have a role in achieving the acquisition of facts by revolution and the acquisition of the concepts of scientific principles..

Due to the fact that scientific knowledge intended to develop the capabilities and skills of individuals in order to meet the requirements of the various aspects of life, as well as the community. Accordingly, the development and developing the capacity to build the skills of human individuals.

The human Plays a distinct role in the knowledge according to the fact that the human has activity n component and social, scientific structure that inherent rights throughout his life.

One of the most important things science gave us security – we figured out how to take care of our physiological needs, as well as our physical needs. We made the most out of science by initially trying to understand how things worked, and then making them work to our advantage; the result is that, besides offering us what we needed, science allowed us to create what we wanted.

. Science allowed us to understand how things work, but also how to make things happen. Social sciences made us understand that, if we want to survive, we need to work together in an organized matter.

**C. Theories of the creation and formation of the universe.**

From an examination of creation as described in the Qur'an, an extremely important general concept emerges, it is possible to compare the six days of creation .

"Your Lord is God who created the heavens and the earth in six days." Qur'an, 7:54

. It refers both to the heavens before the earth and the earth before the heavens, when it talks of creation in general, as in this verse of chapter Taa Haa:

"(God) who created the earth and heavens above." Qur'an, 20:4

These ideas are expressed in chapters Fussilat and al-Anbiyaa:

"God then rose turning towards the heaven when it was smoke" Qur'an, 41:11

According to modern science, the separation process which appears dozens of times in the Qur'an. For example, look at the first chapter of the Qur'an, al-Faatihah:( "Praise be to God, the Lord of the Worlds." Qur'an, 1:1 ). These Qur'anic references are in perfect agreement with modern ideas on the existence of primary nebula (galactic dust), followed by the separation of the elements which resulted in the formation of galaxies and then stars from which the planets were born.

"God is the one who created the heavens, the earth and what is between them..." Qur'an, 25:59

This brief survey of Qur'anic references to creation clearly shows us how modern scientific data and statements in the Qur'an consistently agree on a large number of points.

**Expansion of the Universe**

Chapter ath-Thaariyaat of the Qur'an also seems to allude to one of the most imposing discoveries of modern science, the expansion of the Universe.

"I built the heaven with power and it is I, who am expanding it." Qur'an,51:47 The expansion of the universe was first suggested by the general theory of relativity and is supported by the calculations of astrophysics.

Authority to travel in space can only come from the Creator of the laws which govern movement and space. The whole of this Qur'anic chapter invites humankind to recognize God's beneficence. The universe consists of hundreds of billions of galaxies, each with hundreds of billions of stars. This myth of seven heavens was a common idea prevalent during the time when the Qur'an was first recited. See ye not how Allah has created the seven heavens one above another.Qur'an 71:15

Surely We have adorned the nearest heaven with an adornment, the starsQur'an 37:6

The earth first formed around 9 billion years after the Big Bang. The Qur'an, however, repeats the prevailing middle eastern myth that the earth and universe were formed in six days. We created the heavens and the earth and all between them in Six Days, nor did any sense of weariness touch UsQur'an 50:38

Modern science has proposed that all the elements that make up the earth (Oxygen, Nitrogen, Carbon, Iron, etc.) was originally formed by nucleosynthesis in stars and then expelled into the universe when those stars supernova. He it is Who created for you all that is in the earth. Then turned He to the heaven, and fashioned it as seven heavens. And He is knower of all things. Qur'an 2:29 . But according to this theory, the Universe was formed about 13.8 billion years ago due to arapid expansion from singularity. The earth was formed 4.54 billion years ago from accretion of debris that surrounded the precursor of the Sun.

The next verse clearly says that stars were created after the seven heavens which took two days to create, and after the earth which was created even earlier. This verse suggests that the Sun is flat and can be folded up. The Sun appears as a flat disc from the perspective of a person on earth, but the sun is in fact a near-perfect sphere (oblate spheroid. When the sun (with its spacious light) is folded up Qur'an 81:1

And the moon darkens And the sun and the moon are joined Qur'an 75:8-9

. Many Bedouin people living in Arabia imagined the sky as a large tent covering, similar to the tents they used.

It is He Who made the Earth a couch for you, and the sky a dome.Qur'an 2:22

**Evolution**

. Opinion polls show that the majority of Muslims agree Islam and evolution are not compatible.The Qur'an state that humans were created instantaneously from mud or clay.

We created man from sounding clay, from mud molded into shape;Qur'an 15:26

The Qur'an explicitly states that humans were first created outside of the earth and then only later sent down to live on the planet after being expelled from paradise. "Then did Satan make them slip from the (garden), and get them out of the state (of felicity) in which they had been. We said: 'Get ye down, all (ye people), with enmity between yourselves. On **earth will be your dwelling-place** and your means of livelihood - for a time.'Qur'an 2:36

B. There are several theories to explain the emergence of life and

C. trying to answer the classic question, how life was began the answers such as; the fall of some organic molecules on Earth from comets and planting the land with life by the intelligent beings on advanced planets..

The evolution has contributed to clarify the continuing march of life regardless of the theories put forward, including Lamarck and Darwin. Many theories put forward on the creation of the heavens and the earth that were formed from the old material was scattered in space and the earth were not present before.

Theories of formation of life The most specialized Theories in the formation of life: The fall of some organic molecules on earth from comets.

- General fertilization year (planting the ground by life from the intelligent beings inhabiting advanced planets).
- The emergence of life after eons of consecutive chemical filtration in the sea.
- The emergence of life quickly, after the land formation in brief time.
- The emergence of life through escalation of dusty water droplets to the surface, the collected chemicals turned quickly to life.

These theories are summarized according to the following:-:

Theory of Laplace (nebular theory):

Laplace theory" Nebular theory": The planets revolved around the sun according to widespread nebula across the sky in flaming gas that blocks and revolve around itself. Chambrin and Moulton theory (the theory of planetary particles) It is assumed the presence of a foreign body leads to the occurrence of a series of large explosions in the body of the sun.

Gas theory of the tides( Gaseous tidal theory): This theory assumes the approach of the star to the sun then to resurrection of the tongue of a huge gas-fired toward the opposite side of the star, later formed planets.

- Levon theory: This theory reflect the emergence of planets of the solar system itself when it hit the planet that led to a small external parts flying from the sun..
- Theory of the great cloud of dust: The sun and all the planets was in the form of a huge cloud of gas and dust result in condensation and force of gravity..

• General enrichment “ ground planting life by intelligent beings inhabiting in advanced planets.

..’ In Islamic doctrine, a divine oath signifies the magnitude of importance of a special relation to the Creator, and manifests His majesty and the supreme Truth in a special way.

"Water evaporates from both the aquatic and terrestrial environments as it is heated by the Sun’s energy. The rates of evaporation and precipitation depend on solar energy, as do the patterns of circulation of moisture in the air and currents in the ocean. "The sunlight that we can see represents one group of wavelengths, visible light. Other wavelengths emitted by the sun include x-rays and ultraviolet radiation. X-rays and some ultraviolet light waves are absorbed high in Earth’s atmosphere. Ultraviolet light waves are the rays that can cause sunburn. Most ultraviolet light waves are absorbed by a thicker layer of gas closer to Earth called the ozone layer. the upper stratospheric regions, absorption of ultraviolet light from the Sun breaks down oxygen molecules; recombination of oxygen atoms with O2 molecules into ozone (O3) creates the ozone layer, which shields the lower ecosphere from harmful short-wavelength radiation… since the ozone layer serves as a shield against ultraviolet radiation, which has been found to cause skin cancer.

- **Life theories**. Louis Pasteur (1872) supported wrongly the idea of self-evolution and said and that the objects can arise only from living organisms.

- Then this scientists proved to the world in his experiments that microorganisms that live in the water, and independent living organisms, are set into the water from the outside and breed in it and in the process of fermentation.. And that life is not self-reproduce, but the presence of creative power of protoplasmic material aired in life. Accordingly these theories can be divided into those and other to:Theories of lifeInclude the notion of self- evolution of life..

**Physical theories**

- The materialists confirms that life is generated from the union of simple organic , inorganic compounds on Earth's atmosphere prevailed in a period of time, after which these organic materials united to form of a giant molecules entered into a series of complex interactions and formed protoplasm.

*** Chemical theories**

- Earth began and was free of oxygen but it was composed of water vapor , methane , carbon dioxide and ammonia gas, making molecules before the beginning of life and these gases could be mixed in the laboratory, in special glass and then shelling a spark then -producing amino acids , sugar and nitrogenous bases then absorbed by detonators Clays -clay-rich nickel.

- Then the amino acids can be placed in conditions of very hot and dry to form droplets called pro –cells and organizers of the experiment indicates that the origin of life already out of genetics. Accordingly in the laboratory under environmental conditions can create building blocks and small molecules under similar conditions that prevailed on the surface of thc ground (before the creation of chemical life)

- The process of turning gas into organic compounds (amino acids, proteins) by using the energy from sunlight, ultraviolet rays, which is readily absorbed by methane and water vapor and ammonia. Also Can be used electric discharge, as suggested by the researcher Yuri, as well as Miller, that has designed the organization of an experiment on electric discharge.

-

- It is believed that millions of years had passed before are complex organic materials characterized by complex organic materials such as proteins and nucleic acids in oceans water, where it gathered in such large molecules as blocks and systems working by physico- chemical forces, led to the formation of more complex new structures, and finally the formation of simple living organism as a result of repeated interactions and the ability to self- reproduction.

The process of conversion of gases to organic compounds (amino acids, proteins) could be carried out by using energy of sunlight including UV irradiation and then are readily absorbed by methane and water vapor and ammonia. It can also use electric discharge, as suggested by the researcher Yuri, as well as that, the Miller organizing experience about electric discharge.

*** Miller theory**

Miller theory include the formation of amino acid from the components of the atmosphere that surround the globe, including ammonia, hydrogen, methane and water vapor. The experiment was conducted in a special pot (a closed system of glass and poles made of Tc) .

These ingredients were mixed with water vapor and different gases (methane, ammonia, hydrogen) and exposed to an electric discharge, then passing the gases through the cooling area of liquefied gases. After a week of the reaction observed the presence of organic materials such as amino acids and other compounds.

Miller theory suffered to doubts, especially in:

-It did not explain how the amino nitrile converted to amino acid in the land that contains a high concentration of ammonia conditions (characterized by the Earth's atmosphere at the beginning of the composition being flown and turned by oxidative mechanism which is not known)-Miller did not explain the source of oxygen in the composition of the atmosphere.

**Oparin's theory**

Oparin believed that life was originated on the result of the combing materials to form organic compounds, then aformation system consists of low-organism. And then that the planet may be formed as a result of accumlation of blocks of gases.

Earth's atmosphere was imbued with a variety of gases (methane, ammonia, water vapor, and hydrogen gas) was followed by the escape of many of the gases from the earth's atmosphere and the formation of the field of gravity, however, then continued to leak hydrogen gas of the earth's atmosphere, hydrogen was then drained all of the methane and ammonia forming of carbon dioxide, nitrogen, and then the water was hydrolyzed by light energy to hydrogen and oxygen, leaving oxygen and turning the earth's atmosphere to the shorthand version of the formula oxidizing.

Oparin Says (that all attempts made to produce life from inorganic materials either under normal conditions or laboratory have failed). And the physical and chemical conditions that prevailed on the ground in nature before the emergence of life, in which the complex chemical reactions that led to the emergence of life differ from the circumstances now prevailing.

Oparin also indicates that these organic materials are beginning to accumulate in water leading to the emergence of self-reproductive capacity but could not prove it. That the transformation of non-biological material to the spiritual life not consistent with the mathematical concepts or thermodynamics laws, well as mechanisms of reactions.

**D. Intellectual life of human beings**

The scientific progress made by man is in fact the result of intellectual growth in the presence and the disclosure of the laws governing the universe. In modern times, we find that the old traditional division between mental philosophy that assumes the mind is the source of knowledge and the experimental philosophy, which is the source of knowledge is experience is still great and did not fade.

• The large scale of the mind in human thinking.

• The experiment is  an important tool for the application of mental scale, but not alone.

• The mental trend is supposed not to ignore the role of science in human experience and knowledge.

Experiment of Komeras and his aides after a thorough study of the reaction of Strycker have shown as follows:

- The nitrile amine, intermediate material for interaction be stable in the prescence of ammonia as a result the theory of Miller was subjected to doubts and in particular.
- It did not explain how and the mechanism of a nitrile amine conversion to amino acid in the circumstances which contains a high concentration of ammonia (characterized by the Earth's atmosphere at the beginning of a reducing atmosphere composition being turned into an atmosphere oxidative mechanism is unknown).

Miller did not explain the source composition of oxygen in the atmosphere and was viewed as a missing link.

As noted by many researchers that the hydrogen peroxide was capable to hydrolyze the amino nitrile to the amino acid after passing through an intermediate material called the amino amide and those had proved the following:

- Hydrogen peroxide accelerate the amino acids in the conditions on earth that contain high concentrations of ammonia with access to byproduct oxygen.
- The development of the land, transforming it from pre bio-dynamic environment to bio-environment in accordance with the following scenarios:

- The Earth was created at the beginning of ozone layer protected and that the water molecules exposed to UN radiation and cosmic rays forming of hydrogen peroxide (in a free radical).

-

- Then the hydrogen peroxide molecule structures entered in the atmosphere of most of the planets circle around the sun, with environmental conditions similar to Earth's environment at its formation. And then entered in the reaction of nitryl amine hydrolysis to amino acids then to throw in the air and oxygen gas which is a byproduct of this reaction.

-

- Oxygen molecules accumulated after thousands of years in an earth-air and the amino acid pool in the surrounding water (sea).

-

- Under the high concentration of oxygen, upon arrival at the upper layers of the earth to "Ozonization" carried out of the Ozone Layer, which is protecting of the ground to prevent cosmic rays that has stopped functioning hydrogen peroxide in the atmosphere.

- Then it was converted from the reduced environment that contains methane, hydrogen and ammonia to oxidizing atmosphere containing oxygen and ozone layer protected with a high accumulation of amino acids in the sea and therefore creation of atmosphere suitable for life.

## E. Models of the concepts of mystery of life

There are three philosophical concepts of the world, developed as a result of human intellectual effort, the spiritual concept of the real material and the realistic concept of the divine. These contents could be evaluated and trying to rush to one of them or the formulation of the concept of compromise between them.

The conflict between the divine and the physical manifestation of the conflict between idealism and realism and the philosophical conception of the world by one of two things the ideal and the concept of material does not correspond to reality at all. at the reality.

In the scientific field, there is no divine, and the philosopher, whether divine or material believes in the positive side of science. There is no question of divine, according to philosopher of scientific material, but there are two incompatible and when it was the question of existence beyond nature.

The Divine believe that the world is the fact that just about to rule, outside the experiment and material, deny it believe to be natural causes that revealed by the experiment and spread to the hands of science that is the primary reasons for existence.

The nature of the evidence that can be given by Divine that is mind, not by the direct experiment, unlike the material traditionally regarded in the experiment as defined evidence of the materialism which is their own version.. If we look at a set of basic concepts of life and way of thinking, then could be addressed first to the theory of knowledge and secondly the philosophical conception of life when we study the theory of knowledge. The focus becomes mental dependence of the way of thinking includes knowledge over experiment, as well as a valuable study of human knowledge on the basis of sense mental sense, not physical.

**F. Theory of knowledge and secret of life**

The vision and ratification, which represent the expressing of the perception .The first represent a presence, such as heat, light and sound while the second (ratification), which represents a recognition that, for example heat is an energy comes from the sun and other concepts has dealt with a number of theories of perception, including:

- The theory of recollection

  It is isolation from the material, and we can correct some mistakes of this theory which is that the self is not something that exists but an abstract before the existence of the body, but of the fundamental movement of the material.

- **Mental theory**

  This view put forward by Decarte, Cant, pointed to the existence of two perceptions of first is the sense of (heat, light, taste, etc...) and the second instinct (the human mind has the meanings and perceptions did not emerge from the common, but are fixed at the center of instinct

- **Theory of sensory**

  This theory is based on the experiment and the sense which is the infrastructure of which this theory and the base of the human imagination. Therefore, the theory of knowledge could be used to follow-up to the secret of life and access to it .In this perception (scenario of objected value) of life is an expressing the presence of thing in our brain..

When talking about the mind to explore the mystery of life it is necessary to underscored that the measurement is the first of the human thinking ,Which is in general for being processed cards dealing enrich human thought with energy ,dealing beyond the material. The intellectual track is gradually move from public issues into private and the rationalism does not ignore the role of experiment.

**G. The science and thought in the understanding of life**

After clarification the thinking part of life, we could say the deeper main reason of the universe and the world in general is the reason due in particular, which ends by the sequence of causes and the only question that deserves its generation is due to this reason in particular, which is the first fountain of existence, is it the same material, or something else over the border.

Models of inherited science

Scientific heritage is the most precious legacy of the human to human being and the complete presence of human that provide him with the meanings that make sense for a human to distinguish them from other organisms, however at the Arab and Islamic library of hundreds of men did not lift them up after the dust of oblivion, .

The period after the Islam, which has widened the prospects of Arab scientific movement and the consequences of this period emphasizes inherited a multi-faceted so-called Arab Islamic civilization that include the following:

- Profiles of the scientific legacy.
- Islam and its impact in the scientific tradition.
  - Holy Quran as a model
- Scientific thought in Islam.
- The translation of science.
- Methods and ways of scientific research in the scientific heritage.

## H. Scientific disciplines

The scientific progress made by man is in fact the result of intellectual growth in the presence and the disclosure of the laws governing the universe. In modern times, we find that the old traditional division between mental philosophy that assumes that the mind is the source of knowledge and the experimental philosophy, which is the source of knowledge ..

- The large scale of the mind in human thinking.
- That the experiment as an important tool for the application of mental scale, but not alone.
- The mental trend is supposed to not ignore the role of science in human experience and knowledge.
- Science can be divided into different sections, including the basic (Pure) and applied humanities and basic sciences include the natural sciences with the exception of Engineering and Applied **Science, that** include mathematics, physics, chemistry, life sciences, earth sciences, astronomy and meteorology.
- The humanities include philosophy ,social and other undergraduate majors that have developed with the evolution of the various fields such as engineering, agriculture, pure and social sciences .

- The scientific advances in life sciences, mathematics, computer sciences and other sciences brought about developments and tremendous quality in chemistry and life sciences.
- Sciences were not known during the first half of the last century, and the results of such events and large-scale changes in the structures of study and research approaches, the developed world universities that turned to study and research curricula and certificates have been implemented in different ways.
- The splitting of the many Pure Science such as chemistry formed a new terms such as industrial, medical chemistry.
- The introduction of these terms in the international universities in physics, chemistry and mathematics to prepare graduates for some sectors such as engineering and physics engineering, medical and agricultural .
- This was done in British universities through the expansion of preliminary to benefit from an increase in systematic college teaching and practical opportunities and qualify students to work in various sectors of production.
- Some universities introduced in the world of modern curricula at the level of initial studies, as in Britain and Germany, for example, in the field of chemistry especially during the second half of the twentieth century, dealing with developments chemical industries.
- Some US universities has expanded dramatically in granting bachelor's degree, in the disciplines of chemistry related to medicine, agriculture, engineering , life sciences, physics, education, and others. Mathematics:

It is worth mentioning that the evolution of mathematics inherent in physics, then ripped off mathematics and liberation and began the development Systems abstract using set theory and mathematical logic, and this, using the workers and researchers in mathematics, and ways to create a broader understanding of the economy and population change., adoption of the so-called mathematical modeling introduced then "applied..

In the US and British universities from the beginning of the twentieth century and to the previous years at the end of the twentieth century sections of Applied Mathematics and other pure mathematics has become at the present time for Mathematical Sciences.

Mathematics speeds varying and expanding its uses and trends in the areas of knowledge, including the construction of mathematical models, and collaboration with life scientists and members of the organism and to understand human behavior and medicine to build mathematical models to help man in control of some natural phenomena, and installation constructivist of DNA.
(Catastroph Theory) which can be used to describe some phenomena is continuing and is a multiple features, including applied and philosophical side, it can be adopted in the life and physical sciences:.

Computer:

Mathematics contributed to the evolution of technology such as manufacturing computer with very high computational capabilities, the specialized topics bases in Computer Science, Mathematics robot, mathematics and artificial intelligence, and engineering computing, that the impact of computers not only on mathematics developed (computational science) and the emergence of computational mathematics.,

The mathematics concerned with the study of common properties for sports models and the development of algorithms, but has spread to other sciences.

Then It began the development of computer hardware and adoption of operational programs "Abermjiat" This rapid development has led to the development of computer science in a large number of universities and colleges to form a separate fields of computer science, such as information, education and computer cybernetics , information processing and data processing interfere with a lot of material.

Mmaterials have led to this mating with different fields of knowledge, especially in the field of development and field applications, for example,

"Computational Linguistics" as a result of overlapping of computer science with linguistics.

The most important disciplines in the near future, computer technology and robotics, software and information systems and systems engineering and engineering software and information technology level, and information systems and neural networks and virtual reality and networks of fiber optic and information security, encryption, and data processing and systems of teaching and learning systems design, production and neural networks and systems knowledge.

The Iraqi universities new specializations in the field of computer, such as computer technology and robotics, systems engineering, artificial intelligence, information management and computer security, information systems, and virtual reality, and decision support systems, machine translation neural networks, and the protection of information networks and knowledge of computing and information systems.

**Biology**

The terms of reference for the Life Sciences include

- Immunology.
- Toxicology.
- biological sciences nerve (neuroscience).
- Biomedicine (Biomedical Sciences).
- Human Genetics.
- molecular biology.
- Biochemistry.
- Bioethics (Bioemics).
- research centers.

But modern research centers in the light of future research for life sciences include:

- establishment of a center for computer applications in the life sciences.
- creation of a new nature reserves.
- biodiversity and the protection of national germplasm.
- studies of the aquatic environment and fisheries.
- studies and research center in the field of inherited diseases in Iraq.

**Geology:**

The development of the study of Earth Sciences in the initial stages requires updating the curriculum and general vocabulary of the main themes of Earth Sciences and the adoption of modern books.

Providing topics of modern teaching such as geology, environment, geophysics, engineering, mathematical modeling, software, mining, seismic layers, geology sedimentary, analysis seas, relics of Geology and, soil Geosciences and organic chemistry.

It proposes the introduction of a scientific departments specializing in petroleum geology and Geology and Mineral Resources of rocks and industrial water, geology, resources, environment, geology, engineering, and geology of the sea.

The development of graduate studies and strengthening of Applied curriculum in graduate and topics relating to mineral resources, oil and water, in addition to purely scientific disciplines.

The use of modern research centers such as: Water Research Center, and oil research center Desertification Research Center Engineering Geology Research Center.

The foundations of bioinformatics

Different forms use the methods and tools for a variety of information consistent with the idea of the basic units of D.n.a genes, but it is at the same time vary depending on the sequence of these same units and the extent of its simplicity or complexity.

That the circumstances that prevailed in the incidence of various diseases, and related to these conditions of major change which is to a large extent in increasing trend towards updating of information about human biology, including abnormalities and diseases related example is characterized by the fact that the cell individual contain a third version, it is determined by microscopic examination.

The tremendous progress has been possible to follow changes in the D.n.a including mutations responsible for many genetic diseases. Importantly, however, that recent data is heavily dependent on modern technologies that are meant connected to the researchers of these operations and become one of the fundamentals of bioinformatics of science-related advanced and effective device, hence the importance and seriousness of this means.

A common relations between computer scientists and molecular biology scientists can do to provide the computer world that rules necessary to encrypt the D.n.a and thus progress in the future and are shown adequate important information then to determine the organism interpretation.

Moreover, there is a possibility for the development of the computer using the D.n.a and the use of mathematical rules as symbols When adding a certain amount of the D.n.a all the information in the computer Storage. The idea of a computer that appeared in the 1995 adaptation of the D.n.a afford toa process information.

Finally, the biology is no longer limited to the professionals but entered into computer science and scientists who are beginning to learn this science and participate in the research teams of specialists. Computer life develop, and can be imagined led by computer scientists and biology became note in terms of information which is called bioinformatics and life size computer components billions of times smaller than silicon chip that is characterized by a very huge storage capacity.

From a computer look the the D.n.a and installation of structural represents a system intelligent for the storage of information and computer scientists have become accustomed to deal with the digital binary system to express the alphabet and numbers, symbols and diagnosed directly alphabetical character quadruple in the D.n.a in order to encrypt messages and that all three-relay in the D.n.a.

That the information stored in sophisticated computers evoke different senses dramatically, as actually live researcher within the computer and sees human life is stored and the whole picture of technology modern era necessarily means exploit and develop as the many researchers in the countries that have contributed to the success of the Human Genome Project that developed means and comparisons.

The Human Genome Project (HGP) Applied tool of information for scientists and researchers with vital information on the basis of the genes is crucial to the success of research programs for the purpose of basic human life according to information identity.

The Human Genome Initiative was the first time in 1988 and aimed to find sites about (100.000) gene in the human genome in D.n.a and expresses the 24 pairs of chromosomes, turns into an information content when follow the sequence of rules need to be based on solutions of Computer science , mathematics , statistics and empirical science.

Note that the Computer Science offers mostly contributions to programs and solutions are characterized by the skills led to the invention of access in programming languages and describe the information that implements a specific order and provides methods to describe the complex dynamic processes with a number of software Lines.

After researchers and observers have said that -first century will be the century of biology and analytical power resulting from (HGP) that will be interpreted radically and medical research as it has been:

- Finding the nature of genomes and the nature of its composition and organization in different scientific institutions.

• Acceleration in the implementation of the project, which plans to accomplish during the 15 years, but the technical progress shortened the time to ten years.

• • Finding in genome content to know the type of information or material contained in contact to diagnose all the genes the (100.000) in the D.n.a of human as well as determine the sequences of nearly three billion chemical bases.

• Study the nature of the information stored in the evolution of computer technology and elaborately efficient sequence and the evolution of the tools that contribute to the analysis of information.

• Progress and development in Pure Science, for example, led to the development topics and disciplines and new disciplines and sciences Interfaces new were not known during the first half of the last century. The results of such events and large-scale changes in the structures of study and research approaches, as these developments in the developed world universities turned into curricula study and research.

• The splitting of the terms of reference for many Pure Science in chemistry for example, given the competence of Chemistry in the industry next to the jurisdiction of the Industrial Chemistry, Medical Chemistry.

• These terms of reference have been developed in US universities in physics, chemistry and mathematics to prepare graduates in some sectors, such as chemistry, physics, engineering and medical chemistry.

• Some of the topics were added assistance to many of the terms of reference of Pure Science and Education, including education and literature and Library Services and the use of modern equipment and robots and computers.

• Specialties in British universities through the expansion of preliminary to benefit from an increase in college of teaching systematic and practical opportunities and qualify students to work in various sectors of production.

Computer Science has developed since the discovery of the computer and processing laboratories in a number of US and British universities in the early sixties. Computer science was not at that time have crystallized a separate note itself, similar to science Pure or Applied, but it was growing up in the arms of scientific sections, especially math and departments of electrical engineering departments,.

where the computer materials and methods were not possible to needed to do different sections and sufficiently, also focused Computing materials in two directions One of them has to do with geometric and architectural components of computer software and the other direction, where he teaches a number of programming languages in science and engineering faculties, and used to solve mathematical and computational, engineering .

And the evolution of computers in terms of physical devices (Hardware) and operational software (Software) began expanding scientificLY and theoretical material and scientific developer, rapid expansion led to develop computer science as a fundamental science.
large number of universities and colleges to form separate sections with the knowledge of computer Computer Sscience or science Computer and Information Computer and Information Science or Computer education or Cybernetics or information processing Data refers to one type of information processing.

The expansion of computer education even overlapped articles with a large number of scientific and humanitarian materials and computer materials become part of the curriculum requirements of the various Pure and Applied Sciences and humanitarian materials.

This has resulted in computer science with other sciences knowledge and scientific fields and new, especially in the field of application for example, the development of computational linguistics field of Computational Linguistics as a result of overlap so computer with linguistics or cognitive field between computer science or information science systems and science Management and Operations Research .

Although modern science bioinformatics very aware of the complexity derives its assets and the accountability of non-science, the most important of mathematics and computer science as well as many of the explanations of medical science resulting from human genome studies (gene content) for different objects, and as well as the use of electronic information to understand Genetic network and the development.

## I. Frontier sciences

Using the United States as an example, some of the topics to be discussed are the views of public officials who influence the distribution of research funds, the response of funding agencies and the views of scientists.

Finally, we shall look at the co-evolution of science and society and attempt to draw some conclusions concerning their related future and the implications for the future of technology.

Public officials who are involved in setting or influencing science policy have expressed opinions that indicate that they intend to change the basis for supporting research and development.. the public officials wish to alter somewhat the pattern of funding for science.

Their motivation is to orient research more toward programs that, for example, ensure a stronger economy and improvements in the environment. It is becoming increasingly apparent that those public officials who control public funds, will be reluctant to fund research programs that they consider unrelated to national needs.

Academic disciplines have evolved with the development of sciences and various new disciplines such as engineering, agriculture, science whereas the twentieth century indicates other disciplines such as business management, journalism, information and library science, economics, politics and world affairs were added.

The world witnessed in the twentieth century breakthrough in all fields and scientific trends, so there are no boundaries between different disciplines. For example, medical science requires engineering science and recent tests of modern science depends on the physical, chemical extraction and analysis and also relies on mathematics to lay the groundwork mathematics.

The progress and development in pure science, for example, develop new subjects and disciplines and specialties of science. New interfaces were not known during the first half of the last century. The results of these major changes in curriculum and build up research transformed these developments to the university curricula. Seminars and researches are now carried out in different ways including:

- Bachelor based on the study and theses.
- Some universities in Britain and Germany developed curricula at the level of initial studies that include research and study.
- Divide the present fields such as industrial chemistry, chemistry of life with medical side and other disciplines in the branches of pure science.
- Develop competencies.

These terms of reference have been developed in American universities in physics, chemistry and mathematics, to prepare graduate in some sectors such as engineering, chemistry physics and chemistry, agricultural engineering, and medical studies.

- Adding assistance topics.

Some topics have been added as assistance of many of the terms of reference of pure sciences, including education, literature and library services and use modern machinery and computers.

- Other disciplines (Sandwich)

British universities were carrying out by expanding the initial years of university study for use in increasing opportunities for the systematic teaching and applied for and rehabilitation work in various production sector.

**Chemistry and biology**

Amazing developments have taken place in the chemical sciences particularly during the second half of the century, including implicit and other interfaces. Developments on the implicit content and the vocabulary and mechanisms are known in chemistry and provide improved or new interpretation of events and phenomena and chemical reactions, as a result also of new subjects and disciplines within the science of chemistry itself. These developments have led to the opening of new channels in scientific research and technological innovations such as chemical industries to create new chemicals, or chemical industries, and new techniques.

The developments of the second type of chemical sciences interface had addressed the disciplines of science linking chemical sciences and applied various treatments. These developments have led to the developments of science or the new terms of reference were not known before.

The recent trends in the chemistry of life

Bio- chemistry studies the chemical and physical characteristics of the components of the cell and features of the life systems of the components, as well as the interpretation of what these systems in the cell Biochemistry provided a lot of accomplishments, it has helped to clarify the mechanism of medicine and contributed to the diagnosis and treatment of many diseases.

Biochemistry lasted over the age of a century in different disciplines, some with a study of the materials that make up plant cell and then called the chemistry of plant life, and then which is related the animal cell which is called chemistry of animal life if the human cell is the target.

Chemistry has expanded to clinical biochemistry that includes chemistry of life, becoming a physical, organic and biochemistry and inorganic chemistry as well as nutrition. Interested in chemistry, life functions of the modern systems of life, have contributed to the means of study in the last century with the observation of these systems directly during the work,.

Either at the present time which has changed the picture and it became possible to obtain the most desirable observations by the development of viable technologies (electron microscope, radioactive isotopes, Immunology, spectrum).

The scientists believed at the end of the nineteenth century that it is possible to obtain some information relating to the systems of life, by studying the chemistry of cells and for decades was followed by chemists adopted the chemical methods available and succeeded in obtaining useful developments.

Significant improvements to the technical methods such as the use of chemical isotopes have greatly increased the sensitivity of diagnosis of different types of molecules of life and others, and when it is necessary to separate the components of the chemical reaction through life and is very sensitive, then used deportation electric traditionally.

When the attention has turned physicists, chemists, physicists about the science of life (and perhaps due to the ability of living cells to configure the system, although the laws of physics, emphasizes the universe there is a tendency towards non- attendance) then emerged the technical methods of physical, chemical, physical,.

The progress achieved in the chemistry of life has begun to acknowledge that the livelihood systems containing small particles interested m organic chemistry to study and clarify as well as large molecules called macroscopic particles which are not molecular weights less than 100 million times the mass of one atom of hydrogen..

The importance of macroscopic particles of the life systems in its ability to privacy in life interactions composition of building blocks, and can say clearly that he had made in the past years considerable effort to characterize the annexation of macroscopic particles as well as the reactors that occur between them and the need for advanced methods of separation and purification and characterization of macroscopic particles.

The objective of biochemistry for nearly half a century is to collect and organize interactions that occur in living cells. The motivation for this major effort is that a significant number of the attributes of living cells can be understood through these interactions that are typically characterized by the formation or breaking covalent bonds.

It is been clarified on the liberalization of energy as a result of break chemical transformation processes as well as molecules of life and mutual assembly operations amino acids, sugars and fats to form macroscopic particles.

During the last thirty years clearly demonstrated that the reactors that occur between molecules due to physical, those that are not or break covalent bonds have the same importance of chemical reactions, for example. The organization of chemical reactions (ie, the degree of permitted them to occur) performed by the physical changes that occur in the structure of large molecules, as well as the creation of active centers in these molecules and the resulting interdependence of the non-covalent small molecules, in addition to, many of the qualities of a macroscopic aggregates molecules in cells .

Plurality of molecules of life structure consists of installation of the first structural molecular structures of multiple different types of units place (serial), for example, the sequence of amino acids found in proteins and sequence by chemical analysis. The secondary structural composition involves the formation of a complex three- dimensional structures is called to direct all of the units for multi-particles to other units and is called the secondary structural composition tradition or (body and image) or the status of the foundation structure or backbone of multiple chains.

The forms, which consists of surfaces and different types of these mixed forms, and called on the direction of (position) of side chains relative (amino acids, nucleic acids or bases) triangular structural composition. A lot of multiple molecules of life with each other to be as complex as the structures of several structural units viruses, membranes and capillaries bonds and are usually in one level, where you specify the types of bilateral structures of proteins. On the other hand that includes the alpha carbon to allow for many types of structural combinations.

The two phosphate ester bonds in nucleic acids are subject to sag as well, because the flexible rule and hate water and one level surrounded by a few of so they are usually located one above the other, thus reducing the adhesion of water, and this increases the structural rigidity of installation.

The multi- life linear molecules, which has no free rotation about the bonds, which do not interact aggregates side is called the file is not a random combination structurally specific dimensions or size of distinct wraps by Brownian movement. Size can be measured by the value equal to the rate of rotation of the radius around a point or an axis.

Nucleic acids - the mystery of the mysteries of life

A nucleic acid represent the brain of the cell brain cell with a specific developed program, to be issued through the instructions for that cell fusion and installation of the life and death and plan for the future.

There are two types of nucleic acids (DNA, RNA) both of their differences centered a long chain molecules composed of nucleotides and position of certain forms.

In 1953 Crick and Watson was able, who have previously received the Nobel Prize in 1962 developed a model for the DNA structure, consisting of two strands of units of the four nucleotides arranged in orderly fashion, and every one of them is a multi-helical nucleotides wrapped around a common axis to form the double helix right direction. (Figure1 ).

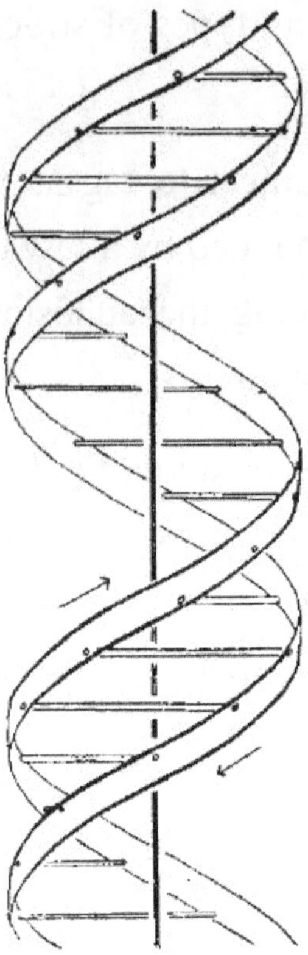

Figure 1

**The DNA double helix showing a right handed or B-helix.**

Indeed, this diagram is the **only** one that actually appears in the article, and one would seek in vain any diagrammatic elaboration of what the molecular structure of DNA is (although components such as deoxyribose or guanine are named as such in the text).

Anyone seeking to repeat Watson and Crick's model building would certainly have to acquire additional molecular data from another sources. Some of that missing information although this only describes the connectivity of the various atoms in a single strand of DNA, and not the two or three dimensional relationships of the (125 in this example) individual atoms.

Note also that this diagram is presented here for visual consumption by a human, who still has to recover additional semantics such as the stereochemistry at the three stereogenic ribose centres, and note carefully that the unit represented must be accompanied by positively charged counter-ions.

**The molecular basis of one strand of DNA, based on the CG bases.**

Armed only with the one diagram actually published, curiosity might lead one to pose a scientific question such as "How did Watson and Crick assign the helix as right rather than left handed".

In other words, on what data did they base that conclusion? in most cases certainly erroneously Coincidentally, similar issues of left or right-handedness were to be found when Pauling presented his α-helix models of proteins. In fact, almost all protein helices exhibit right-handedness .

- that both chains follow right handed helices ...
- because left handed helices can only be constructed by violating permissible van der Waals contacts.
- We are informed that such permissible contacts include the approach of any two hydrogen atoms in the molecule to a distance of **no less** than 2.1Å.
- We are not however informed what the violations might be in a left handed helix that excludes this model. In other words, just how close can two hydrogen atoms separated (for intramolecular contacts) by at least four bonds approach? In fact, distances of ~1.85Å or less have been observed .

In this same full article by Watson and Crick . we are given a table of numerical (polar) coordinates describing the positions of twelve key atoms, but it would have taken a very determined scientist to have used only this combination of information to easily confirm the assertion that a left-handed helix is excluded.

Perhaps the lack of a model with which the reader could experiment might account for the relatively slow recognition of the importance of this article in the immediate years following its publication, and the observation that whilst a physical model of DNA had of course been built, it was only available for viewing (but not modifying) by visiting Cambridge!

One tool that modern chemistry now has at its disposal (which Watson and Crick did not have) are accurate molecular models based on quantum mechanical calculations. Such a molecule is quite a challenge to model, since the computation has to take into account subtle interactions such as dispersion (long range correlation) effects, which are more or less equivalent to the van der Waals contacts referred to by Watson and Crick.

The ionic phosphate groups, the planar bases and how they stack, so-called anomeric effects at the base-sugar connecting C-N bond, hydrogen bonds between both the obvious NH...N and NH...O atoms and less obvious ones such as C-H...O, and not least the capacity to deal self-consistently and accurately with the optimal positions of (at least) 250-254 atoms.

In reality, such models have only very recently become available . To illustrate how this famous article from 1953 could now be published in a journal in 2011, I have taken the liberty of updating the original diagram .

The models explained by x-ray, have two sections through the longitudinal axis of the first of 0,34 nm and the second 3,4 nm.

Model of "Watson & Crick Model" Both Watson and Kirk in 1953 the first specialist in genetics and the second physicist to develop a model structure that represents the structural basis of DNA. Jarkav in the light of studies on the percentage of nitrogen bases, X-ray dimensions, as well as the fact that adenine = thymine and Guanine = Cytosine

Features of the model of Watson and Crick, this is called beta-form and include:

- The DNA is composed of two strands of deoxy nucleotides wrapped through multi-spiral system.
- The nucleotide chains are connected by diester bonds within one strand.
- The bases purines and pyrimidines facing each other, so that in particular, adenine faces thymine and guanine faces cytosine through hydrogen bonds.
- The order of nitrogenous bases one in series vary from other.

- The levels of sugar rings parallel to the axis of and phosphate groups abroad.
- DNA is divide in two parts the first is called water hating (nitrogenous bases), located internally and a second which is externally faced of the surrounding water molecules containing phosphate groups.

The second half of the nineteenth century witnessed a series of discoveries of life such as, a serious (cell) theory at the hands by Matthias in the plant and then Theodorishvan in the animal. Since of plant and animal are composed of cells, that evolution (cell theory) and their development is considering critical stage in the progress of life science similar to atomic energy.

The human Physiology science as one of the branches of the life science that refers to the amazing facts explaining the greatness of the Creator and accuracy of the details and secrets. The digestive system for example (the greatest chemical plant in the world) including, the methods of food analysis of chemical analysis of various surprising and distribution of fairly Safe food distributed to millions of living cells.

If we explore the science of life, we will find another secret of that biggest secrets, the secret of the mysterious life, which fills the moral conscience of mankind, with the concept of divine fear and faith, firmly established in it. The theory of self-regeneration was collapsed at the depth but the unequivocal scientific experiments, demonstrated the invalidity of the theory of self-regeneration. The material basis of life science was examined and then basically spread the idea of elements. The atoms are spread better for the basic materials of the universe and second nature that the elements consist of a central core electrons of the nucleus orbit (negative) and the nucleus contains protons and neutrons. Attempts were made to alter the material to absolute energy, no electric charge.

In other words removing character from element in the light of the theory of relativity of Anstein, where the body mass is relative, not fixed, and increase with the speed according to Anstein equation energy = mass of × square of the speed of light and mass = energy ÷ square of the speed of light.

As a result, the atom, including of protons and electrons are condensed energy. Appeared in various forms and multiple images, whereas materials has been converted into energy and energy to the material.

It follows from the views put forward that the original materials the world-life the reality show one common in various forms, and the physical properties of compounds are accidental such as the liquidity of water is incidental, not self-evidence, since it is consisted of two toms and possible separation these two elements from each othe..

The characteristics of the simple elements themselves are not self-rule but are incidental to the material. That such material characteristics become the light of the above facts incidental, it is encroached to be among the identified energy and philosophically, the presumption of tahe material in the world of life on the top reason capable for denial, as well as of effectiveness.

## Genetics

Mendel discovered the basic principles of heredity and passed him a head later by scientists. As he concluded after mating successive generations of peas plants, that split successor inherits the characteristics of, according to a mathematical formula that could be the secret to life and then called the laws of Mendel. Then genetics born at the beginning of the twentieth century after the principles of, which were designated as Mendelian inheritance. Followed by several changes altered the traditional characters of life science to prolong and settled genetics then rolled up on the basis of discoveries that quoted life science version of the conventional version of the description and classification.

The Darwinian concept, was retreated which relies on the theory of evolution that the changes and the characters that we can get as a result of practice or reaction with the environment can be transferred by inheritance to the descendants. The hypothesis of evolution of species has been the trend due to the mutations, some aspects of the sudden change in the number of cases that called for the assumption that the diversity of animal emerged from mutations, and some of these changes may have inherited.

After that, the transfer life science version of the traditional formula and tradition description, classification, manifestations and modalities of organic evolution and the cell in its entirety to the life science microscopy, which focuses on exploration of the nucleus molecules and chemical structures. It is a mass of spherical material or oval that looks heavier than around it. Then it emerged that in the nucleus, clusters of fine particulate organic form, renamed chromosome or chromosomes that contain the genetic factors mentioned by Mendel each factor was called gene. The cell contain genes and for each type of species there is a special number of chromosomes in every cell of the human body, forty-six (46) chromosomes except in the female reproductive cell (egg) and sperm in the male sperm, each containing two (23) chromosomes. But there is one chromosome in a set of chromosomes of the male sperm determines sex of the fetus generated from the fertilized egg, it may be x or y. Not only the impact of these chromosomes to determine the sex of newborn, but also that genes determine the hereditary characteristics of male and female, then was found that the nucleic acids present in the nucleus of cells issued instructions for their growth and break apart and there are two types, DNA, RNA. Crick and Watson (1962) managed to develop an acceptable model for this structure which is composed of two bands of nucleic acid units (with four bases adenine, thymine, guanine, cytosine) in a corresponding sequential arrangement to RNA and such a model and a specific genetic information is transferred to RNA which controls the composition of proteins.

## Evolution

There are different opinions about the evolution of such as that living organisms in all its forms and types are fixed and does not change, but some of the scientists are not convinced with validity of this opinion, and expressed the possibility of changing living organisms, or that living organisms are not static but in constant change, depending on the prevailing natural conditions. In the following number of views on development with theories that have been presented each gained continuity and the other stopped until:

- Both humans and animals, with a single installation consists of degrees of nature non- living then plants, animals and low organisms and marine animals and finally humans.
- The organisms, which represent high-degree, can not arise from the low-level organism but, it was created on this picture.

.

- Akhavan Safa (brother of safa)

Akhavan Safa pointed to explain the evolution and the doctrine of evolution that the worlds of animals, plants and inanimate are and one separated from each other within a sharp border.

**Lamarck**

Lamarck says that the organic and inorganic world is constantly changing, influenced by environmental factors, mating and the use of organs or neglected. And high living organisms originated from the living organism over a long but slow period and low- living organism in turn originated by self regeneration or from the non living organism in the period of evolution. This evolution depends on use, neglect and transmission of acquired qualities genetically such as the appearance of new organs in the body commensurate with requirements. Contemporary animals are divided into six grades according to the default effect of the Creator.

**Darwin**

Darwin believed according to his theory, that the factors that led to the evolution of living organisms, heterogeneity (contrast), natural selection (survival of the fittest) and artificial election and conflict on stay. In heterogeneity, the differences in members of the same species of plant or animal to the environment and the willingness of the transformation and self-use or neglect of Darwinism consider that the human is the last episode of a series of episodes from the single-cell organisms which have been grappling with the forces of nature for survival. For access to the most complicated forms of combination, life is as convinced by Darwin of the powerful and intelligent, and death for the weak and the survival of the fittest. Through Darwinian launched many new bases of life. In the light of the perceptions of genetics, the inheritance of the individual includes nuclear material of living cells of reproduction and genetic traits due to microscopic precision parts of a gene. In the light of declining Darwinism perception that the changes and attributes acquired by the animal as a result of experience and practice or to interact with the environment or type of food can be transmitted by inheritance to his descendants. And, therefore, researchers turn to that the evolution of species was determined by mutations in some aspects by observing the sudden change. Science has arrived to learn many things, including:

- Genetic Engineering
- Biotechnology
- Communications life

**Genetic Engineering**

Genetic engineering caused major developments in life sciences, including applications in medicine, which includes diagnosis and treatment.

The concept of the genetic engineering and technologies based on multi-splitting of DNA By special enzymes work on specific sites and then linking the pieces formed with DNA from other sources, is then the proliferation of hybrid is able to reproduce on vehicle “Cloning”, including bacteria and viruses which have been used to develop genetic engineering techniques and other such as electropheresis and auto radiography.

This technology can be used for the preparation of special DNA sensors for the purpose of searching for specific genes or specific parts of the DNA technology to clone parts of the DNA in large quantities by the use of the enzyme "PCR".

**Applications of genetic engineering:**

- Gene therapy: It is used in the treatment of some diseases, including the treatment of brain tumors and in the reduction of cholesterol in the blood.
- Genetic mapping in humans: one can look at the matter in the future after the location of many human genes.
- Stimulation the immune system to produce antibodies more efficient and accurate (a vaccine against the virus of hepatitis “B”).
- Production of proteins of medical importance.
- Determination of the nature and location of genes for some genetic diseases.
- Diagnosis of genetic diseases before birth.

**Biotechnology**

Biotechnology caused enormous developments human, including those which were aimed at human use:

- Industrial electronics.
- Products of biotechnology space.
- Environmental treatment
- Extracting of oil by microorganism
- Medicinal plants.

The transfer of chromosomes from one cell to another is of chemical and living concept for the process of cloning, and its backbone. The genetic chromosomal transfer is not considered new, it has been exercised during the implantation of an egg in the uterus of a female in (1978) (the birth of a tube child - Louise).

Specifications and features of the cloning by the cloning is characterized with specific characteristics of chemical features of some new and some old concepts. These specifications are as follows:

- Converting an adult cell (totally grown) to a cell could reproduce without vaccination.
- Converting the reproducing cell to full a creature breeders (replica of the mother).
- Transfer of mature cell grown to fully grown live immature egg in the uterus of another object (the sheep, for example).
- Development of a new principle called the principle of Wilmurt can be expressed as follows: (very specialized mature cell from an animal that awakens static sophisticated genetic information in the chromosome to become a source of a whole new creature).
- This process of reproduction is not a sexual in the traditional sense (without sexual contact between male and female).
- The genetical cell can not be used through challenge and breeding.
- The cloning from adult stem cells facilitate the researcher to wait to see the nature of the thing itself before proceeding to reproduction.

**Features applied to clones (the experiment of Ian's death)**

- Converting a cell of the nipple of the gland of sheep (full growth - extremely) to a cell with a capacity of breeding and without sexual relationship.
- Transfer the mature full cell to fully live the immature egg in the uterus of a sheep.
- The cloned sheep of inherited properties, from the donor mother (the birth of Dolly the sheep).
- The sheep (Dolly) is not the nascent daughter and her mother, can be its twin.

- The sheep of the cell prepared for a clone of the breast, containing all the genes necessary for a complete sheep.

**Old and new cloning**

The cloning technique, distributed according to principles and specifications on several areas, including gene cloning embryonic cloning and cloning by nucleus. Gene cloning refers to the production of similar genes similar to the original gene, and the best example of this, is the process that occurs naturally when giving birth to identical twins, as a result of one split of egg genes and its distribution into two cells identical to the growth of each to produce transferred uniform embryo. But when the separation of embryo's cells from each other then process is referred to by certain embryonic cloning, each cell is growing separately to produce an integrated organism. Researchers have successfully cloned monkeys and frogs from embryonic stem cells match similar to the original. It was easier to deal with embryonic stem cells because they do not have any discrimination that they were not yet have been developed to convert to and brain, muscle and other organs, in the wombs of their mothers. But it turned out with the progress of development, the property of discrimination occur a change in the DNA. The passivity of embryonic cloning researcher stresses the intrinsic character of being unaware of what will emerge.

The third cloning, which represents the third occurrence of mature cells, a researcher can wait to see the nature of the thing with his own eyes before they proceed on reproduction. This represents a new type of reproduction, which is using the nucleus of an adult cell, adopted by Wilmot the specialist knowledge of embryos, which represents technical innovation for the reproduction of fetal sheep from an adult cell, and he was able to clone sheep, genetically modified by human genes to produce factors that deal with clot the blood. The Dolly, of Wilmot considered as the first which was cloned with the latest technology and adult cell cloning technology has international new principles of life, especially relying on a somatic cell, and successful steps can be carried out each and every one has a recipe associated with these principles:

- Dealing with the cell donor (host) intended for cloning and extracted from the membrane cells of a pregnant sheep is characterized by white color and necessary genes needed to form the purpose complete sheep. This step represents a use an adult cell transferred to a cell with the ability to reproduce without sexual mating to become a creature without full sexual intercourse and instead of being subject to embryonic cell fusion.
- An unfertilized egg cell that has been stripped from another type of sheep with a black color in the laboratory and the nucleus was removed, but the cytoplasm remain in the proper position.
- The defined cell donors exposed to famine and by halting its development and to prevent divisions and food resources for a period of ten days and surrendered to a state of sleep. The egg were put near the nucleus of the cell donor and prepared for cloning. Using weak electric firing bursts then nucleus and the egg unite and start acting as fertilized egg.
- Allowing the embryonic cells to grow and divide and then formation of embryo in the pot laboratory, implants in the uterus of sheep with black head. The sheep that was born become and was named an international replica of the white sheep of the cell donor not the same as the black sheep used for the lap.

**Bio communications of life**

Communications carries out number of operations which include successive generations such bioelectrical- communication and chemical, reflects the continuity of generations, which represents the survival of reproduction in different ways. The mainstream way is connection in the process of sexual relationship resulting in the integration of the male cell (sperm) female cell (egg). Reproduction may occur without convergence between male and female, as in the cloning processes, which we have mentioned, or, as occurs in the primitive animals, with single cell in a simple dichotomy, as do animal and amibia, paramseyoum. In the process of sexual convergence the number of chromosomes go back to the original number in all of the cells of the human body and becomes a forty-six chromosomes (23 from the nucleus + 23 from the sperm). Therefore, the characteristics of many genes on the chromosomes are transferred. It is worth mentioning that the vaccination is not described in the same way it may be externally or internally in the fish and some other animals, the sperm and eggs put in the water in which they live. The artificial insemination is carried out by sperm from a male and put in the female's vagina to cause pregnancy for the purpose of the transfer of the qualities required to a large number of females.

The other type of communication named electrical contact is carried out via the nervous system, which includes the central nervous system (brain and nervous system or spinal cord) and peripheral nervous system (nerves of the brain, nerves and spinal cord) nervous system (self-sympathetic and para sympathetic). The brain, which represents the main component of the central nervous system, contains about 12 billion nerve cells that do not divide, and when they die does not change with other cells. The nerve cells for example, connect impulses and the installation of a specific architecture. It is worth noting that electric current flows in a given direction and one inside the neuron, and to understand how we entry into nerve impulses in the nerve cell should be conceived that in all cells there is a difference in voltage between the inner and outer surfaces of the cell membrane surrounding the cell. The chemical communication is carried out by glands that produce hormones by channels "Ductless Glands" then they carry out considerable influence on many of the functions of the body.

There are concepts of life in different formats pave the way to clarify the mysteries of life, including cell and technologies and weaving techniques and giant molecules. The secret of the evolution of life in this area, where many scientists believe that life appeared on earth since about 2000 million years and that the earth has been more than 2000 million a year devoid of life. Some have said that life must have originated from water and others say that the air more than water and is flexible others remember that life is generated from the mud while Hrkulait believed that the origin of the universe comes from the fire, was able to turn to air and then to water and the last change to dry and then back into water, then air to fire. The Ambdoukls has reached to the theory of the four elements; the origin of all beings is earth, water, fire and air. Therefore, the thinking of these theories can be deduced from scientific conceptions and we can continue with it to the path of breeding that have been fabricated intellectually by Alersai first and Pasteur II. Humans have many and great works of intellectual and experimental to underline his perception of this universe, have attempted to link his idea of theoretical and experimental trends. In final his perception it was confirmed that life arises only from life, as well as to the truth of God still plays a significant role. The early stage study of the body carried out by the microscope and chemical analysis and other techniques, then developed later, first has allowed the researcher to see fine details, arrangement of atoms can not know by this technology, while the chemical analysis refers to compounds that make up the body and the components of these compounds.

Later several techniques developed, including electron microscope, where it was possible to obtain precise details of the minutes. To reflect the elements of body composition other technique called X-ray have been used to study, the order of atoms in many of the biological compounds.

The isotope techniques have continued as a useful tool in the search for the secret of life is as well as a means of treatment. Researchers have manufactured hundreds of radioactive isotopes, generated from the non-radioactive elements in nature, including sodium, sulfur, calcium, chlorine, copper, cobalt, gold, iron, mercury and silver. The main uses of these element the process of photosynthesis, as well as follow- up to 14C in the development of a new technology called radio- immunoassay which can be used to determine the concentrations of compounds found in very small quantities, especially hormones.

Used several techniques to separate, including electrical and deportation chromatography. The first has been used to isolate many of the vehicles life and purification, and the first uses and which still constitute the mainstay of the first structural study of structure of proteins and amino acids, thanks to the use of positive ion equivalent.

### J. Living organism and the secret of life

When studying living organisms as secret of life and mysteries dependence of the of life it might be important to consider many concepts such as the idealism that reflect the objects of which they are the realities of life and is found independently of sense perception, and is the way of our thinking for our perception. As a matter of realism philosophy, another question is the position of living objects reason for all the phenomena of existence and the universe, or bypassed to another reason represented the deepest area of the material, another beyond the spiritual and the last living as a reason to click the Spirit a realistic concept of Divine (the divine), and this concept does not mean dispensing with the reasons natural or something to rebel against the facts of sound science, a notion that God is deeper reason for science to explore the wider area, including the continuing nature of living organisms. The Philosopher, whether materialistic or divinely believes the positive side of science, such as exploring the unity of life of organisms with the general knowledge that is not in issue, the scientific issues , divine philosophy materialistic philosopher material. Agasiz introduced in the year (1858) the idea represented by all kinds of organisms created by special acts of erecting force. This opinion is consistent with view of both Al-Razi and Pasteur and settled their minds with that every living being that must be generated by an organism like him. Also, Hermann Erhard Brichter has stated that each living thing is eternal and produce only from the cell.

### The cell as secret of life

The human digestive system which is an arm of the human physiology of man, explain grandeur of the Creator and the accuracy in detail of the multiple and various secrets. The digestive system is sophisticated chemical laboratory having different methods of food analysis then the food will be distributed equitably to millions of cells that make up the human body consideration involving the secret of life and admiration for the cell. These cells are different technologies for tissue engineering and in the digestive system nearly two hundred thousand reaction within 24 hours. Some of which are the heart muscle shrinks and flattens millions of times during the whole year tirelessly, to obtain the necessary energy for thinking and movement and speech, including also the disposal of waste and toxins in the body, looking at the cell that its approach is one of the secrets of life. These secrets are adopted according to serious lay cell; it is one of the discoveries, theory at the hands of Hleydn and Schwan then cell theory, considered a critical stage in the progress of life science, similar to atomic theory in chemistry. The cell in the body of an organism is also similar to personnel in the communities or the living cell act as technique specific to perform a particular job. The cell becomes a plant or accurate chemical plant. The nerve cell act as a system, for example electrochemical transformation of chemical energy to electrical energy and electrical energy to mechanical energy or kinetic energy. There are also some cells manufacture of the of hormones and other life products used as system used defensive attack their products all exotic and cells in the process of purification and filtration and the cells that absorbs. Furthermore arises from a single fertilized cell, tissues are various heterogeneous tissues, the different organs and different functions, the bones, muscles, cartilage and twigs, leather, and the blood vessels. Then a living cell had made specific technologies in particular, for example, nerve cell is electrochemical system that can shift the chemical energy into electrical energy and the last to mechanical or dynamic or may become a cell laboratory or chemical plant carried hundreds of chemical processes complex and there are cells specialized in the manufacture of hormones and the other to produce biological weapons of and cells of the nomination and purification.

It is clear that the tissues that originate from a single fertilized cell then divided into thousands of cells to materialize into the bones and muscles, cartilage, twigs, leather, and the blood vessels, these tissues are formed in the early embryos and mutate into organs and systems in a stand- alone, but integrated in the performance of its functions.

According to the information of originated from the cell that there is strange power lies in living cells. Walker, professor of Plant Physiology say (that components of a cell arranged in a strange way in which life emerged). But researchers still are unable to make blood cell components and accurate knowledge of this so-called the mystery of life.

**Nucleus and secret of life**

There are at the center of the cell mass of material in the spherical or oval clusters of objects in the body and continues to represent the mysteries of life and plans, regulations, and ideas of life.

In the cell of the human body forty-six chromosomes except the egg (cell reproduction female) sperm and egg, each containing 23 chromosomes (half the number in the human cell non-reproductive). The secret of life is due to this process of somatic cell fusion where each chromosome separated into two parts, it becomes in each of the cells of the fission 46 chromosomes. The chromosome (multi-genes), each genetic factor arranged in two strands one received from the mother and the other from the father.

The nucleus synthesize nucleic acids, seen in the central nucleus filamentous structures and spread over its surface of granules of quick- impact dyes include the nuclear network and the network should be clear when they are not in the case of splitting and dividing at smaller and thickening of these lines is called chromosome. The chromosomes in a cell division looks like similar pairs of fixed shape and fixed number for one type of living organisms.

(Russell and Wallace) says that the cell nucleus is not chemical but structure if analyzed and during the processes of analysis of the most secret mysteries of life may be lost.

Molecules of life, which are building the organism

The space of cell is containing the liquid water containing the various ions and compounds with molecular weights of small, medium and macroscopic, and it is possible to measure the ion composition in each cellular organelle, where each one of them has different ionic compositions.

The sodium ion "Na +" ion is the main ion out the cell in which the 140 mM / L is also found positive ion in the fluid cell in the Interior position that the , is potassium "K +" Cation cell procedure. There is magnesium ion "2 + Mg" in all cellular spaces inside and outside but with lower concentrations of sodium, potassium and chloride is "CL-" ,the main negative ion outside the cell, with hydrogen carbonate ions “and” small amounts of phosphate and sulfate, and the proteins carry a negative charge at pH 7,4 in the tissue fluids.

All living cells contain a different chemical components of water 70-90% and 2-5% of inorganic ions such as sodium, potassium, chloride and sulfate, and magnesium ,carbonate molecules of life, as well as small, medium and macroscopic molecule that constitute 8-25%.

It has been proven that all the elements in the periodic table of Mendeleev’s constitute in the composition of the organism divided into small elements and large. The carbon, oxygen, hydrogen and nitrogen constitute 96% of the elements in the cell, while calcium and phosphorus constitute 3% and each of potassium and sulfur, iron, sodium and chlorine 1% There are very small quantities of the elements iodine, magnesium, copper, manganese, cobalt, boron, zinc, fluorine, selenium and molydenom.

The chemical side is concentrated in the molecules of life on the carbon, which constitutes about 50% by weight bio- molecules are characterized by life-covalent bonds, four of which related to carbon stubs and have different angles of particular value from one carbon atom to another in different molecules of life and because of that there are different types of building structures with three-dimensional, these structures contribute to clarify the complexity of the cellular composition with particular reference to its failure, as well as various forms. In addition organic compounds are characterized by free rotation. The tetrahedral bonds emphasizes on individual carbon atom of the very important properties of organic molecules and the presence of four different groups or different atoms connected to carbon and the last become non-symmetrical (a carbon-atom covalently bonded with four different groups) and composed (which every one of which is mirror images of each other) with a symmetric arrangement in space and called isomers mirror of light for the chemical similarity of the interactions but differ in physical properties of the rotation of polarized light.

**Enzymes**

The word of the enzyme was proposed by the researcher (kon) in 1878 and began to study the specificity of the enzymes at the end of 19th century. Fischer introduce in Fischer" in 1894 the idea of the specifity of the enzyme and the relationship between the enzyme and the substrate stereochemically.

The purification studies started on the isolation of enzymes and then purification in 1920 and in 1922 Wilstater has isolated some of enzymes Kdama and Dixon in 1926 extract an enzyme that oxidize Xanthin, and then the research was continued in developing the science of enzymes and its different branches. The chemical nature of enzyme was elucidated as protein, produced by the cells of the body according to the different needs of those cells.

The enzyme as a catalyst differ from inorganic catalysts in the chemical nature, and mode of operation dynamics of interactions (mode of action), the need for special materials called Co-enzymes. The important properties to the living cell, its ability to catalyze complex reactions at temperatures of its environment without these reactions the cell activity occurs slowly, the cell room has thousands of enzymes and each enzyme specially designed to perform a specific reaction, according to a specific rule within the cell and the light can not imagine the existence of this huge number of specialized enzymes in an organism, which provides a way to control the cell chemistry as accurately as possible. "Oswold" has defined the catalyst as accelator of the speed of reactions, which is characterized by the following characteristics:

- It maintains the structure of the enzyme during the chemical reaction but may occur some natural changes in some of the reactions.
- Speeds up the reaction to reach the state of balance "Equilibrium" without prejudicing "Equilibrium Constant" or position, but also effect the speed of the reaction for each of the directions equally and keep the concentrations of various materials in the fixed chemical balance.

- Contribution to access to slowest energy

**Activation Energy**

This factor is characterized by the specificity of its reaction "Reaction Specificity", where there is usually an incentive, one for each reaction and group of close reactions.

Enzymes produced by various cells of the body according to the need for these cells. These stimuli differ from their counterparts the nonorganic catalysts by the chemical nature and natural way of working and the mechanism of the reaction that have driven.

The enzymes which consists of protein are different from non-organic catalyst such as manganese dioxide, platinum, nickel, iron filings, etc, The enzymes are more specialized than non-organic catalysts involved in several reactions and be quite different from each other. The molecular weight of enzymes is large, more than the molecular weight of non-organic factors, the enzymes affected by temperature that increase their reactions.

"Esterases" catalyze the several types of esters in fat and phosphatides, varies its influence from ester to another. These enzymes do not affect one type of compound, but on many types which are similar in chemical composition in terms of the existence of the certain chemical bonds in all.

The enzyme "Lipase", which is one type of esterases specific in the analysis of fat and can not cataltze non-fat, esters. The relative specifity is not an enzyme specialized in the impact on a particular compound, but have specialized in certain chemicals, regardless of the building blocks are related to the chemical bonding.

Enzymes are specific for the space look-alikes "Stereo chemical Specificity" the enzyme is working on a particular compound and does not work in other particular compound, such as "L-amino and oxidase" affect the amino acid L but does not stimulate the conversion of the amino acid D.

"Structural Specificity" some of the enzymes required for the completion of their specialization, the existence of special groups adjacent to, e.g "Carboxy Peptidase" affect on the bond adjacent to the peptide group &COOH free in many complex peptides. And the enzyme "Aminopeptidase" only affects the foreign peptide on the link adjacent to the free amino group of many peptides, affects the internal peptide bonds in the protein molecule.

Specificity of the reaction, the enzyme has the ability to choose one of multiple reactions (specificity of the reaction). The enzyme also stimulate one of the multiple reactions, through the catalysis of the specific substrates. For example, the enzyme "amino acid oxidase" oxidize and remove the amino acid "Oxidative Deamination" and removing $CO_2$ when there is an enzyme "Decarboxylase" and interaction that took place for the third amino acid is the transfer of a amino group "Transamination" through the exchange between a group of ketones in "Oxalaacetic", these three enzyme catalyzed reactions characterize the specificity of the protein part of the enzyme "Apoenzyme" and not to the enzyme assistant (co-enzyme ), in the "Decarboxylation Tnansamination" the need for an enzyme and one assistant "Pyridoxal Phosphate".

**k. Recent scientific trends in Iraq.**

**Chemistry**

Amazing developments occurred in the chemical sciences, especially during the second half of the last century, including implicit and other interfaces. Developments implied addressed the contents and mechanisms of new interpretations of the phenomena and events and chemical reactions, arose because of it also subjects new disciplines in within taught himself chemistry, led these developments as well as to open new channels in the search for science and technology, chemical and construction of new chemical industries, or chemical industries new technologies .

The developments of the second type of chemical interfaces science has dealt with disciplines that bind Chemical Sciences Pure and Applied Sciences and various led these developments to develop terms of reference or a new science were not known before and took these terms of reference, science its proper place in the programs and curricula and plans of study and research. Some of these updated science is still incomplete features at the present time, but the scientific forecasts indicate near the completeness or ingrained with the turn of the century.

And it reflected the effects of these developments both types of implicit and interface to the application of curriculum topics sandwich to work in various scientific and industrial fields. And it took some US universities to expand significantly in the granting of bachelor degrees, but also in graduate studies, in double chemistry disciplines dealt with medicine, agriculture, engineering or biological sciences, physics, education and others.

Attention to scientific research on the level of the initial study in universities to prepare messages also increased next to pass the courses. And the expansion and development of the two concepts in the event in the chemical sciences as well as to the valley bisected some specialties chemical known to specialists or more, and this would develop new terms of reference were not known before.

And it took international universities known enter the computer not only in the subjects of study specialized in chemistry, but also taking the means of clarification is necessary in the chemical t for the purpose of clarification and display structures and chemical structures and arranged stereo chemistry and methods of preparation and diagnosis of some vehicles, for the purposes of types of processors and accounts chemical necessary in the present day.

And the expansion of development in the chemical sciences led to the splitting of some terms of reference chemical known to specialists or more, and this would led to the development of new terms of reference were not known before. Amazing developments occurred in the chemical sciences, especially during the second half of this century, including implicit and other interfaces. Developments implied dealt with content and vocabulary and mechanisms known in chemistry and provide improved or new interpretations of the events and phenomena and chemical reactions. And arise because of it also new subjects and disciplines within taught himself chemistry. These developments also led to the opening of new channels of scientific research in chemical and technology to create a new chemical industries, chemical industries or new technologies.

The developments of the second type of chemical inter-disciplinary science has addressed that connects Chemical Sciences Pure and Applied sciences different. These developments have led to the development of the terms of reference or a new science were not known before. And it took these terms of reference and its proper place in science programs and curricula and plans of study and research in a lot of sober universities of the world. Some of these disciplines or updated science is still incomplete features at the present time, but the scientific forecasts indicate near completeness and ingrained with the beginning of the next century. And it reflected the effects of these developments both types of implicit and interface to the application of the topics Alhtairah curricula in some British universities to increase the rehabilitation of sections of chemistry graduates to work in various scientific and industrial fields. Some US universities and taken to expand significantly in the granting of bachelor degrees, but also in graduate studies, in double chemistry disciplines dealt with medicine, agriculture, engineering or biological sciences, physics, education and others. Attention to scientific research on the level of the initial study in universities has also increased even some British universities have become demanding preliminary study students in chemistry and other sciences to prepare scientific messages next to pass the courses. And the expansion and development of the two concepts in the event in the chemical sciences as well as to the valley bisected some specialties chemical known to specialists or more, and this would develop new terms of reference were not known before. And it took universities in the world known computer enter not only study subjects specialized in chemistry, but also as a means necessary in chemical Altdrissat for the purpose of clarification and display compositions and structures and chemical and arrangements of steric and methods of preparation and diagnosis of some vehicles, for the purposes of conducting types of processors necessary chemical and accounts in the present day. Also have been vaccinated chemistry curriculum in universities in the world in a new Help topics it was not known before and believed it had become a day of themes and supplements the scientific culture

## Some recent trends in chemistry and Bio-chemistry

Of the difficulties faced by the researcher to contain the recent trends in science variety and the limited field of view. Although it can review these trends very briefly and then focus on each other and thus the benefit is based effort to prevent COMPLICATIONS. One area of modern chemistry, for example, as follows: -

**Macroscopic cyclic compounds.**

These vehicles susceptibility to separate chemical elements and drawn with high efficiency, used as models to study the movement of ions across the cell wall and have a high efficiency as anti-life.

• **multiple molecules (polymers).**

therapeutic radiology It has a high ability to connect to the electrical elements and the movement of the medication to the exact location.

• Events interactions transition, to make an impact on increasing the effectiveness of life for the purpose of treatment compounds.

• Use life molecules of life for chip manufacturing computers in the next generation.

• Manufacture of organic chemicals super connectivity.

• The development of a new generation of super-computers and sensitive devices, can be grown within the human body to repair some damaged tissue.

• The development of spectroscopy devices to follow the pre-life interactions.

• Making new chemical molecules for the treatment of diseases (AIDS, cancer). Pave the recent trends in biochemistry

Looking biochemistry in chemical and physical properties of the components of the cell and the features of the systems life of these components, as well as the interpretation of the nature of these regimes in the cell in a minute.

Biochemistry made a lot of achievements, has helped to clarify the mechanism of drugs and contributed to the diagnosis and treatment of many diseases and provided techniques that could be used measuring a lot of compounds in vivo level.

Biochemistry century has different disciplines, some with the study of materials that make up plant cell and called then the chemistry of life and plant which relates to animal cell and if the human cell is intended whether natural or satisfactory called the clinical chemistry has expanded biochemistry, becoming a cover biochemistry Physical and organic chemistry of life and life Inorganic chemistry and nutrition as well as chemistry.

Interested in science biochemistry modern functions of life systems, it means the study of the past century contributed to note these systems directly during their work, either at the present time the picture has changed and it became possible to obtain observations favorite most Tbhra by midwife to the evolution of technology (electron microscope, radioactive isotopes , immune, spectrum).

The progress achieved in the biochemistry has begun to recognize that life systems contain small molecules and clarified as well as the huge molecules called macroscopic particles which are no less molecular weights from 100 million times as much mass hydrogen atom. It Occupies important molecules of macroscopic Life Systems with ability towards life interactions in the composition of the structural units, and we can say clearly that in the past years huge efforts to characterize the annexation of macroscopic molecules as well as the reactors that occur between them and needs so advanced methods of separation, purification and characterization and to obtain information on structural composition of macroscopic molecule.

The aim of biochemistry for almost half a century is to collect and organize the chemical reactions that occur in living cells. The catalyst for this great effort is that a significant number of the qualities of life cells can be understood through these interactions that are usually characterized by the composition or breaking covalent bonds. It has been clarified as a result of the process of liberalization of the energy break-up as well as the chemical transformation of molecules of life and mutual assembly operations ofamino acids and sugars and fats molecules to form macroscopic processes.

Bio-security and the impact of CT on biochemical systems

There are specific levels of biosecurity have been developed in accordance with the rules set by the institutions (scientific, global) radiation enters, including in coordination with various other influences.

Form used radioactive materials in diagnostics, treatment and scientific research regardless of its source, whether a hospital or medical institution or research or the result of diagnostic and and nuclear medicine threat to public health and the environment, the same time by the right doctor in his battle against the disease.

Used, for example, radioactive isotopes in radiographic diagnosis Kmchks the drug field as being the detection and diagnosis of radioactive isotopes in specific steps sequential stages of the process and it was possible to study all the organs in the body radioisotope such as the brain, thyroid, heart, lungs, liver, spleen, kidneys and other.

Although the amount of radioactive isotopes associated diagnostic number used in the small immune radiation test, but it may cause complications radioactive I121 isotope, I131, for example, in the case dealt break down active compound radiation then the body absorbs irradiated concentrates in the thyroid gland, so The thyroid received doses greater of radiation permitted as a result of the random distribution in the body, the controls Biosecurity at the use of these radioactive isotopes used in spite of the small amounts of radioactive chemicals generally constitute a radioactive isotope and several other levels of bio-security can refer to them during the search.

**Some recent trends in mathematics**

Some universities went to the development of Projects and clinics. Mathematical Clinics project to confirm the applications, as well as that combination studies such as mathematics and statistics, mathematics and philosophy, mathematics and economy, mathematics and physics, mathematics and engineering, mathematics, computers, and others have been introduced.

Computing and introduced a way to teach some math and devise new ways of teaching a substitute for the currently prevailing lecture and publish mathematical interest and awareness of mathematical culture. Emphasis was placed on a philosophy of mathematics, multiple directions (axes) such as Applied Mathematics (geometric), athletic life sciences, mathematical economy, financial mathematics, probability, statistics, operations, research and other research mathematics.

As has been confirmed applications of mathematics through public programs, and in particular to give a private sports importance of modeling, I have gone some universities to develop Project and sports clinics Mathematical Clinics to confirm the project applications. Moreover, a combination studies such as Mathematics and Statistics (NBS) has developed, mathematics and philosophy, mathematics and economy, mathematics and physics, mathematics and engineering, mathematics, computers, and others. Computing and introduced a way to teach some math and devise new ways of teaching a substitute for the currently prevailing lecture and publish mathematical interest and awareness of mathematical culture: for graduates.

Emphasis was placed on the unity of mathematics philosophy and the designation of the relevant departments of mathematics on behalf of the Department of Science mathematical This section includes multiple directions (axes) such as Applied Mathematics (geometric), Science sports life, sports economy, financial mathematics, probability and statistics, operations research mathematics and other research.

It was also stressed on the applications of mathematics through public programs, and in particular to give special modeling mathematical significance, has gone some universities to develop a project (Project), clinics mathematical

(Mathematical Clinics) to emphasize the applications as well as it has developed interfaces studies such as Mathematics and Statistics, Mathematics and Philosophy , Mathematics and Economics, Mathematics and Physics, Mathematics and Engineering, Mathematics and Computing, and others. PCs and introduced as a means to teach some math and devise new ways of teaching substitute for the lecture and the prevailing interest in the deployment of mathematical culture and awareness mathematician.

Emphasis on the unity of mathematics and philosophy, includes several directions, such as, applied mathematics, engineering, life science, mathematical economy, financial mathematics, probability and statistics, operations research, mathematical research and other applications have also been reaffirmed through mathematics public programs, In particular, special importance to the modeling of mathematics, has gone some universities to develop a draft "project" and mathematical clinic confirm this applications. Moreover, studies have developed interface such as mathematics statistics, mathematics philosophy, mathematics and the economy, mathematics and physics, mathematics and engineering, mathematics and computers, and others.

Computers have been introduced as a mean to teach some subjects of mathematics and devise new methods of teaching an alternative to the currently prevailing lecture and attention to the employment of culture and mathematical awareness.

The unity of mathematics as a philosophy and the name of mathematical sciences include many aspects such as applied mathematics (engineering) and life sciences, mathematical economy, financial mathematics, probability and statistics, operations research, mathematics research, and others. It was also stressed the applications of mathematics through public programs, and in particular to give special significance to the mathematical modeling, has gone some universities to develop a draft "Project" clinics "Mathematical Clinic". Moreover, studies have developed an interface such as mathematics and statistics, mathematics and philosophy, mathematics and economics, mathematics and physics, mathematics and engineering, mathematics and computers, and others. Computers were introduced as a means to teach some math and attention to promotion of culture and awareness of mathematical mathematician. To provide students with the qualifications can be expressed as follows:

- A sufficient culture to qualify them to understand the difficulties.
- Encouragement self- confidence of all to address the difficult technical issues.
- The amount of uncertainty makes them ask the right questions.
- The amount of persistence makes them continue to search for appropriate responses.
- Discretion to choose what is right.

**Computer Science and Informatics**

Recent trends suggest that the evolution of Computer Science from the computer discovery and manufacturing in the early sixties. Computer science was not then separated itself along the lines of pure or applied science, but they grew up in the embrace of scientific departments, especially departments of mathematics and electrical engineering departments. The computer hardware and curriculum were not possible to sufficiently and necessary for the different compartments and focused the first two- way computer materials is the direction of engineering ties to computer components and architecture, and the second trend is the compiler. Number of programming languages in science and engineering faculties used to solve mathematical problems and mathematical, engineering or for the purposes of modeling and simulation. And the evolution of computers in terms of physical hardware "Hardware" and software operational "Software" to develop computer science as one of the basic science, where it was subjected to the axioms of this science and the philosophy of pure science and applied at the same time. Since the late sixties, introduced independent scientific sections of computer science "Computer Science" or computer science and information "Computer and Information Science" or computer education "Computer Education" or "Cybernetics" or processing of information "Data Processing", or under other refers to one type of information processing.

Development of computer education even intervened materials with a large number of scientific and humanitarian and computer hardware has become part of the curriculum requirements of the various articles of pure science, applied science and humanities. As a result of the marriage of computer science with other fields of science and new scientific knowledge, especially in the application, example such as field of computer linguistics, "Computational Linguistics" as a result of overlap with computer science, linguistics, or a discipline called decision support systems "Decision Support Systems"

The evolution of computer sciences was started since the discovery of computer laboratories and manufacturing in the number of American universities and the British in the early sixties.

The computer science was not as separate discipline as pure science or applied, but had its inception in the arms of scientific sections, especially sections of mathematics and electrical engineering departments.

As computer hardware and curriculum were not enough to be in different department at the beginning. The first computer hardware trend was related to engineering and architectural computer components. The second trend was the software programs, which examines a number of programming languages in science and engineering faculties used to solve the issues of mathematics computational and engineering and statistical purposes or modeling and simulation. The evolution of computers in terms of physical device "Hardware" and "Software" began to widen in scientific, theoretical and practical trends. This expansion has led to accelerated development of computer science as one of the basic sciences.

The large number of universities and colleges established the formation of separate departments for the computer science, such as, computer science, information computer, information science, computer education or cybernetics, or information processing, data processing. The resulting marriage of computer science with other fields of science, new scientific knowledge was created, especially in the field of application. For example, the evolution of computer "Computational Linguistics" as a result of overlapping computer science with the knowledge of language or field of knowledge called "Decision Support Systems" overlap between computer science or science information systems and management science and operations research and statistics.

## Virtual Informational Technology (IT) Knowledge

Virtual community association is characterized by information and the presence of different types of interactions knowledge, information and economic and the presence of e-government.

Therefore, virtual community knowledge society replaces the real values and standards of a new hypothesis occupies a distinct position in the knowledge society assumption; represent a model of high proficiency, difficult to isolate it from reality.

The virtual universities are academic institutions aimed at providing levels of high-quality education for students in their residence places through the world wide web of the internet where these universities work to create an electronic educational structure, meaning that there is no need for classrooms or buildings or student rallies in classrooms or exams. In other words, these universities established ranks of virtual default continue to conduct tests on after the adoption of advanced programs and therefore it requires Iraq to take the experience of virtual universities in view of what can be provided by funds, where low-cost student and regular pressure on universities. The virtual universities create platforms for a variety of electronic and administrative services and create an oasis of virtual science.

It is noteworthy, the idea that emergence of virtual universities depends on the use of e-learning in terms of curriculum, teaching methods and techniques according to the explanatory educational academic standards developed and invested in self-education "self learning" effective and to obtain academic degree.

**Organisms secret life**

When studying living organisms and adopted a fraction of the mysteries of life can be touched on a number of concepts Kalmthalah that reflect the objects that this life is a facts exist independently of feeling and perception, or is the color of our thinking or our perception as far as it comes to realism philosophy can ask another question The position of the organisms on the borders of this philosophy as a sensible Vtkon is the general cause of all the phenomena of existence and the universe or bypassed to another reason the deepest represented by a physical field and another spiritual and last beyond the organism as to the cause over the Spirit is the concept realistic Divine (Divine password) and this concept does not mean dispensing natural causes or rebel against something Facts sound science, a notion that God is a reason he is called to a deeper knowledge of the wider field to explore the nature and continuation of which living organisms.

Valfelsov whether physically or divinely believes in the positive side of science unit life of organisms with the knowledge that is not a scientific issue philosopher in God and another material.

Agassiz was introduced the year 01 858 m) opinion is that every kind of living organisms create special an act of creative power. This view is consistent with the opinion of all Razi and Pasteur and they settled on that every living creature that has to be generated from a living organism like him. Herman also said Erhard Brichter has stated that every neighborhood eternal nor generated only cellificCell of the mysteries of life

If watched for example the gastrointestinal tract in humans, which represents one devices physiology of man, we find the greatness of the Creator and accuracy in the multiple details of the various secrets, for example, we can imagine being a laboratory chemically developed where being different methods of analysis for food, then distributed the food fair distribution to millions of cells which make up the human body and involving the secret of life.

Self astonishment and admiration which adapts according to the place and circumstances where these cells form different geometric textile technologies mixed organic cell, it is in the digestive system nearly two hundred thousand interaction within 24 hours, some of which makes the heart muscle shrinks millions of times and have fun during the whole year tirelessly and how to get Ateltaqh necessary to think, movement and speech. Including what the disposal of waste and toxins within the body looking at the cell which is the approach one of the secrets of life perceived to be adapted according to the requirements of the position and the circumstances and noted it was one serious life, which saw establish cell theory at the hands of Sheldon and Schwann then considered cell theory serious stage in the progress of the science of life Atomic in chemistry, or that the cell in the body of the living object in the communities or the living cell behave especially with specific work ways that can cause a particular job as a technology, it has become a cell as a plant or a chemical plant accurate.

Neural example behave as a system energy into electrical energy and electrical energy into mechanical energy or kinetic energy. As the industry some cells produce hormones and other life defense articles attacking their products every intruder and there are cells in the process of purification and filtration cells secrete and absorb movement and other commands and signals that each cell as a plan of action does not deviate from them.

And originate from one fertilized cell differentiated tissues and different members of different functions and is embodied in the end in the bones, muscles, cartilage and twigs and skins and vessels and the blood.

The private living cell technologies came, for example, nerve Valkhalih system can transform chemical energy into electrical energy and last into mechanical or motor. Or cell as a laboratory or may become a chemical factory in which hundreds of complex chemical processes take place and there are cells specialized in manufacturing and other hormones to produce weapons cells filtration and purification.

It is clear from tissue that arise from the fertilized cell one then divided into thousands of similar cells is embodied in the end in the bones, muscles, cartilage and twigs and skins and vessels and the blood, where these tissues are formed in the early embryos and mutate to organs and systems stand-alone, but they are integrated in the performance of .

It is clear from the information about the cell that there is a strange force managed in living cells, the world (Walker), professor of Velsjh Plant says (that the cell components group in a strange way emerged through life. And still, the researchers are unable to cell making blood of know the exact components and this so-called the mystery of life) ..

In light of perceptions of genetics, that the inheritance of the individual left untouched nuclear material of living cells for reproduction and genetic traits due to microscopic precision parts which genes. In light of this decline perception Darwinian view that the changes and the qualities that earned by the animal as a result of experience and practice, or to interact with the ocean or type of food can be transmitted by inheritance to his descendants. Consequently, the researchers moved to the emergence of species is by mutations by observing some aspects of the sudden change. •

**Genetic Engineering**

Genetic engineering has brought tremendous developments in the life sciences, including applications in medical science, which includes diagnosis and treatment. Old and new in reproduction

Cloning technology is distributed according to the foundations and specifications on several areas, including gene cloning and gene cloning and reproduction nucleus. Gene Cloning refers to the production of identical genes are similar to the original gene, and the best example of this process is what happens naturally when you give birth to twins as a result of similar split genes per egg and distribution

But when the embryo cells separate from each other in a particular phase is called a gene cloning, where they grow each cell separately to produce an integrated organism. Researchers have successfully cloned monkeys and frogs from the cloned gene cells match similar to the original, and is believed to be easier to deal with embryonic stem cells because they do not possess it any they were not have been after for conversion evolved into hair and brain and muscles and other organs..

The third cloning which represents the occurrence of the completed cell growth, the researcher can wait to see the nature of the thing with his eyes before baptizing on reproducible. That this type represents a new cloning and that is using the nucleus of an adult cell and adopted by the world's Wilmot specialist knowledge of the embryos, which represent innovation embryonic technique to clone a sheep from an adult cell.

is also able to clone sheep genetically modified human genes to blood clot factors was the bearing human genes while considered Dolly the sheep which Wilmot the first that have been reproduced with technology the latest and that an adult cell is a reproduction of Oelmut technique for an international cloning of a new life principles, particularly to rely on a physical cell successful and can take place steps each one has a recipe relates to these principles and are the following:

- Dealing with the donor cell (host) intended for reproduction and extracted from the membrane cells to sheep udder white color swatch This cell is characterized necessary to form a complete sheep genes. The move represents the use of an adult cell for the purpose of turning them into a cell with the ability to reproduce without vaccination in order to become a full creature without sexual intercourse (without vaccination) and instead of embryonic cell be capable of docking.

- Took the cell-egg is fertilized after extracted from ewe from another type of color in a black pot laboratory and removed its nucleus with the cytoplasm to keep in the proper position.

- Donor cell exposed to famine by stopping its development and prevent divisions to prevent food resources for a period of ten days and surrender to a state of sleep and put the egg near the stomach donor cell cloning and using electric induction firing bursts electric weak unite the nucleus and the egg nucleus and begin acting fertilized.

- Allow the embryonic cells to grow and divide and form an embryo at the pot laboratory implants in the womb of sheep with black head and become the sheep that was born of an international replica of the ewe white donor cell Idah and named nor similar black which is used to lap the sheep.

The concept of this multi-genetic engineering and technologies based on splitting the D.n.a. By special enzymes called in specific engineering sites and connecting pieces formed with D.n.a from other sources. It is then formed on the multiplication of hybrid vital carrier capable of reproduction Cloning including bacteria and viruses have been used for the development of genetic engineering and other techniques electric and radiographic self.

This technology can be used for the preparation of the private D.n.a sensors for the purpose of search for specific genes or specific parts of the D.n.a as this technology can also clone parts of the D.n.a in large quantities by the private use of the enzyme reactant sequential PCR.

**Applications of genetic engineering:**

- Gene therapy: used in the treatment of some diseases, including the treatment of brain tumors and in reducing the proportion of in the blood.

- Draw a genetic map of the human beings: it can be seen to the topic of the future after he was identified many human genes sites.

- Stimulate the immune system to produce anti-more efficient and accurate (a vaccine against the virus hepatitis B).

- The production of proteins of medical importance.

- To determine the nature and location of genes for some genetic diseases.

- The diagnosis of genetic diseases before birth.

• bio-technology

Caused enormous ecological technology developments mankind, including those that purpose human use:

- Electronic Industries.

- Bio-technology satellite products.

- Environmental processors.

- Oil extraction catalyst - Medicinal plants.

• Cloning

The transfer of chromosomes from one cell to another chemical and life to the process of cloning concept, also represents her spine. The transport process and chromosomal gene that is not considered new, it has been practice during egg implantation in the womb Female year (1978) (the birth of baby Louise ETT).

**Specifications and features of reproduction**

Cloning is characterized by specific chemical specifications of the features of some new and some old concepts. These include the following specifications:

- Convert an adult cell (total growth) cell to reproduce without vaccination.

- Proliferating cell to convert the entire creature (replica of the mother).

- The transfer of fully mature cell growth to a live egg completed their growth in the womb of another object (the sheep, for example).

- The introduction of a new principle called the principle of my life (and death) can be expressed according to the following: (You may very specialized of sophisticated animal cell to awaken the population genetic information to become the source of a new creature full).

- That this process of reproduction is not the nationality of the traditional (no sexual contact between male and female).

- Do not use a cell are genetic fusion and reproduction.

- The cloning of adult cells to facilitate the growth of the researcher to wait to see the nature of the same thing before baptizing on reproducible.

Applied features for reproduction (the experience of Ian's death)

- Conversion of mammary gland cell from the udder ewe (full growth-adult) to the same cell reproduction ability and without vaccination.

- The transfer of the full amount of cell growth and fully alive to complete egg to grow in the womb of a sheep.

- Inherited the sheep cloned donor parent properties for the first hive (the birth of Dolly the sheep).

- The fledgling sheep is not the daughter of her mother and could be twinning.

- The sheep-giving of the cell intended for a clone of her nipples, containing all the genes necessary for the full ewe.

• Life Communications

Life is meant to communications a number of operations that include the succession of generations and contact life electrical and chemical ones. And it expresses the continuity of generations, which represents the survival of the type that is to reproduce a variety of means, and the main way the prevailing male sexual contact with a female in the process result in the merger of the male cell (sperm) female cell (egg).

Reproduction without convergence between male and female as may happen in cloning processes we have mentioned, or as in the initial single-cell animals in a manner as simple dichotomy does animal. In sexual rendezvous process number of chromosomes back to the original number in all body cells of the human becomes forty-six chromosomes (23 from the nucleus +23 of sperm) and therefore many recipes resulting from the genes on the chromosomes move, in fish and some other animals pose sperm and eggs into the water in which the animals live either artificial insemination are put sperm from one male and put it in the vagina for women's events of pregnancy for the purpose of the qualities required to transfer a large number of females.

The other type of connection, called the contact electrode are through the nervous system, which includes the central nervous system (the brain and cord nerve or spinal) and nervous system peripheral (nerves of the brain and nerves of the spinal cord) and the nervous system of self (the sympathetic).

The brain, which represents the main component of the central nervous system contains about (12) one billion neurons that are not divided and when death does not change the other cells. The neurons connect, for example, nerve impulses and with syntactical specific installation. It is worth mentioning that the current flows in one direction inside the nerve cell, and to understand how nerve installment in the nerve cell into force should imagine that in all cells there is a difference in voltage between the inner and outer surfaces of the cell membrane surrounding the cell.

Academic disciplines have evolved with the development of sciences and various new disciplines such as engineering, agriculture, science and total treatments that began in the nineteenth century, whereas the twentieth century indicates other disciplines such as business management, journalism, information and library science, economics, politics and world affairs were added.

Each country has special methods to determine its own disciplinary university and identification numbers, graduate students and the quality.

The world witnessed in the twentieth century breakthrough in all fields and scientific trends, so there are no boundaries between different disciplines. For example, medical science requires engineering science and recent tests of modern science depends on the physical, chemical extraction and analysis and also relies on mathematics to lay the groundwork mathematics.

The progress and development in pure science, for example, develop new subjects and disciplines and specialties of science. New interfaces were not known during the first half of the last century. The results of these major changes in curriculum and build up research transformed these developments to the university curricula. Seminars and researches are now carried out in different ways including:

•Bachelor degree is based on the study and thesis.

•Some universities in Britain and Germany developed curricula at the level of initial studies that include research and study.

•Division of the present fields such as industrial chemistry, chemistry of life with medical side and other disciplines in the branches of pure science.

•Development of competencies.

These terms of reference have been developed in American universities in physics, chemistry and mathematics, to prepare graduate in some sectors such as engineering, chemistry physics and chemistry, agricultural engineering, and medical studies.

•Addition of assistance topics.

Some topics have been added as assistance of many of the terms of reference of pure sciences, including education, literature and library services and use modern machinery and computers.

•Other disciplines (Sandwich)

British universities were carrying out by expanding the initial years of university study for use in increasing opportunities for the systematic teaching and applied for and rehabilitation work in various production sector.

**Psychology**

Is the study of human and animal and regulations gradient in intelligence, and called on the science of animal behavior Ethology a branch of zoology (many animals, unable to move and respond to environmental changes, feed on plants and animals), and divides to a branch of which continue animals Animal Communication and learning Animal neurological knew self represent study of behavior and the mind, thinking and personal to humans and previously was a science that examines the soul, mind and brain function.

Subdivisions and sensations and the device visual on location and the cerebellum and medulla oblongata responsible for the body and breathing balance. There are several schools of psychology, including the and structural and Padua and Freudian psychoanalysis and modern existential analysis and, moral and social of Science

And it represents one of the branches of philosophy that deals with the philosophical and virtual foundations and implications of various sciences (physics, Akemiae, life sciences and humanities and social sciences) and contents that have been roads to it, including:

- scientific theories
- formulation of various scientific methods.
- scientific method.
- the credibility of the scientific arguments.
- modalities of the production of science.

**The philosophical foundations**

Despite these man builds different perceptions in the general philosophies , for example in the ideal object of philosophy it practiced spiritual and is responsible for his actions. It is noteworthy, Plato believes that the man is composed of two parts, one of which belongs to the world of ideals (self) and the other to the world of sense (the body) and stressed on the idea of bilateralism while the realists believe that the human is organic entity and a social behavior and natural philosophy suggest that the human is finest soul and the present is continued development of the future and education is necessary as long as they continue to longevity.

It is intended that basics of philosophical application of theory in the field of education for the purpose of clarifying the educational process and to identify its features, which is also being organized as intellectual activity, depends on the philosophy as a tool to organize and coordinate the educational process.

Educational philosophy is required due to education as a part of human existence totally as well as, science and language.

The concept of education differ according to the kind of philosophy that deals with it, indicates that education is the formulation of the human being announced and well-being, and Plato believed that education is the consistency between the self and body, while Abu Hamid Al- Ghazali believed that education is a priority for humanitarian and spiritual atmosphere. Education leads to higher degree of ideal maturity for children, and natural philosophy refers to the mental preparation of the child. While existential philosophy believes that man is free and subject to the inevitable and philosophy of pragmatism which is the vision of Dewey that suggests that education is a permanent organization of experience and adaption to social reality.

John Dewey, one of the philosophers, who pointed the principles underlying the concept of modern education, including that education is a small community that comes to life and education continues forever, and the curriculum must be accompanied by life and mission of education is to prepare the individual for life.

Among the Arab philosophers, who were interested in the problems of education, are lbn Sina, Al-Ghazali and lbn Khaldoun (1332-1406 AD) whose belief was limited to the educational principles characterized by gradual transition from known to the unknown, and from easy to difficult.

The subject of philosophy is knowledge of the natural facts (relative) standard and the facts (values and ethics) and the philosophy is considered as perceived by Dewey. For example, the general theory of education is characterized by a turning humanity of human beings to the intellectual and others believe the philosophy Celebrities are the leaders of educational, Plato, Jean-Jacques Rousseau, which reflect about the conditions of philosophy of education and work to evaluate and critique the educational process.

As the fields of philosophy and issues of our time metaphysics (metaphysics), or the divine science, knowledge, values and philosophy of education is to apply the positivist approach in the field of human experience which we call education.

**Idealism**

It is linked to the philosophy of Plato, which is piloted by the perception of two worlds, ideals (fixed) and the real world (variable). The community is composed of two layers, one of them think and other works in other words, the first class is linked to the educational framework for the purpose of access to knowledge, which requires the mind, and thus knowledge is a real constant and does not change. The philosophy believes in and basic principles, based on the ideas of absolute belief in the existence of independent real in a perfect world.

**Positivism**

According to this philosophy of Aristotle (384 BC - 322 BC), that there is only one world, that is the real world, characterized by the firm principles and the care of the senses. It is more important than a focus on imagination and consistent with the ideal of the virtues that is fixed. These realistic narrators follows stability approach and this philosophy depends on the continuous attempts to discover the universe and the world and work to understand the existing laws that include all the facts within a stable and consistent world. The, duties placed on the philosophy of realism for Education illustrated as follows:

Finally, the factual findings to multiple convictions, including:

- Establishment a method of experimentation and exchange of scientific and systematic uncertainty (the way to see the existence of God, real life).
- Encouragement the learner to observe natural phenomena in an order system.
- It was called for acts of reason in the analysis and the independence of the senses.
- The school did not distinguish between the world of ideals and spirit and the serious and actual world.
- Elaboration on intellectual reflection.
- Focusing on professional education.
- It was called for linking the educational curricula with provisions of life.

**Pragmatic philosophy "Pragmatism"**

Of the principles of this philosophy, education is life and this philosophy of education linked to community service while giving the learner a degree of freedom. And human beings adapt to the biological environment, the world is relatively constant and constantly changing, the truth is not absolute but changing society and democratic decision making.

Pragmatic philosophy plays key role in the development of teaching methods and improving the traditional method and manner of trial and error. The school of John Dewey is based on the philosophy of pragmatism played historical intellectual revolution against traditional schools, which focused on information and not believes in the values of morality, but according to the relative renewed convictions of society.

The education in the pragmatic philosophy is not a process of transmitting knowledge to students for knowledge, but help them to meet the needs of the social environment and able to raise the forces required by the student, including social attitudes. Perceptions of the pragmatic philosophy depends on building curriculum writing arithmetic means rather than targets, and approach interests in the facts relating to the nature of the child. Pragmatic philosophy stresses on the development of vocational education, natural sciences, while the humanities and languages is of secondary importance.

**Natural philosophy "Naturalism"**

This philosophy believes that man is good by nature and what is the reason for corruption of man is due to the society and its institutions. Education has a target of opportunity for the growth of the natural child. This philosophy also believes that the sensors are sources of education and entry points for the development of thought and not the role of the stock of knowledge. The natural philosophy calls for the reference to the educational activities that are consistent with the original natural Laws. Rousseau (1712 - 1778) formulate the ideas of natural philosophy, the role of nature in the development of children is illustrated in terms of:

- Release the child from school and classroom activities.
- Meet the needs of the child and the need for tilting the obstacles facing it, and in accordance with environmental requirements.

Rousseau was one of the first to call for self-learning and the nature is one of the most important educational principles that are consistent with, and a woman was found to please men. This philosophy advocated the need to respect individual differences and the protection of children from social pressures.

The first natural philosopher, according to Hellenic tradition, was Thales of Miletus, who flourished in the 6th century BCE. We know of him only through later accounts, for nothing he wrote has survived. He is supposed to have predicted a solar eclipse in 585 BCE and to have invented the formal study of geometry in his demonstration of the bisecting of a circle by its diameter. Most importantly, he tried to explain all observed natural phenomena in terms of the changes of a single substance, water, which can be seen to exist in solid, liquid, and gaseous states. What for Thales guaranteed the regularity and rationality of the world was the innate divinity in all things that directed them to their divinely appointed ends. From these ideas there emerged two characteristics of classical Greek science.

The presence of the elements only guaranteed the presence of their qualities in various proportions. What was not accounted for was the form these elements took, which served to differentiate natural objects from one another. The problem of form was first attacked systematically by the philosopher and cult leader Pythagoras in the 6th century BCE. Legend has it that Pythagoras became convinced of the primacy of number when he realized that the musical notes produced by a monochord were in simple ratio to the length of the string. Qualities (tones) were reduced to quantities (numbers in integral ratios). Thus was born mathematical physics, for this discovery provided the essential bridge between the world of physical experience and that of numerical relationships. Number provided the answer to the question of the origin of forms and qualities.

## ARISTOTLE AND ARCHIMEDES

Aristotle's biological works provided the framework for the science until the time of Charles Darwin. In physics, teleology is not so obvious, and Aristotle had to impose it on the cosmos. From Plato, his teacher, he inherited the theological proposition that the heavenly bodies (stars and planets) are literally divine and, as such, perfect. They could, therefore, move only in perfect, eternal, unchanging motion, which, by Plato's definition, meant perfect circles. The Earth, being obviously not divine, and inert, was at the centre.

His mathematical demonstration of the law of the lever was as exact as a Euclidean proof in geometry. Similarly, his work on hydrostatics introduced and developed the method whereby physical characteristics, in this case specific gravity, which Archimedes discovered, are given mathematical shape and then manipulated by mathematical methods to yield mathematical conclusions that can be translated back into physical terms.

In one major area the Aristotelian and the Archimedean approaches were forced into a rather inconvenient marriage. Astronomy was the dominant physical sciencethroughout antiquity, but it had never been successfully reduced to a coherent system.

# Chapter Two

# Techniques in science

G. **Chromatography**
H. **Electrophoresis**
I. **Electrofocusing**

**J. Sedimentation**

**K. Spectral techniques**

**L. Diagnostic Imaging**

**G .Labelingwithradioactivity**

**H. Autoradiography**

**I. Membrane filtration and screening "Membrane Filtration and dialysis"**

**J. Protein engineering**

**K. Immobilibzed enzyme technology**

**L. Human Genome Project**

**M. Bio- engineering**

**N. Immunoassay**

**O. Biotechnology**

## A. Chromatography

It is used in the separation and diagnosis of many chemical compounds, biological, etc. There are many types of chomotography including adsorption and ion exchange retail and gel filtration and other technical methods for the purpose of use paper and thin layer and gas chromatography. Choromtography include multiple ways that all based on the separation of compounds based on the difference in migration through the passage in the center of force, as well as the tendency to face hard "Stationary phase" for central transgenerational and face the hard nature solid or gaseous, or liquid depends on the tendency of various materials to hard to face multiple methods such as adsorption "Adsorption" ion exchange "Ion exchange" may include all kinds of chromotohraphy these methods, some of them.

**Ion- exchange chromatography**

Based on the tendency of the ions or molecules to materials other than for mobile and non- soluble, which owns the distinct shipments, or molecules that carry one or more of the positive charge exchange with the positive charge associated with the Ionia face, the mobile Resins "Resins" with a negative charge is called this ion exchange process with the positive charge "Cation exchange" and reverse ion exchange is called a negative charge. Examples of the reciprocals of the ion non- animated "Immobile Ion Exchange" that are used in chemical research.

Polystyrene where will attend the multi- way styrene polymerization contributory "Copolymerization" with composite "Divinyl benzene", which adds to the styrene chains cross- shaped multi- written and added then aggregates the active ions altering the chemical composition of the original units that can be prepared for example, styrene resin that contains strong acid groups such as $SO_3H$ hold process of "Styrene- Divinyl Benzene sulfonation". In the same way can be prepared that contains the totals as strong as $NR^{3+}$, or weak acid groups such as the COOH groups or grass- roots groups such as $NH^{3+}$ and types of preparation depends on the ion concentration reciprocals composite "Divinyl benzene" the amount of strings cross referred to the number listed after the name of the resin, such as "8X" Dowex 50, which contains 8% of the "Divinyl benzene".

**Gel-filtration**

The type of chromatography by gel filtration on the difference in the movement of compounds during the gels with regular pores partially used for the purpose of the way in a column filled with from one type of granulated gel filtration.

The pores is able to expel particles with, partial weights more than 10000 if we, for example at the top of the column a small scale solution of dissolved protein and molecular weight of 70,000 with ammonium sulfate the following happens:

- Protein molecules expelled from the pores of granulated gel filtration.
- The migration of protein size start "void volum" outside the granules column.
- Interference ions $NH^{+4}$ and $SO^{-4}$ small pores of the gel granules nomination so there is the amount of liquid required for the expulsion of these molecules outside the pores.

The granules are gel filtration where proteins are separated by major united ammonium by successive periods of time and be dependent on the size of the separation- free liquid, gel filtration is very important ways often used to separate proteins from salts "Desalting".

Electrophoresis is applied techniques migration electric technology application. It is accurate in the middle of an insulator as a result the electric field moves in a minute steady pace, and can then measured the balance between the electric power Eq, and the light of that we get the equation.

The gel gromoatography is the primary means of separation and purification of various enzymes and proteins, as well as the fragmentation of nucleic acids and proteins in the treatment, especially when quantitative diagnosis of some human diseases, as well as by the method of the exchange of tritium to test the protein structure or the structure of DNA and a study of the link between proteins and small molecules. The thin layer chromatography is mainly used for amino acids, polysaccharides and simple sugars, fat and various steroids and other small molecules. The ion exchange chromatography applications, including chromatography cellulose DNA, to purify proteins associated with DNA to separate in general the Bio-compounds according to molecular weights. The affinity chromatography is used to purify the enzymes and antibodies and transport proteins, membrane proteins and the chips and sugary proteins and the separation of animal cells in particular.

**B. Electrophoresis**

Most of the Bio- polymers with the electrical charge is transmitted in the electric field, which have the advantage to classify macroscopic particles and measuring the molecular weight and discrimination and diagnosis of amino acid changes with charges the components without charge, and vice versa. The migration of the electric voltage low- lying useless to separate small molecules such as amino acids, either in the case of the migration of high voltage electricity will become more rapid separation of the amino acids, for example and using the electric gel migration and multi- acrylamide is the last for the time being one of the best prevailing circles to separate proteins and molecules with the addition of the compound. Dodecyl sulfate becomes possible to measure the molecular weights and nucleic acids and proteins, and at times used for this purpose the Agarose gel. The separation of DNA with a single when the use of acrylamide and Agarose. The use of other electrical relay with a multi- acrylamide gel to separate the circular DNA which access through the gel. There are electric displacement by the immune system, which can separate the materials that have the same movement with vertical orientation or by gromotography followed by deportation cuts.

Starch gel used for the first time in the migration electricity and usually consists of' starch pastes, potato starch which burned grain thermally and after putting them in the organizer and the creation of the gel horizontally, as noted in the shape of the sample in a small part which consists of pieces of gel using a razor blade and sealed with wax is usually the part or material lubricating and then begin operation and pass the voltage specified.

Migration of the electric type of the SDS gel, it is possible to calculate the molecular weights of most proteins, the measurement of movement in transition in the multi- acrylamide gel, which contains "(SDS) Sodium Dodecyl Sulfate". In the neutral pH and concentration of 1% of "0.1, SDS" of mercaptothenal. Most proteins associated with multiple strings SDS and analyzed sulphide bilateral ties by "Mercaptoethanol". The result is damaged to the bilateral structure of proteins.

The complexes consisting of units of high protein and the SDS and imposes the existence and status of helical and random act of proteins of this method, as if with a regular form, have an equal proportion of the charge / mass, owing to the percentage of the amount of the SDS associated with each unit by weight.

Using samples of unknown molecular weight of two known molecular weights can then calculate the molecular weight of the sample unknown degree of accuracy between 5% - 10%, and this is certainly one of the most well-known methods and used these days to estimate the molecular weight of subordinate units. It is noted that the way the SDS gel can be used to measure the probability of presence of aggregates SDS. When an electrical relay to a series of proteins of known molecular weights by gel column first separated into a series of packets and then draw distance movement of the samples against the logarithm of molecular weight together there is a straight line.

**C.Electrofocusing**

proteins meaning they contain both negatively charged groups and positively charged, depending on the pH is positively charged when the pH and negative charge in the event that the pH is high. In addition every pH in which the zero-charge and the so- called point equivalent to the movement. For a mixture of proteins, it may point equivalent to multiple sites called points equivalent, which consists of a pH gradient in a column containing the tube- negative and positive. A wide range of points equal to the charge and the various compositions is a mixture of polymers of carboxylic acids.

**Immunological methods**

Provided practical methods to test the immune small quantities of non-radioactive materials in complex mixtures. The developed immunologic tests, which are used to estimate small quantities of antigenic non- radioactive compounds in a mixture of large numbers and quantities of miscellaneous materials testing immune- ray equivalent to or greater than the sensitivity tests of natural chromatography. That the purpose of radio-immunoassay test is to assess the basic particles:

- Those that were not labeled by radioactivity within the body by proper specific activity or without adequate labeling of other compounds.
- Those do not know the identity, it could interact jointly and thus to compete with the antigen known.

There are many substances that are measured in the test radioimmunoassay, including the hormones and pharmaceutical agents, vitamins, and factors assigned and materials in the blood and viral antigens (viral) nucleic acids and nucleotides. In some cases, there is no acceptable method for the preparation of labeled antigen, which is permitted by measuring these materials using immuno-radio metric method, which can be measured by an unknown amount of antigen directly, through its combination with specie labeled. Examples of immunological methods that used in the biological tests of:

- Diagnosis and weakening the various types of bacteria by agglutination.
- Diagnosis of (viruses) by inhibition of virus generated by agglutination of red blood cells especially the antibody present in the serum.
- Diagnosis of gonadotrophin in the urine of pregnant women by testing the inhibition test and complements.
- Measurement of making DNA in phage infected by complement fixation.
- Diagnosis of relations between proteins by specific reaction.

- Diagnosis of tumors.
- Testing materials of clinical importance in children.

**D.Sedimentation**

Sedimentation techniques is common in these days, which is used for the purpose of characterization of macroscopic particles. And using the appropriate type of these methods that can be measured by molecular weight, density macroscopic form of the molecule. It can also measure changes in these transactions and use one as a basis for separation of components of the mixture prepared for analytical purposes and preparations. Moreover, with the modern equipment measurements to make of high- speed centrifuge most of who has benefit. The main thing that occurs in high- speed centrifugal the movement of the minutes by which can measured the distribution of focus during the centrifuge tube several times and called the measurement during the movement of molecules through the center of the centrifuge, account for the molecular weight.

**E.Spectral techniques**

In order to simplify the regulations of life systems, spectral techniques are used to study the structure of many of the synthetic compounds with important skill of life, including protein, nucleic acids and others. Moreover spectral techniques are used in the follow- up of chemical reactions of life and employment regulations of life in the human body.

The techniques continued often its development in medical diagnosis, which can be linked to any disease, accurate diagnosis of the variables affecting the livelihood systems and chemical structure of various compounds, or monitor the status of one or more of the fabric that differ in normal and pathological cases. Molecules absorb light and length of waves that are absorbed and the efficiency of this absorption that are dependent on both the structure of the molecule and its surroundings, making it a useful tool for absorption spectroscopy characterization of small macroscopic particles.

- Spectrum of ultraviolet and visible, the measurement of absorbance in the ultraviolet and visible for several purposes, including measurement of the unknown substance test of some chemical reactions and diagnosis of materials and determine the structural parameters of the macroscopic molecules and follow- up the transition the coil for DNA, example is the double helix as well as the titration of pH spectrum of proteins and to identify features some of the proteins by way of solvent perturbation and by difference spectroscopy and identifying revealed the linking of small molecules to proteins and the union of protein - protein and solvent disorder of nucleic acids.
- Applications of infrared in the Biochemistry applications of modern infrared for diagnosis of mutual hydrogen in proteins and diagnosis of the number of hydrogen bonds and aggregates and measuring the common effective group and broken through the process of metamorphosis and diagnosis of tautorserism forms by and the union between small molecules such as riboflavin and protein. In addition the identification the carboxyl groups in proteins as well as determining the status of hydrogen bonds in proteins and multi peptides to measure the direction of groups.
- Raman spectrum "Raman Spectroscopy": of the important applications of the Raman spectrum in the compounds life, study the mechanisms of differs automerism structure and different kinds of amino acids and discriminate adenosine mono phosphate and triple phosphate and their ionic forms in solution as well as the diagnosis of helix, beta related structure random coil of amino acids and determine the number of

disulfide bonds in proteins and determine the number of double bases in DNA.

- Fluorescence Spectroscopy: there are two types of materials used in the fluorescence analysis of macroscopic molecules; the first contained the same molecules and macroscopic materials fluorescent foreign insert. For example, there are three types of fluor of proteins self tryptophan, tyrosine, phenylalanine due to the fluoresce of the proteins resulted from these compounds, which could be used to study the changing perceptions of the enzyme by the positive correlation factor, assistant recipes enzyme active center and studies on the metamorphosis of the protein and the location of the tryptophars in the enzyme. In many cases the material could be added the fluor in the molecules composed as is being considered, either by chemical duplication or by simple correlation method. The name of the method which will reacted by adding particles when analyzed by outer fluorescence. There are several requirements for materials and when the use of external fluorescence:
    - The material is strongly linked in a prime location.
    - The fluorescence must be very sensitive.
    - Should not affect the macroscopic features of the molecule that studied.

**IR spectra**

Measurement of the infrared spectrum between 4000 $cm^{-1}$ in the upper left end and 925 $cm^{-1}$ in the lower left end of the various groups (methyl, carbonyl, amide …etc) and also to the totals effective adsorption characteristic frequencies to the group from certain areas and this range can be diagnosed by several groups of effective frequency and make the infrared fast methods and reliable.

The infrared spectroscopy in terms of a preliminary is not different from spectrometer for the visible and ultraviolet, where the device usually consists of three main sections:

- The source of radiation the body which is heated to 1500 to 1800 Kelvin "K".
- Radial Beam Analyzer, which is used to test wavelength.

**Raman spectrum**

Small portion located in the infrared in a non- extended with a frequency shifts, called (Raman scattering) and the reason for frequency is due to the high level of vibration by the addition of seismic energy to the molecule to electromagnetic wave optics.

In the process of scattering, the light is usually an irritating scattering (scattering) to the high level of vibration and lose energy and the frequency is decreased, on the other hand, when scattering center is located in the highest level of vibration (chattering in advance with the solvent molecules) can be transferred to the seismic capacity of incident light. Raman spectroscopy uses the regular sources of the laser beam and light scattered through the Raman spectrometer using condenser of electro- optical and consists mostly of light scattered from the baseline formed either by absorption. The emission bred lines that compose a weaker Raman spectrum at lower and higher energy. And thus increasing the frequency and the use of normal temperature, where there is a vibrant and molecules other than the shaky, causing a decrease in frequency more normal, so the test fiber Raman vibrational transitions, such as infrared.

The Raman spectrum is not widespread and can be used to detect some effective groups. Some applications, including proof of structure for various kinds of amino acids, which is accompanied by distinct spectral changes in the support groups and carboxylic acids.

**Fluorescence**

It is possible that light energy absorb only when the molecule move from the lower energy to the top, such transitions in the diagram with vertical lines. When the molecule is initially irritating, it represent the excess energy that will be shaky as energy molecule in one of the levels of seismic and seismic energy appears as heat as a result of collision with solvent molecules (when the inflammatory molecule in solution and reduce the molecular, level to the lower vibration for S1).

In many cases, material could be added to the partial been studied either by duplication or by chemical simple correlation.

**Devices used to measure the fluorescence**

A beam of light with high intensity it passess during the X- ray Beam Analyzer to choose the wavelength to the cause of irritation (eg, wavelength is absorbed by the material enough brilliant), after going through that light Irritated through cell containing the sample. To avoid detection the incident light is restored then the fact that emitted in all directions so that it can shine.

**Dispersion of the optical rotation spectrum and spectra of circular birefringence "CD, ORD"**

A set of techniques that serve to know the status of the molecule or macroscopic and interactions, including absorption spectrum provides useful information of this kind, but that the study of absorbency of polarized light in a spectrum of optical rotation dispersion ORD and spectra of circular birefringence "CD" The method for measuring adoption of the wavelength on the viability of active rotation of polarized light and absorb the difference excellence polarized light and the direction right hand and the left. The physical basis for each of the "ORD" and the "CD" are identical, and are in fact different to address the challenge of polarized light with optically active molecules. And cause to contain a very large part of the molecules of life on active duty visually Hence, the "ORD" and the "CD" has too many applications due to the fact that the spectrum of "ORD" and the "CD" of proteins and nucleic acids. The resulting from the spatial asymmetry of the components of the amino acids and nucleotides sequentially, but for the macroscopic particles it states each of the "ORD" and the "CD" in the structural studies of proteins and nucleic acids and proteins.

The pens and solids are used for this purpose at times, but the solutions are used in most of the time measurements for the "ORD" and the "CD".

Solution is placed in a container called a cell, and the device consists of the light source changes then the wavelength, and a system for polarization of the light, and system for measuring the polarization after the passage of light through the cell.

That the way to know the secondary structure of protein based on the measurement of the curves of CD, ORD experimental multi- peptide. As for the proteins they include measurements of the three main forms, alpha helix, beta form and the coil random. It was found that the spectrum of proteins and the type that gives the same due to the impact of side chains on the rotary power by force, as well as peptide that occurs at times because of the hydrogen between two amino acids and peptides that have multiple heterogeneous long chain of the same strength turnover in each component, such as those owned by small cascade. The side chains of phenylalanine and tyrosine and histidine and tryptophan to the spectrum of CD when they are in certain situations and the sulphide bridges give two CD bands.

Recent developments in the application of circular birefringence CD and optical rotation dispersion ORD advantage of these modern methods to know the status of the molecule or macroscopic and the interactions between them, as well as the structure of poly peptide and it is believed that the key of the structural knowledge of the secondary structure of protein based on the ORD curves and the CD. Furthermore, it is believed to use the test of changes on the status of the proteins by CD and study the changes in the structure of enzymes caused by the substrate, inhibiters that are related to enzymes, which can be illustrated by the CD spectrum for a variety of enzymes when the enzymes interact with the substrate and inhibitors of enzymes to help, as accompanying the metamorphosis of protein changes in the CD due to the loss of the structure of the alpha and beta and increase the spectrum of the components of the coil at random.

**Nuclear Magnetic Resanace**

NMR magnet is a spectral method that can provide sufficient information about the structure of bio-multicellular molecules and the interactions that occur between molecules, as well as molecular motion.

NMR spectrum depends on:

- Displacement of chemical "Chemical Shift".
- Fixed double "Coupling Constant".

Accordingly it used for the following applications:

- Diagnosis of chemical structures.
- Theoretical studies on the chemical tautomerism of displacement.
- Studies on the tautomerism composition.
- The dynamic characterization of chemicals.
- Characterization of the spatial structures of chemicals.
- Effect of solvents.

Technological developments of nuclear magnetic resonance technical developments have taken place in nuclear magnetic resonance study after the hydrogen nuclei (protons) and the nucleus of fluorine with high sensitivity and where the introduced technology transfer. Fourer the use of flashes of measuring the sensitivity of nuclei such as carbon -13 Bio- molecule such as proteins require nuclear magnetic resonance equipment with high frequency 300Hz and more, where it is difficult to analyze a device at low frequency (60Hz) or average (220Hz). New equipments of nuclear magnetic resonance frequency of

600Hz, these machines analyze many of the protein molecules and those of other techniques with the use of two-dimensional and three- dimensions.

New devices of NMR manufactured magnet after the entry of two-dimensional superconducting magnets, where the magnet save at -270 degree using liquid helium and liquid nitrogen to maintain the temperature of the helium and prevent it from evaporation. To distinguish between one dimensions and two notes that the spectrum of NMR magnet with a one- dimensional recorded displacement and pairing constant on the same axis while the two dimensional displacement is recorded the on two different directions. The three-dimensional recording the two perpendicular to each other and are getting a nice Dox- spectrum. The most important uses of the latest is in the field of proteins.

**Applications of NMR spectrum**

NMH magnet is a spectral method that can provide sufficient information about the structure of molecules and macroscopic bio interactions that occur between molecules, which can calculate the arrangement of atoms in the spectrum, and the hydrogen atoms (difficult in a segregation analysis of X-ray diffraction can be identified on site by this spectrum), It is spread in the macroscopic particles, can also test different atoms (phosphorus, carbon, nitrogen and hydrogen) separately. These can be applied in determining the spectrum of protein structure and enzymatic study of active centers and link small molecules to proteins.

In general, there is a difficulty in analyzing the macroscopic particles, the presence of large numbers of lines of spectrum analysis, which distinguishes itself lead to difficult to diagnose due to the large numbers of potential tackles of each atom, where the border of the spectrum characterized by complexity theory. As well as would be possible to study chemical reactions and the effect of drugs on the disease and to study the changes that occur to the water molecules in living cells.

There is great potential for the study of chemical reactions using a spectrum of protein and phosphorus in living cells has made a technical developments in this study. The nuclei of other important life such as sodium, potassium, calcium, iron, cobalt and nitrogen. Examples of phosphorus- use spectral follow- up of the heart and life processes within the heart and to stop the interactions of life organic phosphates components as evidence of changes related to energy. The specialists in the field of NMR spectrum nucleus of sodium, potassium and cesium due to the presence of sodium and potassium in the human body and play significant roles in the interactions that occur in the cell and pressure in the maintenance of fluid balance inside and outside the cells and the role of sodium in the transmission of nerve signals through the cells and the neurological contract and it can be completed as well as there are free and linked and contribute to solutions in the process of blood clotting and visual processes and muscle contraction, and here we began to record NMR spectrum of the nuclei of calcium and magnesium as well as nucleus of silicon, tin, nickel and other industrial significance.

It is played a great interest in oxygen -17 and carbon -13 as well as which gives information on the spatial structures and the use of nuclear magnetic resonance technique to the nucleus of carbon that is determined and these structures in detail. As for the oil industry played important roles in this technique and to identify the percentage of items to articles in different sections of asphalt, as well as technology that helps NMR magnet to determine the oil compound and petroleum products after the cracker. Also used this issue to determine the quantities of oil in different kinds of grain linked to liquid plant cells without crushing and grinding grain.

The techniques of NMR magnet in the other areas are numerous and to identify the proportion of oils and fats, saturated fats in commercial fish, and set fat dairy products, cocoa and coffee industry and set the percentage of alcohol in alcoholic beverages. As well as to set the chemical structures of pesticides and chemicals in dyes and determine the toxicity and a lot of industrial materials. In the pharmaceutical industry it has been used a lot for confirmation of the purity of drugs that produced and the appointment of any impurities in pharmaceuticals.

A nuclear magnetic resonance other spectral method that can provide sufficient information about the structure of polymers and living on the interactions that occur between molecules as well as on molecular movement. The multiplicity of benefits and return to:

- The possibility of calculating the order of the atoms of the NMR spectrum because it theoretically can provide information for the purpose of this account.
- The hydrogen atoms (difficult in a resolution of analysis of X-rays), but it is possible to determine its location by magnetic resonance image.

Nuclear magnet line up in parallel magnetic field and the direction of high-energy or nuclear magnet lines up opposite field and can make a transition from two energy absorption quantity and energy of electromagnetic radiation and the appropriate energy to the heart of the nuclear magnet equipped with radio frequency range if we use the magnitude of around 10,000 chaos, but we see the resonance signals that nuclear magnetic protons that requires: Posted regular magnetic and the receiver of radio frequency. In practice, the necessary changes in the magnetic field bring the protons in the majority of chemical structures to the resonance is ten parts per million only. To changes in the magnetic properties of some nuclei, particularly hydrogen nuclei (protons) and other nuclei such as $C^{13}$, $P^{31}$, $F^{19}$,$N^{15}$ and can be identified, as well as calculate the number of hydrogen in conditions of various electronic and more clear that can detect changes in the direction of nuclear in strong magnetic fields.

And can the chose in the different atoms (phosphorus, carbon, nitrogen, hydrogen) separately. And nuclear magnetic be very successful to determine the structure, when the use of small molecules. But when using macromolecules, it is difficult to identify large numbers of lines of the spectrum, as well as analytical distinguishing itself leading to the difficulty of diagnosing produced because of the line, but among the range of possibilities for every atom of tackles. Using nuclear magnetic resonance, which will be important in the future to help and solve the problems of chemical and will be the difference between

NMR and other methods when studying the macroscopic particles, where also be clarified the limited NMR characterized by complexity in the theory and devices that have been noted that many of these device had improved..

Nuclear magnetic requires resonance spectroscopy of the radio waves of the circular disk of the magnetic field and the absorption of radio waves recorded. The sample in the tube between the B- polarization of the magnet A load (104- 105 Chaos) describes the spiral shape in a plane perpendicular to the electric field of the magnet that surrounds the sample. The sender broadcasts reluctant high and constant (roughly 108 or circuits per second).

And the others in the magnet or in the frequency of radio waves in part cause no change in magnet strength, either the change in the frequency of radio waves.

Nuclear magnetic resonance capability to provide much information about protein structure, where the constants of the spectrum sensitive to changes in both the arrangement and status, for example, in the absence of any fusion or federation is a complex spectrum of peptide or protein which is the sum of the spectrum of protein components, which are amino acids.

Spectrum indicates the enzyme in natural forms and (random coil), which can be calculated from the amino acids of the protein, it is clear that there are significant differences between the spectrum which is calculated so that, it is noticeable, for example, noted the occurrence of poles that occurs when the large chemical proton in amino acid. But in the end, or carboxylic acids, or when the neighboring nitrogen atom to the bond, also a small composed displacement, when the proton adjacent to the carbon atom of the peptide bond, or carbon or nitrogen atom of the amino acid closest 2, also generated a small displacement, when carbon or nitrogen atom in the amino acid closest, in addition to this, when the protein status of normal displacement, spoke of the distinct chemical protons some acids in the form of security.

**F.Diagnostic Imaging**

In medical diagnosis it is adopted mainly on the knowledge of diagnostic imaging technology spectrum, including the use of X-rays and gamma rays from, which is characterized by being electromagnetic radiation ionizing radiation, then began to think about using the term of this non- ionizing radiation infrared or microwave radiation and technical NMR magnet. The examples of spectral techniques used in diagnostic imaging:

- X- rays
- Gamma- ray
- Ultrasound
- Infrared
- Anti electric tissue
- Visual mechanisms

**X-ray**

The oldest techniques that is used in diagnosis and therefore will not focus on the importance of being where they were getting on the first pictorial representation of various tissues obtain after the development that is built on a limited computer assistance.

**Gamma rays**

The purpose of gamma- ray is the imaging profile then it was developed as computer- assisted also in the eighties which was called "ECT" and was then developed using imaging "Postiron emission tomoyraphy (PET)" where the radiation of tissue is carried out by position (positively charged) and thus can get a picture to clarify the life processes of the tissues that carry electrons and draw.

**Ultrasonography**

The speed of these waves are characterized by being less of electromagnetic waves, which provides an opportunity to measure the fetus as well as during the stages of development in the womb, added to that the fact that this technique is based on the fact that the X- ray is not ionized therefore it is not a preferred use in diagnostic imaging.

**Nuclear magnetic resonance imaging**

Despite this technology it is old, but it was then developed for the purpose of medical diagnostic imaging has gone from the seventies, where the nuclei of atoms is measured by the disposal of certain substances found in different body tissues. The criterion for the disposal of these seizures depends on the radio pulses that are similar to the frequency in the field of outer- core magnet and thus to obtain a diagnostic can be used.

In the medical applications for the purification of nuclear magnetic resonance imaging to obtain imagery of infarction that occurs in some parts of the brain and important developments in this area the integration of multiple techniques and access to advanced apparatus for nuclear resonance imaging, including the "TMR" and "MRI".

It is important experiments that experiments are used the magnet resonance imaging of kidney transplantation, which was filmed nearby parts of the kidney and then infected the interactions that take place within the body after transplantation and efficiency of the cultivated parts. As well as imaging of tumors within the liver and liver imaging at the time of myocardial fibrosis or within, as possible, filming parts of the stomach and colon and to identify tumors. It was also to obtain information about stroke and is believed to imagery obtained of cancerous tumors of the brain were more pronounced than the use of X- ray.

It can be measured by any inflation occurs as a result of heart disease, and can also study the problems of the heart due to the presence of any obstruction or infarction in one of the blood vessels and could also portray the evolution of stroke, heart attack and its impact on the heart.

**A nuclear magnetic resonance imaging "MRI"**

This device is used which was created as a result of the development in the technology of magnetic resonance spectrum by the registration of spectra of life processes taking place within the animal body where the magnet- making with full slot by placing the human within the magnet and thus these devices provide a complete picture of the part which is conceived, and the advantage of the fact that this device magnetic field is not harmful, and the microwave radiation used is not harmful too.

It is possible through this device to study the effects of ongoing parts of the human body while taking a particular medicine can also be follow-up of the various core elements and sequentially, as well as to study the changes occurring stereoisomers of chemicals inside the cell as a result with other molecules.

**New Technologies Electron Microscope**

Electron microscope is using a torrent or stream of electrons, where the wavelength is too shoat then we can get on the ability of the analysis is very high. Extent of segregation (analysis) of the optical microscope and is an estimate of 2000 and this is not enough to see parts of the cell, viruses, and macroscopic particles, but the use of electron microscope segregation less than the uranium atom (in approximate) in special circumstances. It is clear that electron microscopes large and complex and expensive operation that is similar to the foundations of the optical microscope, which reveal the sale electrons emanating from the source mail (metallic thread with a high degree preheated in vacuum) and reveals extensive by lenses and lens-body grows electromagnetic diffraction and finally drop the image as by the final lens of the projector.

To see the image on the screen is up brilliantly by the lens or can be scanned to imagine, as we mentioned earlier, the high segregation ability of this microscope to enable the researcher to view more details when you enlarge the optical microscope, the exact address. When the materials to be examined too thick for the passage of electrons then therefore requires the creation of a thin section, and it must be the sample used for the purpose of this solid and cut easily.

**G.Labeling with radioactivity**

Require a lot of chemical analysis revealed small amounts of material with amount of concentrations $10^{-4}$ – $10^{-6}$ molari therefore it requires the development of other ways to respond to the concentration of low- lying, such as the development of experimental methods by radioactive to solve many of the other problems that might face them. Some of these methods that could be used by dual- labeling for follow-up of two similar materials formed at various times by pulse method for follow- up fugitive substance at a time after the configuration without interference of other material. An example is the use of radioactive materials in the chemistry of life:

- Choose a material that resides on small concentrations, which are difficult to measure by direct chemical methods.
- Distinguishing similar molecules in different chemical sites.
- Analysis of mixtures that are very complex, which can not be done by various conventional chemical methods. Including:
  - Enzyme interactions (DNA polymerase).
  - Measurement of molecular weight of the DNA by labeling the final group.
  - Diagnosis particle by settling with the anti body.
  - Protein purification, which does not have a chemical test.

The isotopic properties will make the labeled compound more easily identifiable. For example, the radioiodine – labeled thyroxine molecules can be identified and quantified easily by virtue of their radioactivity .

The use of isotopes, both stable and radioactive, has proved great body of information in the medical scince. Stable isotopes are non radioactive and are suitable for use as tracers in humans. Especially infants children and pregnant women, stable isotopes have also been used in the quantative analysis of various substances in recent years.

Radioactivity measurements depends on the ability of radionuclides to produce ionized or excited atoms within the detector . Two basic types of radiation detectors are in common use: gas ionization ans scintillation .Radioisotopes allow the detection of minute quantities and differenate physically between substances.

The use of radioactive isotopes in biochemistry and clinical chemistry has proved us with a wealth of information about biological processes, that offers such as adiverse range of applications, using enzyme assays, biochemical pathways of synthesis and degradation, analysis of biomolecules, measurement of antibodies, binding and transport studies .

The use of radionuclide in nuclear medicine began when Frederick proescher published the paper entitled the use of radium for therapy of various diseases (1). Early experimental and diagnostic applications were performed with naturally occurring radionuclides, then the radioisotope with physical short half loves have become increasingly popular for imaging applications .

The first commercially available radioisotope generator was the $^{132}$Tc-$^{132}$I, (5) several other generators (such as Mo-$^{99m}$Tc, $^{68}$GE-68Ga, $^{113}$Sn-$^{113}$In, $^{87}$Y-$^{87}$mSr….etc) subsequently evolved . These generators must meet certain physical basic criteria to be useful. It should. It should be simple and convenient to operate, radiation must be adequately shielded------- yield adaughter product of high purity in terms of both radioactivity stable contaminants during every clution throughout the life of the generator, the product should be in a chemical form suitable for use with amininmum of additional chemical or physical manipulation, lastly the radioactive yield of the daughter product during each elution should be high. Labeled compounds either be used in biochemical research --- routine medical diagnosis were carried out in vivo for medical diagnosis such as those labeled with gamma emitting isotopes to permit detection external to the patient, (12) but those labeled with beta-emitting isotopes such as: $^{14}C$, $^{3}H$, $^{35}S$ and $^{32}P$ were principally used in biochemical research .

There are various methods which were used for preparing labeled compounds such as of the followings:

1- isotope exchange reactions, in which one or more atoms in the molecule, exchange with atoms of the same element and of different mass, these atoms may be radioactive or stable isotopes, according to the following:

$$AX^* + BX \rightarrow BX^* + AX$$

The compound BX under certain reaction conditions will exchange its X atom (s) with the compound AX* where X* atom (s) is an isotopic form of the element X. awide range of compounds labeled with different stable or radioactive isotopes are prepared by exchange methods, which have the advantage that they can normally be carred out on a small chemical scale. An example is the preparation of urea C, .

$$CO(NH_2)_2 + {}^{14}CO_2 \rightarrow {}^{14}CO(NH_2)_2 + CO$$

2- chemical synthesis in volves the construction of complex moleculry from simple isotopically labeled intermediates, yields are usually expre as a percentage radiochemical yield.

For example, the preparation of carboxyl-labelled fatty acids by reaction with the corresponding grignard reagent or acetic anhydride- $^{14}C$, steroids $^{14}C$ and amino acids-$^{14}C$ .

3- biochemical methods: these include different procedures such as enzymatic synthesis which isvery similar to chemical synthesis in that such aconversion usually occurs without any change in the specificity of the labeling or the molar specific activity (16). Total biosynthetic methods are normally of value only when microorganisms are employed, but the production of uniformly labeled carbohydrates by photosynthesis in detached leaves is an exception to this .

4- recoil labeling this method depend on the ability of recoil atom produced in a nuclear reaction to form a stable bond with an organic (or an inorganic) compound. For example, if an organic compound is mixed with a lithium carbonate or chloride and irradiated in anuclear reactor at fixed neutron flux, tritium compounds are produced by the recoiling "tritons" from the nuclear reaction $^{6}Li(n, \alpha)^{3}H$

One of the most radioisotopes used in clinical application in both cases as pure radioisotope or in labeled compounds is technetium-99m, which have a short half-life about six hours, with a predominate single photon gama emission having an energy of 140 kev . $^{99m}Tc$-labelled compounds are diagnostic imaging agents used in the field of unclear medicine to visualize tissue anatomical structures and metabolic disorders. After interavenous administration $^{99m}Tc$ or it labeled compounds localized in specific target organ or tissue, can then be imaged using stable instrument .

## TECHNETIUM CHEMISTRY

Technetium is not a naturally abundant element, some of its properties were produced by mendeleev in 1869, who called it ekamangabese and gave it the symbol (EM) . After world war II, Perrier and segre gave element 43 the name technetium as the first artificial element .

To give rise to multiple oxidation states an forms coordination complexes with avariety of inorganic and organic ligands . The chemistry of Tc in its I-V oxidation states was sureveyd by davison and jones, many of the thermodynamically stable Tc-complexes have oxygen bound to it like TcO4, $TcO_2.2H_2O$, Tc-O-Ligand (27-29). Tc V and IV complexes are knwn to have the ligand coordinated to the Tc alone, i.e Tc, $Tc_2Ln$....etc, the polyncuclear formation of Tc species is usually minimized by keeping the Tc-concentration as low as possible, to prevent the formation of Tc complexes with more than one oxidation state, efforts are made to control the reducing agent and the reaction conditions,.

**H.Autoradiography**

This method is used to detect and locate radioactive materials in the cells or tissue for example, and so the molecule itself and is done by the impact of radiation emanating from radioactive materials or emulsions of photographic plates specially designed for radiation imaging device self-motivate, where silver halides grains, located in the emulsion as a result of the dissolution of radioactive materials in the sample, and the emission of radiation, including activation and work output reduction as indicators minutes for the site radiological effectiveness. And signaling models resulting from the grains chemically and radiation efficiency in the presence of structures that are in contact with these granules and the microscope can be obtained from the resulting image on the two types of information at the site of radioactive materials and the quantity of a radiation of as the amount of silver particles is directly proportional to the severity of radiation present.

Of the modern applications of this technology as follows:

- Measurement the number of molecules of DNA in bacteria phage.
- Measurement the number of secondary units of the chromosomes.
- Double vision in the DNA molecule of bacteria.

**I.Membrane filtration and screening "Membrane Filtration and dialysis"**

(Like cheesecloth that has been used to separate the serum from the leaky). The cheesecloth filter extracts used in textiles. Then use the cards instead of porous fabrics for the purpose of controlling the size of chips and then create filters made up of cells, or glass yarn, either the softest materials they include sorting tubes membrane, which allows the passage of small molecules and ions. But keep the particles and macroscopic aggregates macroscopic particles. It is called the membrane tubes stitches the contrary, and that have the ability to separate the macroscopic particles from small.

### J.Protein engineering

It is the technique that allows the installation of structural proteins desired in order to build a clone- mediated DNA "Cloned DNA". There is no relationship between the latter and engineering of proteins used, including the building of protein functionally, chemically and physically.

The DNA could be modified by two ways using:

- Mutagenic in private venues.
- Switch sections of the nucleotides.

The protein engineering include modify the structure with protein mediated by genetic engineering and most protein engineering is carried out currently in the field of enzymes, either to speed up its response to the incentive or to become more receptive to acid and heat.

Example: "Cloning" the cDNA for the receptor of "acetyl choline receptor" facilitated the technology which is called site directed mutagensis for getting sequences skilled "Deletions" or substituting some of the amino acids in an additional unit "subunits" of the receptor and then it can test these changes on the functional aspect, and are also defined as follows:

There are many examples of this type of modification for production of complex of organic compound that have catalytic activity have of it chemically synthesized for example the myoglobin of which associated with oxygen, but it docs not have catalytic activity. This Bio- molecule with three complexes of ruthenium "ruthenium" carrier of the electron through the surface of the histidines components generate a complex that has the ability to reduce oxygen and the oxidation of the natural ascorbate.

The construction of DNA contributed significantly to the development to the stage of protein engineering to construct proteins that do not exist in nature. The technique has evolved to the point can modifies the gene by an engineering to change the protein in a predictable and have to improve some functional characteristics such as:

- No. transformation "turnover number".
- Static Km of substrate specific.
- Thermostability.
- temperature optimum.
- Stability and activity in non-aqueous solvents.
- Privacy of interaction and substrate "Specificity".
- Requirements of co- factors.
- Protease resistance.
- Allosteric regulation.
- Molecular weight and composition of the structural unit "Sub- unit structure".

And for engineering the protein molecule, it is clearly necessary to ensure a series of rules relating to major synthetic building blocks of proteins that recipe as desired. After seeing the structural composition of protein crystals, it is then possible to diagnose those areas in which it occurs possible modifications to improve the catalytic molecule, protein, and this is done to modify the sequence of amino acids in the protein.

**Major modifications protein**

The use of site- directed mutagenesis determined then what is aimed to, because the change in one base in the gene result in a change in the sequence of amino acids in the protein, which in turn improve the protein in question. Large modifications in proteins by removing the "delete" section mediated by enzymes or by the unequivocal chemical structure of part of the gene. In this way, the production of spare "klenow fragment" "DNA polymerase" free of analytical activity, also can add sequence of amino acids through docking to improve the stability of proteins made in E. coli and finally can collect or part of a fusion gene or the whole of all or part of the other, thereby generating new proteins.

Determination the general features of the installation of the structural protein. Protein engineering based on the availability of information on the district and synthetic building blocks that are obtained from the methods of X-ray diffraction and nuclear magnetic resonance two- way "Two dimensiona nuclear magnetic resonance NMR" and the latter is the alternative method in the future. Many researchers expect success in engineering of proteins "Protein Engineering", especially after the great progress which has been in embryonic technique, where each protein is produced by genetic conditions of its own machine of the cell consisting of enzymes when they become three characters of the genetic material and arranged in advance and checked that then wrap as a specimen to be specific proteins effectively.

When you know the rules that allow the protein to form belts wrapped can then change the genetic information of proteins and identified so that it works in another way as soon as a large and powerful grants stability, and thus can benefit economically from the proteins of the broad areas of application by micro- organisms and can be more clear: for example, improved production of proteins (new physical properties and functional).

Important notes that are related to protein engineering is to clarify the potential relationship of proteins, where the protein for example, a specimen 15-amino acid. There are 103× 3 possible sequence of these acids is larger than the number of atoms that make up technical enzymes immibolized onboard, the development of these enzymes are restricted or limited to a solid surface to be in constant contact with the foundation to which the article in the mobile phase "mobile phase". It is clear from this that there is a possibility to use the many pathways that retains its effectiveness.

Technical features of immibolized enzymes"

- Prevent the entry of the immibolized enzyme in the mobile phase.
- The product is characterized by being cleansed of the enzyme and does not accumulate.

- Using the enzyme for long.

The globe, despite the lack of clear understanding of the rules that govern protein engineering, but the equipment contribute to give some suggestions on how to achieve a stereo structure of the protein. In this area one can not expect for example bacterial cell to produce human that differs in form of human protein.

**K.Immobilibzed enzyme technology**

At present, there are important industrial applications of immibolized enzyme technology represented by the following enzymes:

- Glucose isomerase.
- Aminoacylase.
- Penicillin acylase.
- Lactase.

The latter has been "Immobilized" on the particles of silica. It is used to convert the lactose in whey to glucose and galactose.

Applications to include of immibolized in the future as follows:

- Use enzyme "Cholinesterase" for the purpose of pesticide detection "Pesticides" and watching the inhibition of this enzyme either by the method of electrical "Calorimetrically electrochemical" or by the color method.
- Other enzymes that may be used in the same method in order to detect toxic chemicals, the enzymc "Carbonic anhydrase" is very sensitive to low concentrations of chlorinated hydrocarbons from low- lying "Chlorinated hydrocarbon" and "Hexokinase" to "Chlordane".
- Immobilized diisopropyl phosphor fluoridate extracted from the nerve cells.

**General aspects of enzymes immibolization**

This process is intended as we mention it to determine kinetics of enzymes, as yell as cells that characterized by (desorption) on the surface such as fibers gels, etc., also can be used as phenomenon shooting accordingly.

Advantages of the immiblization process are the followings:

- Finding the status of enzymes similar to those found within cells and tissues.
- Prolonging the period of use and has repeatedly given to the survival of catalytic activity and stability.
- Use appropriate concentrations and may be high for the purpose of increasing the speed of the reaction, given the focus to fit with the speed in specific circumstances of the reaction.
- Contributing of the immibolization process to facilitate the purification process of related to products of reaction.
- The use of multiple systems from the fermentation (continuous and open).
- Reducing energy consumption and cost.

The immibolization methods are numerous, including:

- Chemical methods: they are similar to affinity chromatography such as use the covalent and casual.
- Physical methods: such as packaging inside a capsule adsorption and shooting.

As for choosing the appropriate method to be immibolized are determined according to the specific bases represented by measurement of activity, stability, so it must be taken into account the business side that is, have used with less expensive. And choose the easiest method because they are all tough and stay away from hazardous substances to human health, and the technical side is important in the selection process since there is a special mechanical pressure during the operation.

The immibolization cells vary from cell since it is being more of enzymatic system builders with the installation of diverse chemical content, therefore, requires that the appropriate modalities, simple and stay away from these that require to use extreme circumstances. It also requires that to taken into account the number of cells to be immibolized so the method must be a convenient and linking cells are good and avoid the use of hazardous materials. The characters of the immibolized cells are numerous advantages including the use of small amounts of carbon and energy sources and re- use of cells, so it is possible separate the growth phase from production phase, where it is possible control the fermentation. Immibolization depends on the type of cells, microbial cell reduce the size of the manufacturing process and thus reduce the cost of the production process. The Eukaryotic cells which are characterized as specialized capable of limited division of which are specific plant or animal cells and preferred to be immibolized, particularly those that are separated as any single and are generally used for the purpose of the immiboliaztion of adsorbed on the hollow fiber.

**Enzyme Technology**

The biotechnology is considered as one of the technical life in science and engineering. It was one of the enzymatic technology trends that have grown with the technology of life, despite being preceded by technical life, keeping in mind that enzymes from an engineering standpoint is a special case of the factors that have qualities such as privacy.

Bio- systems are used in critical periods in history to get the desired chemical conversions such as transformations of like milk to cheese and fermenting of liquids that contain sugar to alcohol, but such research trends have changed during the evolution of Biotechnology with the fact that these processes such as cheese, bread and alcohol industry still very important.

The history of enzymatic techniques started with the developments that have emerged a number of chemical transformations using the tissue of life, which include, for example hydrogen peroxide decomposition and degradation of starch to sugar and digestion of proteins.

**L.Human Genome Project**

The initiative of (HGP) came the first time in 1988 and aimed at finding the sites of some 100000 human gene in DNA and the (HGP) expresses 24 pairs of the human of chromosomes, is turning into information content when it follow the sequence of rules need to be resolved based on computer science, mathematics, statistics and experimental sciences. Note that the computer science often provided in contributions in programs and solutions that are characterized by the skills that led to the invention of language access code information described the performs a particular order and provides methods to describe complex biological processes by the number of code rather than their natural language with hundreds of pages. Then the researchers and observers said that the twentieth century be the century of biology and analytical power resulting from the HGP that will explain drastically all life and medical research as it was:

- Research on the nature of genomes and the nature of the composition and organization of various scientific institutions.
- Acceleration in the implementation of the project, which was planned to complete within 15 years of technical progress, but then shortened the time to ten years.
- The search of the genome to find the type of information or material contained in the communication as well as identifying the sequence about three billion chemical bases.
- Study the nature of the information stored in the computer and the evolution of elaborate by efficient techniques of and sequence evolution in the tools that contribute to the analysis of information.

- Study the effects to be set in the community and to what extent this can be achieved, and the type of response.

In spite of all reported studies and research conducted by methods and techniques in various vital information as well as numerous writings and published in this area, there are still many other fields and various study and research. Some of these fields has not been touched so far in the country, especially the human genome projects, and areas to attract the attention of researchers, but it's mostly a few problems, mostly dealing with partial or subsidiary. **Genome**

The human chromosomes is 46 consisting of strips of the double helical DNA wrapped circumvent complex shapes of helix, normal and high and consists of the DNA with four units of high repeated synthetic (nucleotides incomplete oxygen), each of which consists of three components: nitrogenous base and sugar phosphate penta and not organic.

There are four types of nitrogenous bases and symbolized by TCGA arranged in pairs along the stretch of DNA and the numbers of secondary units in the DNA molecule. Approximately $3 \times 10^7$ base pairs in each of the cells of the human body, and the length of DNA equal $8 \times 10^3$ times the distance between the earth and the moon and bigger than the distance between the earth and the sun 300 times (the length of all the DNA, In the body $2 \times 10^{10}$ km).

The gene (one gene) constitutes a piece of DNA it consists of a large number of secondary units and in the molecular weight of the gene is $600 \times 10^3$ and is a very long string of four characters, and each character represents a nitrogenous base. There are usually genes in the nucleus of the cell and molecular genetic consists of two bands are linked together by special bonds, each other on some twisting spiral and there is peace on the same wrapped that consists of a sequence of nitrogenous bases, or nucleotides that contain the bases arranged in a manner different from the gene to another and then discriminate organism from the other because of all the genes governing cell functions, guidance on ways making a specific protein or another compound with medical importance, hence a single gene responsible for the general one recipe and therefore we find that the qualities beauty, shapes and colors that each one of them result of a single gene or the number of genes. Genes are transmitted from parents to offspring by mating the structural change in terms of affected and then the subsequent processes of making many of the compound causing the disease, which may be cancerous or always defect organisms.

The genome is intended to aggregate the DNA (genes) of the bacterial cell example, contain about 200- 300 gene, while the genome of a human cell include the thousand times as much as the genes found in bacterial cell 200.000 to 300.000 and the organization of these genes depends on the number, in the chromosomes of the cell Eukaryotic (human cell) is more complex than the primitive cell nucleus (bacteria).

The genetic map represents the order of genes (genes) within the cell chromosomes and that this arrangement within the human chromosomes is more complicated than other organisms. Thus, the process of discovering how to arrange these genes, given the sheer number and complexity associated with variation built and responsibility to control complex cellular functions. Hence the decoding process and diagnosis and scheduling of full human genome by genetic map and the preliminary draft of a preliminary genetic blueprint of human genes and the previous process is equivalent to a significant scientific breakthrough, scientific achievements made during the twentieth century, including the discovery of penicillin and landing on the lunar surface and use a computer and other discoveries. And according to this perception announced 26.6.2000, the end of the main phase of the Human Genome Project, which represents the first achievement in the twenty- century atheist and the development of the draft map is almost complete and a preliminary blueprint for a human gene content of the human genome and was named the human genome. It had been prepared jointly by both the research centers m the United States, Britain, Japan, France, Germany, China and other countries with long experience in genetics and genetic engineering, funded by 18 countries.

**Specifications of the human genome (Genetic human map)**

This map is characterized thoroughly without gaps up 99.9% and 97% of the components of the human genome, has been decoded and 85% of the sequence and gene order has been tabulated and analyzed. The rest of the map requires additional time to accomplish. The map we have opened a new era of molecular deal with the situation of life, has made it clear that in a manner distinct and different, Dr. Ahmed Zewail Nobel Laureate in nanotechnology, said that the molecules are arranged genes that act in the movement of one can not detect them, but that can be pursued with sophisticated and sensitive to femto second.

**Futurism of human genetic**

The discovery of the human genome and complete the approximate locations and sequencing of this large number of genes input to future developments are to:

- A new look for the human body.
- To find new ways of treating diseases (gene therapy developed) such as AIDS, cancer and heart disease.
- Correcting the genetic errors.
- Organ transplantation.
- To address the social and ethical consequences.
- Sustaining life.

The battle against cancer, AIDS and other incurable diseases are ready to locate the responsible genes for these diseases that have discovered the map and according to that can solve the problems of treating these chronic diseases and decrypt secrets, and the negative aspects of this discovery is immoral exploitation of the future such as racial discrimination in accordance with the genetic composition and control of human qualities and it requires the issuance of special legislation related to human rights and prevent the future destruction of rights (death of persons with disabilities and life of the insane, for example).

## Gene therapy

Thousands of diseases due to the presence of genes responsible for the appearance and many of them dangerous to humans and non- treatment or cure and concerns and applications of biotechnology to find what is known as gene therapy, which is either by bringing the damaged gene or gene intact repair defective gene. This could be done through the intervention to repair the gene in somatic cells, or by intervening in the cell construction.

In order to spread the use of gene therapy it must be certain that the expiry date and free from damage and researchers that must be able to transport techniques and control gene expression in the correct and consistent, and should not obscure the international success of the many risks carried by this treatment.

- The genetic balance of any human being is the only thing that can not be replaced but must be preserved and transferred to the generations of while it is possible.
- Here we must make sure that it can allow gene therapy in somatic cells, and prevention must be in the manipulation because of its many negative consequences both in terms of genetic or moral.
- Gene therapy in somatic cell only affects the individual patient treat him, while affecting.
- Gene therapy in cells on the construction of successive generations.

## Recent progress in the field of gene therapy

It has become possible to do some practice in the field of genetic medicine with the evolution of technology DNA "Recombinant DNA" as has been addressed most of the problems related to the production and disposal of genes and to consider their ability to modify objects, and the laboratory tests on animals proved that non- genetic, genetic medicine can be successful. It has been treated several human cells in inherited tissue laboratory for the use of retroviral vector. And proved the possibility of a peace process, the introduction of white blood that had been genetically engineered in the patient. It was also carried out several recent clinical tests for human genetic medicine, culminating in successfully treating a patient in the loss of immune complex advanced emergency resulting from the adenosine.

It seems that gene therapy is the only way to cure genetic diseases or chronic (such as cancer, Acquired Immune Deficiency Syndrome). In this case, there is an objective one is to improve the health status or save the life of a patient work of the highly desirable, and then there is no difference between the unit body and the unity of the genes. Gene therapy holds great danger is the use of this technology in order to improve the human race; can we change the balance of genetic risk for the human species? This may have dire consequences especially on the reduction of formal diversity. This method, which promises also carry the hopes of many fears. Do we have verified that all the risks in the long run, where would we be? What are our borders? Where the border between the correction of what eugenics? The achievement of this modern way of thinking must be accompanied by a profound and moral debate.

**General characteristics of the human gene therapy**

It was discovered more than two thousand genetic disease; all affect the genetic information in the patient and move it to the next generation. In order to restore the natural functions of the victim there are two ways in gene therapy.

- Gene therapy in somatic cell.
- Gene therapy in the cell construction.

In each of these methods a special set of scientific and ethical considerations. Construction cells are sperm cells and egg cells, which include the rest of the cells in vivo somatic cells. The gene therapy of somatic cells in the introduction of DNA in this type of cells so the added gene in the progeny of the patient on the contrary, affects gene therapy in cells on the construction rights in the early stages of embryo development.

## Gene therapy in somatic cell

Before embarking on any attempt of human gene therapy it must first determine the exact mutation that leads to a particular disease. This information is not available currently, but in terms of a small number of diseases, but current advances in genetic engineering point of what will happen soon on detailed analysis of the genetic and second it should be identified on the type of mutation affected the cell in the body and their genetic transformation. Only current clinical trials to treat somatic cells, which are based on the introduction of a gene in somatic cells of a small child or a young man. Thus, cells not exposed to structural change, which prevents the transmission of the gene to offspring and this, is something which makes having a person who benefited from the treatment is always vulnerable to this disease. The structural gene therapy did not apply to humans, as it was rejected on moral grounds. But a team of researchers are currently considering the possibility of its application in the treatment of incurable genetic disease and it comes in all cases. The treatment depends on the addendum, any patient that the gene dysfunction and genetic causes of the disease will not heal or replaced, but added to the cell intact copy of it and this method of treatment do not apply except in cases of genetic diseases caused by genes elected. In general, gene expression is not valid only in a given tissue can be determined in different ways in experiments conducted on the living body, determines the quality of the input method the target tissue. Incomes through the trachea transfer of genes into the pulmonary epithelium, and injection in the liver gene transfer to liver tissue, either tumor injection in the objects is transferred genes into tumor cells also contributes to the carrier of the virus in determining the target tissue.

## Treatment in the cells construction

Cells contribute to construction in the genetic heritage of successive generations. Gene therapy through the cell affect the construction of genetic stock to his descendants too, and then the sum of the genetic traits of humanity. The majority of scientists that may not be morally any attempts of this kind of treatment, another group believes that the gene therapy in the cell construction is the only way to eliminate genetic diseases suffered by millions of people.

**Different methods of gene transfer**

- Viruses: Retro virus, including the only to ensure that the genes transmitted via cellular divisions. We therefore consider these viruses are the most successful means of gene transfer in the laboratory, where they allow in principle a final treatment of genetic diseases.
- Chemical methods: There are numerous studies on the possible use of fat bodies and compounds multi positively charged.
- Physical methods: The target cells is ejected shells are small technisten DNA speed due to electrical discharge or explosion compressed gas.
- Intramuscular injection: Move DNA the intentions of cells around the injection site but does not merge with it, but stay for extending from a few weeks to a few months to form ring.

To deliver a specific gene or gene fragment to a cell, you should test the appropriate carrier depending on the cell type and the type of genetic defect, but for the gene replacement method, i.e. a shift in the form of direct mutated gene in position, it can not be used because most of the known vector vaccination consecutive DNA unrelated to semiconductor gives better results. For the benefit of the continuous attempts at gene therapy of the progress in the area of expanded bone marrow transplants to restore the functions of blood cells when infected with a genetic disorder, and used most of these attempts retrograde viral vectors.

New ways to design a treatment using cells of the organism

- Cell culture installed: is converted to the infected cells in organs or in the culture the appropriate gene, and then build a specific structure and in the infected tissues. Thus, cells expressing the gene and provide the onboard product withdrawal.

**Diseases which are currently subjected to genetic manipulation**

- Cancer: The application of gene therapy in cancer is not now aims to correct genetic defects in somatic cells, but sought to allow the introduction of genes to eliminate them.
- Neurological disease: This treatment is used to reduce nerve damage that accompanies Parkinson's and Alzheimer's disease and to enable the infected neurons to recover their function.
- Acquired Immune Deficiency Syndrome: There are many techniques under the anti- vaccination testing, such as installation and self- inoculation cultures for the primary fiber cells that contain transience's carry code of viral proteins and immunization procedure by genes which are to block viral reproduction.
- Gene therapy of developed AIDS, the first attempt of gene therapy in humans (1990), while the enzyme that remove amin of adenosine make the child is not capable of performing the immune response to resist infections after isolation of lymphocytes from patients, and then insert a normal gent for an enzyme that remove the amine through virus vector.

The treatment of patients through the provision of the necessary genes are still the idea compelling. For those who remain in front of researchers in basic science and conscience, much to be done before gene therapy to succeed.

**M.Bio- engineering**

There are a number of scientific developments resulted from the diving in the world of molecules to push medicine forward through the discovery of technical of recombinant DNA (engineering life) and this new knowledge has led to the understanding of the causes of the disease that has eluded science until now, and thus to find new treatments to them. Engineering of life had an impact on medicine borders these have become easier with the forgotten youth of this important scientific field. The reality is that James Watson and Francis Crick did not reach a structural installation with a double helix molecule of DNA. And then it was identified the gene (genes), which manages the production of individual proteins, and then we obtained the tools of partial strong, and in the early seventies researchers began snapped genes of the DNA. One of the species and planting it in DNA another kind for the manufacture of new molecules and in a few years researchers were able to transfer these genes and to produce objects that are within during the eighties and became a human gene transfer to many microscopic organisms and bacteria turning them into factories for medically useful proteins.

After it has been cloned of human genes in the micro- organisms for a number of hormones, including growth hormones and insulin in human as well as bacteria many of the genes responsible for human proteins with diagnostic value was produced at the level of marketing. It is noteworthy that human insulin is derived from living with diabetes, and also for the development of techniques for the production of antibodies "monoclonal antibodies". Many applications, there is a steady increase in the use of enzymes in the diagnosis and treatment as well as in planting (farming) tissues and cells, "Tissue and cell transplantation" and that the development of engineering of life is still in the young stage, but there have major impacts on medicine and industry is synergy between electronic systems, electrical and life- component electrons so- called life "Bioelectronics" and electrochemistry of life "Bioelectrochemistry". Then there have been the following design of a number of devices depending on what is stated in the above examples include "Glucose monitors" for the purposes of medical sensors and nerve gases for medical purposes and sensors nerve gases "Nerve gas sensors" to military uses. Based on sensors that have been most developed in the present time to reveal the exact products enzymatic activity mediated by the traditional pole "Conventional" where is the install (restricted) "Immobilization" new approaches that lead to devices with more sensitivity that depends on the movement of electrons between the direct- polarization and the redox centers protein "Protein redox centers" In brief the enzymes, which is based on the sensor depend on the medical sensor "Glucose sensor" and other sensors that measure chemicals in blood such as immune sensors include the electronic life "Bioelectronic immunosensors", which was commercially manufactured during the current decade, are measured in a large number of materials in the fluid of life, causing a revolution in the diagnosis, in addition to the incremental progress that has been happening as a result the development of a wide range of models "Sensors", which depends on the synergy between micro- organisms substantiated grants stability, "Immobilized and Stabilized". Finally, various data indicate that the microbiology of life through the engineering involved in the medical field in the production:

- Antibiotics.
- Vitamins.
- Nucleotides.
- Hormones.
- Enzymes.
- Vaccines.
- Antibodies.

The progress that accompanied the engineering of life has affected in particular the daily practice of doctors, because of the speed that accompanied the evolution of knowledge and techniques in the laboratory and hence to the industrial production and then patient care. The expression of human insulin gene in bacteria E. coli, for example, has been studied in 1979 and that this insulin, with the original engineering- life of "Recombinant DNA" has been tested by volunteers with non- diabetes "non- diabetic" in 1950 and clinical trials that have been in patients with diabetes began in 1981.

The attention of most doctors on the applications of modern life engineering in medicine, which tend to be very important in areas which have helped to revolutionize the diagnosis, treatment and understanding of many diseases, and examples of this therapeutically important protein, which was manufactured by engineered mediated microbiologist, microbiology, applications of single origin

"Monoclonal antibodies", enzymes and others that arise out of uniform origin from lymphoid cells, where used in:

- Treatment of cancer.
- Diagnosis of many diseases.

Pharmaceutical industry, pharmaceutical companies have been choosing some clinically significant produced cheaply, such as insulin, which was previously mentioned, and which treats patients with diabetes and extensive use of interferon for the treatment of many diseases, including cancer.

The Bio- engineering worked towards a second method by increasing the secretion of microbiology by called anti- life penicillin produced in fungi, and the third trend in the medical field that is the development of drugs already in the nature and turn them into centers of drugs more effectively.

Containing anti- bacterial drugs that have contributed to engineering life and developed vaccines, hormones, vitamins and antibiotics and life for the purpose of producing these materials from micro- organisms after it was restricted to human and animal cells.

Hormones are the most advanced in terms of the accuracy of the technique used and the large economic returns through the engineering of life and led to great successes through the production of materials likes of the hormones which are stimulated, and stimulating the flesh wounds and the growth of the affected nerves that affect the sense of pain. The success of engineering in the provision of life- hormones of the study and treatment has been a boom due to technical difficulties in extraction, which vaccine and growth hormones as well as the instigator of the secretion of pituitary adreno "ACTH" used to treat infections and diseases is used to treat wounds, burns, and stunting and release thyroid hormones pituitary as well as insulin used to treat diabetes, where possible transmission of their genes to bacteria.

Production of hormones is mediated by microbiology research center in the fields of engineering life in general and genetic engineering in particular, where microbiologists used to convert steroids and the production of hormones from the human body can not produce in sufficient quantities. Then it was grown in importance after the custom of cortisone and its derivatives and their effective role in the treatment of arthritis, which draw many medical companies of steroids from plants, animals and chemical methods of trying to turn them into other steroid prescriptions. The methods of microbiology steroids is turning quickly but with less degree, there is in the addition of specialized microorganisms capable converting steroids quickly. There is also the addition of specialized microorganisms is added hydroxyl group of any carbon atom present in the steroid. There are also some working to add hydrogen to steroids or withdrawal of hydrogen or oxidation or separate pools of chemical side effects. Using growth hormone that is released from the pituitary gland for the treatment of dwarfism find the hormone extracted from the animals be in a non-pure from, but according the production of this hormone is preferred to be extrated from microbiology such as the production from the bacteria E. coli after treatment genetically.

The plant hormones have been possible to produce from fungi, especially those produced from rice, as it is known that plant hormones industry is still expensive despite their limitations. In addition, there are a large number of proteins found in the blood such as the factors that contribute to coagulation missing by patients with haemorrhage as well as the albumin found in serum. These materials have been contributed to the development of production by engineering life in medicine (drugs).

The pharmaceutical industry, which includes anti- bacterial drugs, vitamins, vaccines and hormones of the biggest industries that relied on engineering techniques of life for the purpose of producing these materials from microbiology.

**Medical applications of Bio- engineering**

There are many faces, can be addressed when studying the medical applications of bio- engineering after the gene was designed, including:

- Production of therapeutic: include hormones, such as somatostatin insulin, interferon and anti- biotic, where it was initially isolate the hormone somatostain for regulating secretion of growth hormone from the pituitary gland in the traditional way that requires half a million sheep brains to produce 5-10 mg of this material.

- Treatment many of the genetic diseases: the treatment of many genetic diseases possible to treat many genetic diseases due to loss of protein production remedying these proteins from bacteria, and the examples of this case the planting and production of large amounts of genes to produce hemoglobin, which decreases in "Thalassemia" through the introduction of genes responsible for hemoglobin the patient's bone marrow, and then returned the cells to the patient.

- Diagnosis of a number of diseases before birth: the fetus diagnosed in the prenatal stage, through the identifying the defects in a specific gene that causes the disease, such as some "Gamma- Globuinemia" and the disease lest Nhin as well as Tay- Sachs "Tay- Sachs".

- There has been progress in some areas of medical engineering technology due to the recombinant DNA such as "cloning" the human insulin gene as well as growth hormone and its expression in bacteria that has been marketing of human insulin derived from microbiology and used for the treatment of patients with diabetes in addition to:

  - Production of interferon by a large clone human genes in microorganisms.
  - The development of production techniques and monoclonal antibodies and their uses.
  - The increase in the use of enzymes for the diagnosis and treatment in instilling the cells and tissues "tissue and cell transplantation".
  - Treatment of many diseases of genetic mediation by protein that being lost, which can be mediated by production of bacteria.
  - Diagnosis of diseases before birth by identifying the defect in a gene or several genes.

Turning to the relationship between engineering, medicine, is taken into account the following things:

- Mutant cells and the cells unmodified organisms and their products such as antibiotic cellular life and plants, as well as other life transitions "Bioconversions".
- Modified cells "Modified cells" and their products to ensure that objects Monoclonal "Monoclonal antibodies" of the following uses:
  - Immunological Studies.
  - Immunohistochemistry.
  - Tissue typing for trans- plantation.
  - Diagnosis and monitoring of malignancy.
  - Preparation of medicinal products with a "Prepartion of medically important products".
- Recombinant DNA technology and its use for the production of insulin, interferon and growth hormone and vaccines "Vaccines" and enzymes.
- The application of Bio- engineering techniques of molecular genetics and techniques diagnosis recombinant DNA in the diagnosis and (pathological) human disease:
  - Patriarchal diagnosis of genetic diseases.
  - Effects of genetic diseases on the specie disease.
- Features of the future.

It is believed to that Bio- engineering represented by "Clinical biotechnology" has begun in the application management and industrial production of penicillin in 1940 that the success of the full insulin has created a growing demand for medicine (drugs).

The production of penicillin by fermentation and used in the treatment of diseases using the Bio- engineering problems that has been accompanied by the emergence of side effects and put some Bio- engineering solutions , and the problem of production has been developed through genetic improvement producing strains and control the components of the center other conditions contribute to the process of fermentation.

## Bio- engineering and cancer

Bio- engineering has succeeded results in the field of cancer better than other diseases, as shown in the eighties that the main thing vs. cancer is a change in the genes (genes) from an engineering standpoint.

It was clear from the following entries in the relationship between the Bio-engineering and cancer.

- Through analysis of a group of viruses called regressive "Retroviruses", which cause cancer in animals, as a number of these viruses carrying cancer- causing genes or tumor genes "Oncogenes". It appears that the retroviruses that cause cancer have been captured from the normal gene, cell, and one animal and made it part of their own genetic material. The retroviral infection of new cells in the later planted with genetic material, leading to the transformation of healthy cells into cancerous cells.
- The researchers show that DNA extracted from human tumors can shift the cancer cells to cancer cells in test tubes. Or that a specific gene in a human cell that can transform sound cell into the tumor cell and a tumor- causing gene for bladder cancer in humans and called "ras" almost identical to the viral gene, a causing tumors in mice.
- The gene tumor is often due to the mutant or increase in production and there is general consensus about the fact that any of the original tumor gene mutations may be some inherited mutations. The studies of funmor contribute to inherited breast or ovarian caner, the physician may be able to use that gene to assess the patient's condition and prospects and to provide more effective treatment for patients who have multiple copies of inherited suspicious. Harold Varmus and Michael Bishop has concluded that "Lancogen" the legacies of the genes responsible for causing cancer.

**Bio- Engineering and AIDS**

To understand the relationship between Bio- engineering and the AIDS requires a study of the topic in two cases:

- How should the immune system to destroy virus: the defense forces resulting from the immune system to attack multi- directional and of different media for the virus (a specific target) to:
  - Phagocyte and other cells relevant to specific viral antibodies are chewing.
  - These cells installed in the grooves on proteins known as antigens of human white blood cells.
  - Construction immune complexes on the surface of cells identified by a type of white blood cells (T-help) "Helper T".
  - The recipients are on the T- cells help identify the peptide superficial "epitopr", associated with divide, and secrete small proteins that stimulate and activate T-cells and the toxic or lethal trait.
  - The killer T cells directly attack infected cells and fragmentation of viral particles and peptides associated with molecules of antigens of human white blood cells, when identified by toxic T cells by antigenic recipients on the surface of infected cells and destroy them by producing more of them.
  - The B- cells recognize the antigen norepinephrine viral surfaces as a prelude to their destruction.
  - Immune response and the virus "HIV" contribute the immune steps in defense against the virus "HIV", where they are:
    - Invasion of the virus of T- lymphocytes and cells assistance, followed by cloning and increase the virus and help decrease the number of cells, death, and loss of infected T cells.
    - Launch of viral particles from the cell membrane of T cells after being wounded by the T cells and B- toxic responses to be dispatched a strong defense which resulted in killing infected cells,

viruses, and thus is determined by the breeding assistance and reference cells to a normal level.

- A high level of virus gradually with the decline in the number of cells to help patients and reflects the so- called phase of AIDS when the number of cells less than 150 assistance cell in the blood followed by a rise in the level of virus with the decline of the immune system.

**Monoclonal antibodies**

The areas of application for the production of these antibodies where the potential for many therapeutic and diagnostic enormous, including:

- Treatment of patients with leukemia and production of specific antibody alien objects on the cancerous blood cells, leading to the union of antibodies with and removed from the bloodstream.
- Accepting the objects of a transplanted organ which are used Monoclonal antibodies or clone in the development of the body accept a transplanted organ such as the kidney.
- Birth control through private industry specific antibody to proteins found in human sperm.
- Determining the sex of the fetus through a special antibody to sperm of own unwanted sex.
- Models are highly sensitive and privacy are being used as opposites, and a single origin and widely high sensitivity and privacy in early screening for malignant tumors by using specific proteins associated antigen and the presence of tumor presence.
- Determining the levels of hormones in the body and used Monoclonal antibodies to determine the levels of hormones in the body and determine the effectiveness of the glands.
- Search for the presence of some drugs in the body tissue and blood used Monoclonal antibodies in the search for the presence of some drugs

in the body tissue and blood to prevent the occurrence of cases of poisoning or addiction.

- Diagnosis of crimes using Monoclonal antibodies in the search also in the diagnosis of crimes. The food industry also used Monoclonal antibodies in the field of food industries, especially in the diagnosis and determination of the purity of food, processed meat, and free of unwanted substances and preventing fraud in this area.

Of the significant developments that have taken place for Immunology and molecular biology and biochemistry and the discovery of antibodies and the creation of a single origin "The Monoclonal antibodies" is characterized by privacy "Specificity" and sustainability of production, "Immortality" huge quantities "Large ruantities" and high purity "High Purity" for periods of a very long time.

However, these antibodies Monoclonal antibodies created by the multiple origin (clone) the molecular composition and effectiveness. Studies have shown that the use and applications of antibodies only be successful to detect very small quantities of tumor functions that can be used in early diagnosis of many tumors and by diagnosing the effectiveness of these antibodies could be argued that a large proportion of blood diseases can be categorized.

The advantage of imaging the immune flashlight as we have mentioned that the blue single antibodies prepared in the body of a patient associated antigen, surface of cancer cells without other cells and sputtering when labeling these antibodies with radioactive isotope, it can locate the radioactive iodine, for example by gamma cameras and thus can be located and the size of cancerous tumors, including colon, ovarian and skin cancer. The unilateral clone in addressing some of the tumors where it can be linked to medicine as well as radioactive materials to these antibodies, such as chronic leukemia and thyroid cancer lymphoma and colon cancer has been found that these antibodies injected intravenously is grappling with the tumor cells and selectively and is disposed of, where became can direct these drugs directly to tumors by linking them to the catalytic antibodies to these tumors. Used monoclonal antibodies to treat cancer when there is a high toxic concentrations in the tumor. It could also be linked Monoclonal antibodies radioactive isotope and alive in the body of a cancer patient at which time the radioactive material to the site of the tumor and therefore within the cancer cells and it crashed. There are many researches addressing the use of monoclonal antibodies in the early diagnosis of the body rejecting the case of the tissues and the transplanted organs as well as a lot of studies on the use of these antibodies in the treatment of the case of rejection.

**Some applications objects Monoclonal**

Improving the sensitivity of the current immune for tests or tests new Histocompatibility

Fibronnectin
Blood groups Antigens
Sperm antigen
Interleukins IL
Interferons
Progesterone gastrin
Blood clotting factors
Estrogen
Human growth hormone

Monoclonal antibodies has clear impact and important role in clinical medicine before developing the "hybridoma technology", which provides heterogeneous objects "Homogenous antibodies". The research carried out by each of the "Kohler & Milstein" in the early seventies, created a method used for the manufacture of the anti body homogenized with a quantity of non-specific proliferation applied at large.

The researchers "Kohler" and Mlesstin have participated in the production of monoclonal antibodies, which is derived from specific tissue culture which is called hybridoma, where the latter's has the ability to produce one type of antibodies but does not produce more. This is done by crossbreeding or mating types of cells, the first is produce the antibody and the second for the growth of cancer cells have the ability to reproduce. And then treated with hybrid that has to be the formation of antibodies, where antibodies are produced for this body alone, and perhaps it carries the qualities of cancer, the production of antibodies is very large quantities. It is possible in the light of the use of a composition for the manufacture of an unknown antigen monoclonal each part, and then used these antibodies to probe the chemical composition of the real knowledge of the unknown substance.

Monoclonal antibody can be used for treating patients with cancer of the blood through the manufacture of these antibodies is specific to the alien objects on the cancerous blood cells united for the purpose of removal from the blood stream, and used these antibodies for early detection of the presence of tumor cells through the tests that require purity too high to measure the presence of proteins associated with its existence of these tumors and their locations in particular antigen- mediated tumor.

These antibodies are used in determining the levels of hormones in the body and to determine the endocrine events are also used in the search for the presence of certain drugs in the blood and tissues because of the poisoning have also been introduced in the diagnosis of bacteria in the development of the transfer of the body of a transplanted organ, in particular kidney.

**N.Immunoassay**

The development in many areas of clinical medicine by "Yallow & Berson", which developed a radio- immunoassay technique "RIA" for the purpose of measuring the concentrations of very low- lying materials and the offer of "displacement" antigen, which is marked by radiation from the body own by adding increasing concentrations of antigen, the record is marked by radiation and this applies well in the science of hormones, as the hormone levels in the bloodstream always be very low, making it difficult to measure ways of life and conventional chemical. There are many hormones that can be measured easily and quickly test by radiation immuno assay including, prolactin, which was found to be associated with spinal tumor glandular "anterior pituitary gland tumour" and more permanent with symptoms by menstrual "Menstrual disturbance". In fact, the measurement become a part essential to the tests of modern futility "infertility".

**Measurement by radio- immunoassay and other methods of link**

There are three methods of test for the purpose of measuring the materials of life:

- Biological assays.
- Binding assays.
- Physical chemical assays.

There are also two types of tests in cases of binding namely:

- Test the link "Ligand".
- Tests of Binder.

The test of linking section are all kinds of bands that can be used, namely:

- Cell receptors.
- Circulating binding protein.
- Antibody.

The types of acquisitions "Tracers" used are included:

- Particle.
- Fluorescent.
- Enzyme.
- Isotope.

**Applications of the main principles of radio- immunoassay test**

Radio-immunoassay test depends on the competition between the antigen, which is labeled and non- labeled sites on the anti body component complexes ratio on the amount of antigen without radioactivity.

### Immunohistochemistry

Using this antibodies tagged for the diagnosis and distribution of antigens by the optical microscope or electron microscope in tissue sections and sandwitch technique used widely. The unlabeled antibody is placed on the section washed with labeled antibody increases and the layer enzyme linked or marked by "fluorescence" against immune "Immunoglobuin" and can therefore be signaling antigen under examination.

This technique in applied research (Bio and medical research) Examples using antibodies "antisera ordinary polyclonal" on the "Topographical mapping" of the various types of cells in tissues such as "islets of langerhans".

The immuno tissue chemistry containing various anti blood serum describes beta cells of containing insulin, located in the central mass of the island "islet" cells while the A cell that secrete glucacon in the peripheral side linking them to the cells D secreting "somatostatin". An examples on these the use of monoclonal antibody in the clinical diagnosis of tissue "histopathological" of the disease when the test of the "Lymph nodes biopsy" where it help in the classification of a certain type of lymphoma "lymphoid tumour" (e.g Hodgkin's disease and various types of lymphoma "tymphoma").

### Histocompatibility antigens

These antigens are characterized by being glycoprotein's present on the cells, especially white blood cells in humans and is then called human white blood cells, "Human leukocyte antigen" (HLAs). Carrying this antigen genes of the immune response status, where they control antigens were present in different tissues in the body, in addition to genes, there are mismatch humoral immune response "Humoral" and cellular "Cellular".

**Immunodeficiency**

Immunodeficiency resulting from the lack of a genetic condition in the inability to create an immune cell, or one of its outputs and symptoms of the disease- causing immune deficiency commensurate with the degree of destitution and accompany him.

One of the examples on the case the disease resulting from this deficiency (AIDS), "Acquired immunodeficiency syndrome" (AIDS) or acquired immune deficiency syndrome and was attributable to a virus of the type of regression "reterovirus" with a tendency to lymphatic cells "T" in humans, called "Human T cell lymphocyte".

This virus has several methods to spread such as blood and mucus and the interface is accompanied by injury to the virus to many diseases, so called on the situation of the disease and syndrome of one disease. The assumption is based on the perception of geometric depends on logic that a collapse of the immune system caused by AIDS, produced by relationship engineering between HIV (human immunodeficiency) in the body of the infected and the immune system where it is can do the following:

- Cloning of the virus rapidly that destroy large number of cells of the system as there are two types of tests in cases of a binding.
- Faces a viral reproduction for many years through every defensive response to prevent the virus from reproducing.
- An imbalance in the latter for the benefit of the virus "HIV" event leading to AIDS.
- It can show new geometric forms of the virus as a result of mutations that be able avoid the defense forces of the body in some way, and confuse the immune system, which enable many patients to stay healthy for many years, finally collapse due to the boom continued, speaking of the virus.

**Preparation of medically important products**

The use of monoclonal antibodies the purpose of purification and preparation of medical materials which is represented by the task done by the "Secher, Burks" and that covalently bound to anti- body unilateral origin against assigned to drive in "Sepharose" and therefore can refine 5000 times.

There are a number of the production of insulin- led company "Eli Lilly & Co", which used the Bio- engineering, including recombinant DNA technique as a base for the manufacture of human insulin. The production process has been carried out by "Lilly" in cooperation with "Genentech Inc." According to the following steps:

- Determination the sequence of DNA From the known sequence of amino acids m insulin.
- Chemical structure of genes for the series "A" and Series "B" of insulin, contains each and every one of them in the codon methionine at the end "O".
- Each gene enter in 2 mentioned in the beta- 2 gene "B-galactosidase" of the plasmids, which are the same within the" E. coil".
- Because of the fact that the bacteria had grown in the medium that contains the galactose and not glucose that urges enzyme B- galatosidase and then with a series of insulin "A" or "B" linked with methionine.
- After the breakdown of bacteria, treated with cyanogens bromide "CNBr", that breakdown the proteins at the site where the methionin is present.
- Purification of the two strings "A& B" and then returned to their union with the natural production of insulin, the two strings.

The bacteria do not carry enzymes that change the pro insulin to insulin through manufacture the lofty strings "A, B" in the bacteria followed by purification separate chains bilateral, ties with the sulfide.

It is noteworthy that human insulin that is produced by E. coli was tested by healthy human volunteers and with diabetes (and there were no adverse that there is capacity similar to purified pork or with decrease the blood glucose when injected under the skin or injected inside the vein.

The trials have been carried out on human and compared with animal insulin producer from the pancreas of pork produced then was shown similar effects. This was in the hospital, "Guys" in London and Osaka in Japan.

The department of food and drugs in the United States the U.S. approved marketing of insulin produced by the micro- organism, which called the "Humalin", also got the same thing by Britain in the same year 1982.

**Growth hormones**

These hormones are used in the treatment of growth disorders in children, dwarfism some cases of infertility in women and because of the high cost of treatment using these hormones and the difficulty of obtaining and the need for a large number of the pituitary gland the high cost of the hormone and the likelihood of infection with viruses and expected such as uses future growth of the tissues, and the flesh wounds after surgical operations the flesh of fractures and the assist once in the treatment o burns, sores and the used in the study of malignant diseases.

Researchers have made efforts to extract it and its production in bacteria by bioengineering after the successful transfer of genes responsible for hormone production from human cells to bacteria which was done in 1979.

The length of human growth hormone, 191 amino acid with a molecular weight of 2200 excreted by the pituitary gland, secreted of from the front gland of the longitudinal growth of the structure which means the isolation from the pituitary gland.

The pharmaceutical kebi have cooperated with the company "Genetech" for the production of growth hormone from E. coli using Bio- engineering of life recombinant DNA techniques.

The technical difficulties have been overcome in the collection of pituitary glands as well as the cost that accompany and other problems, including lack of access to one type of the hormone, but a mixture composed of several different forms in the installation of structural and molecular weight than did patients with antibodies inhibiting their production.

**Interferon**

The interferon is extracted from the cells infected by virus and studies have shown that these cells, the immune system stimulates the production of this hormone during the infection by virus.

Then overlap with the later injuries to the work therefore it is called "Interferon" used to treat the casualties.

The cost of purification of interferon is very high and extracted from white blood cells, and the other efforts made to develop the production method for human interferon non- blood through tissue culture (artificially), and then methods developed to produce interferon from bacteria, it is done successfully in 1980 and it became clear that there several species produce a number of genes.

Interferon has been isolated in 1957 and considered at that time the first line of defense against attack by viruses, used to treat many viruses diseases and including:

- Cold.
- Hepatitis.
- Cancer Diseases.

Because of the interferon ability to prevent abnormal complications of the cells studies have shown that the immune system of these cells motivated during virus infection leading to secrete very active material is to overlap with the later injuries to the work.

The interferon's a family of proteins, discovered as a result of infection which flows into the cells by viruses "Virally infected ceils" are characterized by the following characteristics:

- Antivural in other cells.
- Inhibition of "Cellular proliferation" "anti cancer drug".
- Internalization of the immune system "Modulation of the immune system".

It is possible to re- classify the life can be interferon's to the following types:

- α- alpha- interferon leucocyte "α leucocyte interferon".
- β- beta, interferon cell Fiber "fibroblast interferon".
- Gama of lymphocytes, immune interferon "immune interferon", "Lymphocytes T".

Leucocyte interferon reduces the spread of a vesicular composition "vesical formation" and responds to fear of infection of the liver "Hepatits B" as well as in various malignancies "malignancies" such as breast cancer spread "metastatic breast cancer" and non- Hodgkin's lymphoma "Non- Hodgkin's lymphoma" and osteoma flesh "Osteosarcoma" and malignant melanoma "Malignant melanoma".

Research has shown that there are about 20 types of interferon, which produces a number of genes mediated for the purpose of genes engineering more effective against viruses or against tumors.

**Production of interferon's**

- Using 50000 liters of human blood, produced from 0.1 g of pure interferon for the treatment of acute viral diseases.
- A culture producing cells and white blood cells of some healthy donors and encouraging the virus "Sandi Virus" for 24 hours and then

isolated interferon- mediated by centrifuge in the Central Laboratory of Public Health in Helsinki, France and the United States.

- DNA recombinant technology.

**O.Biotechnology**

The concepts of traditional genetics known for thousands of years and then evolved into the development of Mendel's laws, famous, and then changed, according to the progress of various technologies and discoveries.

The increased in gene progress in terms of chemicals, then was reached as to how the work of the gene at the molecular level with the adoption of methods of biochemistry, rather than traditional methods in the interpretation of genetics, which paved the way to the evolution of the concept of genetic engineering.

The high and low living organism from units can only be seen with a microscope, a cell, which contains the kernel and the last, which includes chromatin materials that turn into chromosomes (chromosomes). Studies show that both the egg and sperm contain half the number of chromosomes (chromosomes) and therefore half the number of genes in human egg contains, 23 chromosomes. Chromosomes chemically composed of proteins and the "DNA". Studies have indicated that the gene is made up of sections from the chemical DNA. The latter consists of chemically according to the double helix model of Watson and Kirk.

That has been proven that DNA is the genetic material according to the following:

- The amount of DNA constant in all cells of the individual, regardless of the quality of tissue that make up the member.
- The DNA ability to configure a mirror image of himself during the division.
- The DNA characterized to contain all the genetic information in the order of succession of base nitrogen.

## Genetic engineering

Genetic engineering is intertwined with the vitality and technology based on several basic sciences such as cell science, genetics, biochemistry, physics, and others.

The content of genetic engineering the human ability to control the mechanisms of gene transfer from one cell to another and how to express them within the cell for the future.

To understand the genetic engineering practice to be done the following:

- Isolation of DNA of the object which is meant the transfer of its genetic material.
- Cut the DNA to the sections that each end section to a particular gene.
- Identify the gene required between these parties.
- Ensure the presence of a carrier "vector" suitable for gene transferred in order to carry the gene of the object to the donor organism.

Discoveries that paved the way to genetic engineering:

- Carrier "vector".
- Types of bacteria contain a small chromosome called "Plasmid".
- Restriction enzymes DNA, you cut it off at specific sites.
- Ligases close the gap left by the restriction enzyme.
- Select a succession bases in the DNA Sequencing.
- Synthesis of pieces of DNA "Oligonucleotide synthesis" for the purpose of identifying genes within the cell for the purpose of diagnosis of many genetic diseases and led to begin the implementation of the Human Genoma Project.
- Using "Probe" in the processes determining the existence of gene and diagnosis and genetic makeup of the individual.

The genetic engineering since its birth in the seventies of this century a lot of fear, it is double- edged sword usable for good or evil and see the use in the prevention of disease or treatment, whether genetic surgery that change the genes with other as well as another gene in the filing of another object to obtain large quantities of secretion this gene for use as a drug for some diseases. After the success of the possibility of transferring genes from one cell to another, there were some concerns, including the following:

- The possibility of introducing genes that are synthesized toxic material within the cells of bacteria and make them so harmful effect.
- The introduction of parts of DNA Tumor Virus in another virus bacteria.
- Disable the genetic diversity, where the plants or animals that were subject to genetic engineering are usually homogeneous, making it vulnerable to bacterial and viral diseases and others.

Scientists played down such fears, the development of standards and controls to reduce the risk of manipulation of genes and these terms:

The issue of genetic engineering, genetic since its birth in the seventies of the last century a lot of fear, they double- edged sword usable for good or evil and see the use in the prevention of disease or treatment, whether surgery, genetic change since intra gene in the cells of the patient and the gene in the filing of another object for large quantities of secretion of this gene for use as a medicine for certain diseases while preventing the use of genetic engineering on sex cells "Germ Cells" for the legitimacy of the dangers.

And taking into account that it also may not be used in genetic engineering purposes, evil and aggressive, or to overcome the genetic barrier between the different races of creatures in order to create objects out of shape, mixed curiosity.

It is not permissible use of genetic engineering policy to alter the genetic structure in so- called improvement of the human race and any attempt to tamper with the genetic character or human intervention suited to individual responsibility is legally.

The prospects and risks of genetic engineering (genetic) after the success of the possibility of transferring genes from one cell to another, there were some concerns, including the following:

- The possibility of introducing genes are synthesized toxic substances within the cells of bacteria and make them so harmful effect.
- The introduction of parts of DNA tumor virus in another virus in bacteria, with the spread of these viruses and bacteria spread of the disease.
- Loss or interruption of genetic diversity, as the plants or animals that were subject to genetic engineering are usually homogeneous, making it vulnerable to bacterial and viral diseases and others.
- It is known that human intestine contains different types of bacteria the bacteria can thus be dealt with genetic engineering techniques to live in the human intestine and increase the chances of the spread of diseases and epidemics.

However, the scientists played down such fears, the development of standards and controls to reduce the risk of manipulation of genes, some of these precautions: Controls for the design of laboratories and security measures to prevent the spread or leakage of bacteria and viruses treatment.

Although the benefits of biotechnology in many areas such as medicine, agriculture, industry and conservation of the environment there is increasing data show that diversity of thought began to intervene in the subjects continued to reduce the time from the jurisdiction of social thought and reason. The evaluation process of biotechnology has not yet started, does not believe that the limited studies on the impact of bad to launch microorganisms genetically engineered has no real value. These studies resulting from the overlap of politics in science and nothing to do with the problems faced by our communities.

**Milestones in the history of biotechnology**

1- 1952 Genetic material (carrier of genetic information)
2- 1953 and Watson and Kirk define the installation of DNA.
3- 1958 repeating half DNA.
4- 1968 link the sections by the specific enzyme.
5- 1970 discovery of the enzyme antiretroviral RNA.
6- 1973 building hybrid molecules of DNA.
7- 1977 technical determination of the bases of DNA.
8- 1985 interaction sequence of polymers (PCR) of exaggerating the genes.
9- 1986 the beginning of the Human Genome Project.
10-1990's first test of gene therapy (U.S. National Institute of Health).
11- 1999 cloning.
12- 2000 human genome.

The Biotechnology representing biology in the modern design to the mosaic of knowledge reflects its roots in genetics and evolutionary biology and molecular biology, and biochemistry and other sciences. That include biotechnology, technical applications that use biological systems, living organisms or their components or products to modify products or processes vital for specific purposes, and therefore include many operations used in agriculture and the food industry, and in the narrow sense means of biotechnology processes for nucleic acid technology and molecular biology, technical applications for breeding, and this involves modifying the gene transfer and the definition and identification of nucleic acids. And the use of biotechnology has led to many benefits, both in the production of food or medicine or in the environmental field at the same time understands the Food and Agriculture Organization, about the attention on the potential danger that may result from the use of biotechnology, namely, the impact on human health and the impact on environment and therefore must be wary of prospects for the transfer of toxic compounds from one organism to another or creating a new toxic substances or allergy- causing compounds transfer from one organism to another.

The potential of biotechnology for military purposes lies in its ability to address the qualities of DNA in bacteria and viruses. The reference to the problem of which takes use of biotechnology is as a warning to the result of this technology developments, and will be of biotechnology in the future or the role may be more than that of any other technological progress in determining the quality of political and economic relations between the nations.

There are sufficient data now indicate that reproductive technologies are increasing in the intensity of the interaction between science and politics (a proliferation of unregulated medicine across the Western world a strong impact on traditional communities and resulted from studies dealing with rich local delivery and obstetrics, infertility and reproductive technologies).

These issues raised by biotechnology as a result they have become at the forefront of basic research and applied consistently reached new levels of progress and complexity and can be far- reaching impact and positions that require scientific, political, moral and social. These issues vary with varying impact and could be referred to some of them:

- Cloning.
- Human Genome.
- Gene therapy.
- Map of the protein.
- Food and genetically modified organisms.
- Advanced technologies (nanotechnology).
- Vital information.
- Monoclonal antibodies.
- Biotechnology, medicine, agriculture.
- The discovery of disease- causing genes.
- Forensic – DNA, Fingerprinting.
- Biotechnology and biosafety.
- Biotechnology and environmental balance.
- Scientific strategies of biotechnology.
- Research and development and stops future.

- The role of education and training in biotechnology development.
- The twenty- first century is the century of Biotechnology vitality and prospects and challenges.
- Technical aspects of genetic engineering.
- Bioinformatics

It can focus on hot issues of the following:

- Cloning.
- Human genome.
- Gene therapy.
- Genetically modified food.
- Ethics.
- Genetic engineering and the internet.
- Biological weapons.

**Cloning**

Cloning techniques are distributed according to principles and specifications of the theory and practical on several areas, genetic cloning and reproduction by nucleus. Gene cloning refers to the production of similar genes resembling the original, and the best example of this when giving birth to identical twins after a split genes of one egg and its distribution into two similar cells and grow each of them to produce a separate identical fetus. But when the cell of the embryo is sperated in the process is referred to by a particular genetic cloning, and it is growing each cell separately to produce an integrated organism researchers have successfully cloned monkeys and frogs of the cloned embryonic stem cells, similar to the original match, and it believes that it is easy to deal with embryonic stem cells because they are not discriminated by the (were not an evolutionary has turned to hair and brain and muscles and other organs) is in the wombs of their mothers, the negative aspects of cloning, the genetic test will be based in is not what will the world of him.

The third cloning occurs from mature cells, the researcher can wait to see the nature of the thing with his own eyes before they proceed on reproduction. This type of cloning is new and are using the nucleus of an adult cell adopted by Wilmot (embryologist) to clone a sheep from an adult cell, as well as it has managed to clone sheep, genetically modified with genes to produce human factors and blood clot of the most famous that carry human genes. Dolly, the most famous reproduced onganism that carry human genies while the cloned Dolly the first to reproduce the new technology. International principles for the reproduction of new life based on the cloning of a somatic cell is a mature steps each one of them have the status associated with these principles and is the following:

- The use of a cell donor (host) intended for reproduction of cells extracted from the membrane to view white pregnant sheep featuring the genes needed to form a complete sheep for the purpose of conversion to a cell capable of regeneration without preproduction and become a creature without full sexual intercourse (without vaccination) and instead of being cell embryonic viable fusion.
- Using an egg cell is extracted from sheep attached to another type of black- headed in the pot tester removed while retaining the nucleus with cytoplasm to put it right.
- The cell donors to famine and a halt to development and divisions, and preventing food resources for a period of ten days and give in to the state of sleep.
- The removed egg from the receptor cell near the nucleus of the cell donor prepared for cloning by using electric induction, small electrical firing bursts per nucleus and the egg unite and begin to act as fertilize.
- Allowing the egg genetic cells for growth and division and the formation embryo in the laboratory, implants in the uterus of sheep with black head and become the sheep that was born and was named an international replica of the white sheep of the cell donor and not the same as that used for the black bosom.

**Fingerprint**

That science is progressing dramatically in the current year, so that it can be made in the last quarter of the last century, equivalent to human progress in its long history as a whole. In the field of genetics offers this impressive progress of science and builds on the many hopes in the future of human. While the human in a state of surprise and astonishment, which inherited the technology to adapt the results of gene. The scientists that discover some of the problems to show us the genes which was later named the fingerprint genetic fingerprinting. What are these? What are the issues that can be resolved, and is unable to traditional means of forensic medicine to find a solution? Genes that carry genetic message from one generation to another and guide the activities of each cell is a giant molecule may be converted to resemble the strings, called DNA that contains all the genetic traits, from eye color to the smallest structures in the body. They are the result of genetics in human cells, for 23 pairs of chromosomes in the nucleus of the cell, the chromosomes constitute from DNA. Proteins play an important role in maintaining the structure of the genetic material that lead to reveal the full individual. The finger printing began until 1984, when the geneticist deploy at the University of Leicester in London search of genetic material that may be repeated several times and re- sequences itself incomprehensible represented in the length and location, this research reach in one year that this sequences characteristic of each individual and can not be similar between the two except in cases of identical twins only, and the potential similarity of fingerprints between one person and another one trillion, making it impossible similarity, that was found that these differences are unique to each person just like fingerprints and therefore called the fingerprint genes. Dr. Alec has recorded his discovery in 1985 and named them the name of the sequences of the human person as defined and as a means of identifying a person through the passages approach sometimes called DNA fingerprint "DNA Typing". The genetic fingerprint known through the courts, although had spent time in the detection through forensic medicine, where possible knowledge of this fingerprint to identify the mutilated bodies and tracking children and missing soldiers, as it can mark genes to identify the person until the bulbs of the hair that has been cleared of many of the defendants by identifying the genetic fingerprint of murder, rape and revealed the true perpetrator of the crime, had a genetic fingerprint of the word on the issue of genealogy polarities of a number of issues to prove paternity, rape, and calculates the ratio of the distinction between individuals using fingerprint genes found that this ratio up to about 1: 300 million people, there is one person with the same genetic fingerprint was also found that fingerprint genes inherited according to Mendel's laws of genetics. It has also found that the fingerprint genes vary according to geographical patterns of the genes in the peoples of the world, for example, is different from Asian (Mongolian or yellow race) for the Africans. For the identification of genetic fingerprint requires a small sample of tissue that can be drawn DNA including, for example:

- A sample of blood in the case to prove filiation.
- A sample of sperm in the case of rape.
- A piece of skin under the nails or hair roots of the body in case of death after resisting the aggressor.
- Blood or semen frozen or dry is on the crime scene.
- A sample of saliva.

**Ethical implications**

That this issue is distinct attention to enrich the scientific research related to this section of the scientific specialization, which is still growing and evolving as a number of questions that put precision together constitute the social issues and scientific issues that require a unique answer response send in self- certainty and a sense of security and assurance. The religions have confirmed the ethics of researcher and research ethics and both sides of the same coin, a search should be moving to the reconstruction and development and preservation of the environment that God created it so well, the search if deviated from their destination and good career development research is not useful and must be liberated with the production and consumption together.

The importance of this subject is first that it does not affect the religious, but very cautiously and in accordance with insights and analysis are limited, and the cure of genetic testing and abortion, infertility and human eugenics and other topics related to the needs of the Muslim scholars to discuss and study and comparison with the fundamentals of the faith and purposes of the law. If we were not the courage and wisdom to show the religious scientific opinion on these issues inherent in our daily lives, will remain controversial among the various currents and contradictory beliefs, which reflects negatively on our future generations and directly affects our faith, one way or another.

**Ethical considerations**

The gene therapy in somatic cells aims to treat serious diseases, and the possibility of morally acceptable. The gene therapy in cells construction remains a subject of controversy with regard to cell construction cells and with regard there are a number of questions.

- Do we have the right to change the genome of an unborn child?
- Who has the right to approve?
- Are we encouraging the introduction of genes (such as growth hormone) to improve the quality of embryos? Any non-therapeutic uses.

Despite the many considerations of discussions on gene therapy technology, millions of people with one of the different types of genetic disease, they hope to apply these technological developments soon in the attempts to mankind and in the absence of other types of treatment, then should allow the growth and development of gene therapy in somatic cells, under the supervision of bio-security. Which include preventive measures that should be adopted to reduce the need for gene therapy in Muslim societies:

- Interests in genetic counseling in public and private hospitals to help people to absorb health education on genetic diseases and to take the necessary measures.

- Promoting genetic studies (epidemiological) in families and tribes that carry infected gene and this makes it easier for genetic counseling, as well as gene therapy.
- Do not marry relatives, particularly when it is in the family ancestors are infected with diseases and hereditary.

**Medical consideration**

It is not preferred to attempts to human gene therapy in the absence of a broad scientific background able to understand the nature of genetic disease and molecular consequences. On the other hand it must be used human gene therapy techniques in the framework of a particular lead to unwanted hard impact and to restore normal cellular function in a person's life that continue throughout the future. It also must be gene expression regulated outside the original a manner as to improve the patient's condition without damaging the cells or the person the future.

The use of viral vectors in human gene therapy is critical concern due to the ability of these vectors to the initiate the particles conditions infected virus that may spread to neighboring cells, or to others in the community.

The treatment of structural cell may cause damage that occur in future generations, and may lead to correct the composition of the affected gene mutations. The remaining operations targeting stay primitive and with non-controlled roads.

**Security considerations**

There are potential dangers from the use of gene transfer by retrogressive virus but it did not cause any minor damage in humans. The National Institutes of Health has described in the United States malignant T cells in monkeys, but discovered later that these resulted from contamination of the carrier virus.

## Religious considerations

- God created man in the best stature and with generosity to other creatures, and tampering with components of the human being and subjected to tests of genetic engineering without a goal is incompatible with the dignity that God bestowed on humans will read on him, "We have honored the sons of Adam".
- Islam is a religion of science and knowledge as stated in the verse, "Are those equal who know and those who do not know", which is not forbidden to the human mind in the field of scientific research and useful genetic engineering in its various aspects in addition to knowledge.
- Everyone has the right to respect dignity and rights whatever genetic characteristics.
- It not conduct any research or carry out any treatment or diagnosis of the genome, of any person unless conducting rigorous and prior assessment of dangers and potential benefits associated with these activities with a commitment to the provisions of law and ethics of this matter and, if not beneficial to health and direct benefit to him. It should respect the right of every person to decide whether he wants or does not want to take note of the results of any examination or genetic consequences.
- All genetic diagnoses, preservation or preparation for the purposes of scientific research or for any genetic examination or its consequences are confidential.
- It is not permissible to offer any person for any form of discrimination based on genetic characteristics, which shall be liable or result of reducing the fundamental rights and freedoms and violating the dignity.
- No research on the human genome or any of this research, particularly in the fields of biology, genetics and medicine, should prevail over the observance of human rights and fundamental freedoms and human dignity of any individual or group of individuals.

- Publication of books should be to simplify scientific information about genetics and genetic engineering to raise awareness and strengthening on the subject.
- The introduction of genetic engineering into the curriculum at different stages and in local media.

**Protein map "Proteome"**

The term "Proteome" which appeared in 1994 to the total pool of proteins present in each cell type the amount of a hundred trillion in each individual and the total proteins produced by cells of the body during different life stages.

After discovering the human genome, which includes (full content of genes (genes) in the amount of the 34 thousands people only, and not one hundred thousand. I think scientists for a long time) and also all the genes inherent in the cells of the body at the present time highlights the important question, what the protein content of these cells.

The type of each protein has to be known as a result of these cells and what function each protein and then what order of these proteins. Asking this question came after attrition rationale the concept of the genome and its consumption and is not enough to know as responsible for stimulating cells to produce the kinds of protein, but only requires the knowledge of the situation in its entirety in routine cases of disease and natural and in accordance with these questions and answers on the back of proteome.

Proteome contains information more complicated and the secrets of the genome is more dangerous than those found in the genome and extensive knowledge and synthetic for more than a million different types of proteins. The concept of proteome is known later human proteome is doing now by scientists and they hope that these will be the beginning of the main achievements under this project, despite the severe difficulties faced by these scientists, in excess of those related to the human genome. The analysis (cell proteome), reached some of the researchers in 2000 to build automated device called the molecular scanner "Molecular Scanner", which is carried out by measuring by mass spectrometer, from which tens of thousands of known proteins in a single day, and at the speed of more than ten times what was known before.

These researchers also managed to build a million boosted the analysis of protein per day to build bigger infrastructure database proteomics mankind. The draft of human proteome or other whereupon many of the laboratories and big budgets and international companies such as the famous, hybrgenics, Clera Genomio, different research directions, the analysis of three- dimensional protein structure and interactions between proteins, which performed many of the key characteristics of the human proteome would pay off represented by the following:

- Specification of fungus or yeast proteome with a single- cell, the first that has been done in the world of proteome.
- This project change from how to design drugs in the near future.
- The appearance of the so- called science and technology human proteome, which will focus on the conversion of most of the drugs manufactured by genetic engineering and biotechnology.

**Genetically Modified organisms (GMO)**

Biotechnology, which contains the processes of nucleic acid technology and molecular biology to separate the specific gene from one organism and transferred to a particular object of another district called gene transfer technology, "Transgenic" or may be called the genetic change or genetic modification "Gene modification" and called on the living modified organisms has been applied this technique recently on agricultural crops in the recent developments in genetics, which is also hot topics in it. A number of genetic modifications on some common food organisms, the addition of a specific gene or several genes, for example when carrying out genetic modification of wheat plant is usually a small percentage due to the fact that this plant has about 80.000 genes. The process of genetic modification is possible in practice so as not to become genetically modified plant to another object or to plant malicious, but maintains the general attributes of the amendment with relative injury.

According to some voices of opposition to the process of genetic modification, that could cause damage to humans and the environment, including poisoning or allergies.

The number of countries including the United States of America, Canada and China that will produce genetically modified crops, including soybeans, corn, flax, beets, potatoes in different proportions. From a technical point of view alone, there are a number of benefits of genetic modification of crops to convert to regular crops resistant to pesticides and weed, disease and insects and reduce pesticide use and increase productivity and improve the nutritional value of crops, and make it more a shift of the circumstances, including salinity, drought and an increase in the quality of the crops for use in food as well as in withstand the transport and storage and make crops resistant to pesticides, insects or insect resistance or both groups. The genetic modification was still in use in plant breeding has a significant impact in providing food for humans and methods that have been used traditionally to improve the crops, but they are not specific or accurate results of modern genetic modification in which the change is unknown in most cases, things such as crops and breeding plant breeding and mutations.

**Bio- safety of genetically modified organisms**

The international convention indicate on genetically modified organisms to the need for special tests on GMOs for fear of the potential impacts on human health and the environment of these tests is the requirements of the United Nations Environment Program and the Organization for Economic Cooperation and Development.

The bio- safety grounds due to the danger potential that result from using this technology mismatch impact on human health, animals and impact on the environment, there is the potential transfer of toxic compounds from one object to another or create new toxins and stresses the Food and Agriculture Organization of the necessity of conducting these tests and countries of this technique. Voices opposition to genetically modified organisms after the production of genetically modified crops for food within a specific industry, voices opposition to it were carried out adoption these voices opposition to genetic modification of agricultural crops on a number of reasons, including:

- Exaggeration to talk about the bio- safety of agricultural crops genetically modified organisms.
- Bias the producers of GM crops.
- An evaluation of the safety of this vital crop locally in the producing countries.
- The introduction of these foods in the world trade conflicts because the majority of exporters are the countries that grow these crops.
- Evaluation of bio- safety in the importing countries.
- Lack of uniformity in laws relating to the products of such countries that imported and produced.

The positions of countries and international bodies of genetically modified crops countries vary in their positions on genetically modified crops. Some producers and other consumers, but they all agree on the specification is limited information on the food label of the product of GM. The European Union is marketing a number of modified food while China and Australia are sowing genetically modified crops and Japan imported crops and genetically modified food and shopping, many of which, while South Africa imports some food container material from genetically modified crops.

**International standards for the product of genetically modified**

There is much debate about the GM product and how to dispose of it in terms of safety and health, and where the consumer's right to choose. The EU believes that the consumer's right to know the chemical composition of food, depending on the nature of the product. Protein must know that while the components for oil and sugar is not necessary with the definition of the product. It is worth mentioning that there was a project for an international standard developed by the international body that develops the specifications for the food. Adoption this standard on the principle of similar semi- finished between GM and the current food intake is limited and there is no justification to identify him and did not know that there must be full use of the terms of use, installation, and the source.

**Advanced technologies**

That the era of advanced technologies "High Technologies" or high-technology "Super Technologies" in which we live the last three decades of the twentieth century, the era in which we do not know how many decades it will take, representing a number of scientific areas and new technology comes on top of these technologies, laser and fiber- optic and space technology, new materials, pharmaceuticals, chemicals, minute nanotechnology, and finally biotechnology and genetic engineering.

The forthcoming technical applications that are difficult to know the extent of today and its impact on humanity can be viewed as the era of advanced technologies as the following day when mankind as a whole interconnected network giant relies on a wide range of communications satellites such as radio waves and X- ray laser, so that every part of the ground contact one of the satellites in the moment and will be available electricity in remote areas with farms, genetically engineered to convert sunlight into carbon and then to the crude stream, and can then run all the equipment and facilities, communications equipment, including satellite and the Internet.

The future applications of these technologies will be radical changes in the forms of life activities and practices relevant to the interests of individuals, groups and the process of coordination between these advanced technologies is a strategic way to bring about a surge in operations research and industrial beginnings began to appear, for example, a draft genome and bioinformatics. We will try in this article and subsequent articles offer examples of advanced technologies.

**Femto**

It means the number 15 and the chemistry of femto, to understand the reasons that lead to some chemical reactions without the other, one of the achievements made at the end of the twentieth century and the efforts of the world that have emerged Ahmed Zewail, who won the Nobel Prize in Chemistry in 1999 and showing the possibility of seeing how to move the atoms within molecules during chemical reactions using laser technology and the rapid use of a new standard of time is Alfmto seconds ($10^{15}$ seconds). Zewail has been used pulses of laser beam of a partial vacuum in the middle of materials to study the chemistry of high- speed stages of the transition, working within the Alfmto seconds be managed after the suddenness of molecules in the interim period and then became a pioneer of so- called Alfmto chemistry using laser technology (laser Alfmto) camera and a very fast, sophisticated and very accurate to portray the ongoing chemical reaction between the molecules in three- dimensional image Alfmto time in its three dimensions, not one dimension only.

Finally, what scientist do is to identify cases of transition of chemical reactions as broken links and new links up, and the development of new chemistry carried the name Alfmto result of invention, or a new laser called laser Alfmto or laser technology and through rapid as we were filmed for the moment the chemical reaction within the atoms in the process of only one part of a billion a second, and therefore this technology and its owner, Dr. Zewail laser secrets complex world characterized by inventing something new the properties of new energy and knowledge of the movement of particles from birth or docking to know what was happening in record time is a million billionth of a second the proportion of this period to the second equivalent of one per second span of time to 22 million years ago.

To reach Dr. Ahmed Zewail of the use of laser microscopy to clarify the picture may have been the most difficult times in less than two and thanks to the time factor has been developed to see things, whether internal or external speed and one millionth of a billionth of a second.

The features of Applied Chemistry Alfmto side is represented as medical, industrial and agricultural in nature and changes in the human body, such as treatment of diseases such as cancer, diabetes, a cell can be imaged in the human body, and according to that disease can be determined in the light of the nature of these cells. That laser Alfmto which has been utilized for imaging the moment of the chemical reaction within the atoms in the cell process is not only is one of the thousand billion from the second.

## Nanotechnology

nanotechnology, the technology that deals nanometer scale (metric unit of measurement), which is manipulating atoms for the manufacture of automatic equipment and information does not extend far a handful of atoms, and then anything can be made by micro- physics is quite different. The first part of the term to reflect the unit of measure (nano= $190^9$ meters)

the term nanotechnology in Engines of Creation written in 1986 and said the possibility of seeing the future of the armies of machinery hidden carry oxygen and nutrients and waste, and manufacture of atomic- sized machines called complexes pads that hold individual atoms. It was found that the control of the maize one and move freely and easily from attributes of nanotechnology.

This technique showed high density in the form of recent innovations in many of the global scientific publications. Including the mutations responsible for many genetic diseases and therefore provide in the future and relevance of information is indicated task to determine the organism is an information system for the manufacture of proteins and other compounds in spite of the inability to resolve the issue of structural triangular shape of proteins, despite the existence of mathematical models serve the purpose.

A fruitful field of nanotechnology research in many parts of the world and in government labs, commercial and academic, have emerged, according to the products on this technique such as sewing pants of fiber and manufacture of precision tennis balls retain flexibility. In the near future, computers will appear smaller tubes made of carbon atom chips represent the atomic scale wires and high strength to build elevator to space, and plants that will manufacture computers minute integrated directly with the human brain to increase intelligence. It is clear from the examples mentioned that nanotechnology is a technology that will change very little minutes every aspect of human life and giving people the ability to control the material, and this technique is the most important applications of medical treatment of human beings through the introduction of precision instruments within the cells to repair infected objects from within or for the diagnosis of patients as well as some developments on the mechanisms of control cells. The first medical use of this technology has been developed a device implanted in the body that may manufacture.

The energy comes biofuels in the cell and the purpose of this engine is the integration of machines in living systems fully, and use some of the scholars of this technology to produce nano- bombs to kill cancer cells. A team of other customers of this technology to produce nano- bombs to kill cancer cells. A team of scientists of any other industry, the crew of siliceous teeth not larger than the size of the cell that can swallow the red blood cells and re- launched into the bloodstream, either antibiotics nano-particles "Nano biotics", which new types of antibiotics contribute to solving the problem of resistance of some types of bacteria to drugs, as well as the modified bacteria organisms, are converging nano- tube micro- rings 2.5 nm diameter amino acids and small hole walls of bacteria are infectious.

Researchers believe that the future of medicine is moving towards nanotechnology that will change medicine, as future devices that will work within the human body to diagnose many diseases and treatment. Russian scientists in the field of quantitative light and laser physics to reach a new discovery has been called the needles agency, a new type of X- ray beam, or special characteristics, as containing the elements of nano- any electronic material on subatomic particles that do not exceed the measurements of nanometer dimensions . It has also developed the first computer chip companies auction that could contribute to increase the power of computers and a reduction, while reducing the amount of energy consumed by the chip is composed of cylindrical molecules of carbon atoms in diameter than a billion to a part of the linker carbon (smaller than a hair a hundred thousand times).

**Bioinformatics**

The modern scientist characterized by the advantages and new versions is in the lab trying to understand the practical environment in which he lives and solve puzzles and formulates symbolic formats for a network of information and communication for the purpose of creating and responding to inquiries, questions of science resulting from a novel link between health and genetics.

The question that presents itself is to us after the introduction on the topic of bioinformatics. The term is relatively new view and bio-informatics is also new. As is the case for many of the terms introduced in modern science, scientists have so far failed to agree on a single definition of the term. It may therefore suffice here to say that the (computing life) is the process whereby the relationships between technologies, biotechnology and computer technology, for the purpose of exchanging information and experiences and the related transfer of ideas and information for the purpose of understanding of life and death of organisms and also represents the integration of mathematics, statistics, computing and life sciences for the purpose of organizing (computing life) and analysis and interpretation. In spite of informatics is modern science that he knew very complex rooted in and accountable to the most important multi-disciplinary mathematics and computer science as well as many explanations of the science or medical studies resulting from the human genome (genetic content) for different organism as well as the use of electronic informatics to understand Genetic Network Models and the development of three- dimensional views of the complex molecules of the computer.

The basis of bioinformatics has taken different forms and the use of methods and tools for a variety of information consistent with the idea of the basic units of DNA. The gene, but at the same time vary according to the sequence of these units themselves and the simplicity or complexity. The conditions and circumstances that prevailed in the incidence of different diseases and conditions associated with this aspect of the major change is largely in the increasing trend towards the modernization of information about human biology, including chromosomal abnormalities and related diseases for example, inherent "Down" is characterized by the fact that cell individual contains a third copy of chromosome 21, and is determine by microscopic examination, but tremendous progrcss has been possible to monitor changes in the DNA including the mutations responsible for many genetic diseases. The important thing anyway is that recent data are heavily dependent on modern technologies that are designed to connect researchers, these operations become aware of the basics of bioinformatics associated with sophisticated and effective, hence the importance and seriousness of these means which undoubtedly provide the benefit greatly from this knowledge and its applications in health and genetics.

The inter- relationship between computer scientists and molecular biologists can computer world help to provide interpretation of the rules and the necessary to encrypt the DNA and therefore provide in the future and relevance of information indicated the task to determine the organism. Moreover, there is a possibility for the development of computer using DNA and the use of mathematical symbols rules when you're adding a certain amount of DNA that can store all the information in the computer world. The idea of computer life which appeared in the 1995 adaptation of DNA affords. For information processing finally, the biology is no longer limited to specialists, but it entered the computer science and scientists, who began learning these sciences and participate in the research teams of specialists to develop computer- life, as it can be envisaged that biology has become a note in terms of information called bioinformatics and the size of the smallest components of the computer- life billions of times the size of silicon chip is that characterized by very huge storage capacity and speed of processing to resolve some complex issues.

From the standpoint of the computer, the DNA and its installation of structural system is an intelligent and sober for the storage and dissemination of information and computer scientists had been accustomed to dealing with a binary digital expression of the alphabet, numbers and symbols, and diagnosed a direct letter of the alphabet of letter DNA. In order to encrypt messages, and all three relay in DNA is an information system for the manufacture of proteins and other compounds in spite of the inability to resolve the issue of structural triangular shape of proteins, despite the existence of mathematical models that serve the purpose. Due to the fact that the storage of information in sophisticated computers to evoke a dramatically different senses, as a researcher actually lives inside a computer, a view of human life, all stored images of modern technology and not necessarily be tapped and developed. As several researchers in countries that have contributed to the success of the development of this human means and comparisons, as were studied in the genetic makeup of living non- human colon fruit flies.

Informatics characterized the modern world and the benefits of new formulas to make it aware in the lab trying to understand the scientific environment in which they live and solve puzzles and formulates symbolic formats for a network of information and communication for the purpose of creating and responding to inquiries, questions about science resulting from a novel link between health and genetics.

The question that presents itself to us after the boot is on the subject of vital information, and there is no reason to enter into a maze of tariffs, the term is relatively new view of childhood, and bio- informatics is also new. As is the case for many of the terms introduced in modern science, scientists would agree so far on a single definition of the term. It may therefore suffice here to say that the (bio- informatics) is the process whereby the relationships between technologies, biotechnology and computer technology, for the purpose of exchanging information and experiences and the related transfer of ideas and information for the purpose of understanding of life and death of organisms and also represents the integration of mathematics, statistics, computing and life sciences for the purpose of organizing analysis.

In spite of modern of bio- informatics science, it is extremely complex, rooted in and accountable to the most important multi-disciplinary mathematics and science, as well as many explanations of the science or medical studies resulting from the human genome (genetic content) for different objects, and as well as the use of information electronic network to understand the genetic and the development of three- dimensional models of complex molecules.

If vital information science, took different forms, and use the methods and tools for a variety of information consistent with the idea of the basic units of DNA the gene, but at the same time vary according to the sequence of these units themselves and the extent of its simplicity or complexity. The conditions and circumstances that prevailed in the incidence of different diseases and conditions associated with this aspect of the major change is largely in the increasing trend towards the modernization of information about human biology, including chromosomal abnormalities and disease- related example, down syndrome characterized by "Down" the fact that cell individual contains a third version of chromosome 21, is determined by microscopic examination, but tremendous progress has been possible to monitor changes in the DNA including the mutations responsible for many genetic diseases. The important thing anyway is that recent data are heavily dependent on modern technologies that are designed to connect researchers, these operations become aware of the basics of bioinformatics and linked with a sophisticated and effective, hence the importance and seriousness of these provide the means by which undoubtedly benefit greatly from this knowledge and provide its applications in health and genetics.

The inter- relationship between scientists and molecular biologists can offer a update knowledge, interpretation of the rules and necessary to encrypt the DNA and therefore provide in the future and relevance of information is indicated task to determine the organism. Moreover, there is a possibility to develop the use of DNA and the use of mathematical symbols rules when adding a certain amount of DNA that appeared in the 1995 adaptation of DNA afford to address the NMC and finally, biology is no longer limited to specialists, but it came to the attention and scientists are beginning to learn this science and to participate in the research teams, of specialists to develop life, as it can be envisaged that biology is located in the heart of a paradigm shift led by scientists and biology became the so- called informational note.

Vital components and the size of vital billions of tunes smaller than the size of silicon chip and is characterized by very huge storage capacity and speed of processing to resolve some complex issues. From the standpoint of DNA and installation of structural system is an intelligent and sober for the storage and dissemination of information that scientists have accustomed to dealing with binary digital expression of the alphabet, numbers and symbols, and diagnosed a direct letter of the alphabet in the four DNA. In order to encrypt messages, and all three relay in DNA is an information system for the manufacture of proteins and other compounds in spite of the inability to resolve the issue of structural triangular shape of proteins, despite the existence of mathematical models serve the purpose due to the fact that the storage of information in the accounts of sophisticated evoke a dramatically different senses, where a researcher is already in the arithmetic of human life, finds its stock, all images of modern technology and not necessarily be tapped and developed, as several researchers in countries that have contributed to the success of the Human Genome Project, the development of this means and comparisons, as were studied in the genetic makeup of living non- human colon fruit flies.

The Human Genome Project "Human Genome project (HGP)" administration major applications of information for scientists, researchers, and vital information on the grounds that genes are not only symbols indicate certain things, and use in critical information is crucial to the success of research programs for the purpose of documenting the basic identity of human life on according to information emanating from that managed many of the pharmaceutical companies or Biotechnology by new medicines to improve human health.

The Human Genome Initiative was the first time in 1988 and aimed at finding the sites of some 100000 in the human gene DNA and the human genome reflects the 24 pairs of chromosomes, is turning into information content when you follow the sequence of rules need to be resolved based on computer science and mathematics, statistics and experimental sciences. Note that the computer science often provided in contributions in programs and solutions are characterized by the skills that led to the invention of language access code information is described that performs a particular order and provides methods to describe complex biological processes a number of lines of code instead of hundreds of pages described in nature. Then the researchers and observers said that the twentieth century atheist and be the century of biology and analytical power resulting from (HGP) will explain all in a drastic life and medical research, as was done.

Research on the nature of genomes and the nature of the composition and organization of various scientific institutions. Acceleration in the implementation of the project, which was planned to complete within 15 years of technical progress, but shortened the time to ten years.

Search the content of the genome to know what kind of information or material contained in the communication and diagnosis of all the genes strictly speaking, the number 100000 in the amount of DNA human as well as identifying sequences around three billion chemical bases. Study the nature of information storage in the computer and the evolution of elaborate techniques of efficient and sequence evolution that contribute to the analysis of information.

Study the effects to be set in the community and to what extent this can be achieved, and the type of response. In spite of all reported studies and research conducted by methods and techniques in various vital information as well as numerous writings in this area, there are still many other fields, and a variety of study and research. Some of these fields has not been touched so far in the country, especially the human genome projects, areas and areas to attract the attention of researchers, but it's mostly a few problems, mostly dealing with partial or subsidiary.

**Diagnostic**

Securing the different types of health services, preventive, curative and rehabilitation of the basic necessities of the individual and society is part of the economic and social development, the Ministry of Health before the embargo the country and the terms of reference of modern medical equipment and refurbished medical equipment, such that the health services in the country in all its aspects to the stage of qualitative and quantitative development admittedly many of the specialized agencies and international experts. That the imposition of the embargo has negatively impacted on the level of health services and spare the necessary medical supplies such as vaccines, medicines and laboratory solutions. In spite of that medicines and medical supplies but not prohibited under UN resolutions, but that the need for medicines and medical supplies that have increased due to the deteriorating state of health, environmental and food, which led to the emergence of many diseases, and chronic diseases and malnutrition. For example, Iraq was clean and free of cholera; the disease returned and appeared again significantly in 1991. The scarcity of essential drugs and lack of availability of the required quantities led to the deterioration of the situation of citizens suffering from chronic diseases such as sugar and heart disease, hypertension, epilepsy, kidney failure and cancer diseases.

The laboratory tests were not no better than drugs, because the lack of laboratory materials and equipment used to conduct those tests and lack of maintenance and sustaining them available because of the acute shortage of spare parts and failing to be delivered to Iraq as well as the lack of diagnostic kits necessary to conduct examinations and laboratory tests, all that reflected negatively on the number of tests performed annually following table shows the percentage decline in the monthly Madal to prepare laboratory tests compared to 1989.

Impediments to the implementation of the diagnostic kits (negatives)

- The lack of some raw materials necessary for the completion of diagnostic kits, including:
  - Chemicals.
  - Other essentials.
  - Hardware.
- Difficulties in meeting the needs of the researcher:
  - Chemicals.
  - Services in the local market.
- Continuing attempts to obtain materials and devices from other outlets outside the country led to:
  - The survival of the need for quite a few of the resources vernacular.
- The cost of scientific resources, stationery and print the necessary reports to parents of high.
  - The difficulty in obtaining journals and literature of modern world.
- The area of examination and evaluation
  - Assigning one for the purpose of examination and evaluation.
  - Not possible to give a certificate of inspection for some diagnostic kits for the following reasons the amount of material sent

for testing are limited, instabilities of some materials, lack of some modern techniques and equipment.

- Unable to implement a number of these numbers.

- A steady increase in prices of materials and equipment and the cost of sustaining an impact on services:
  - Estimates of the prices offered by the researchers.
  - A number of research and is now paragraph of materials and devices is the amount greater than what they have as much time of signing the contract.
- Lack of standard materials and solutions for a number of diagnostic as reference material for the purpose of comparison scanned materials and productive and the lack of number of standard delay in conducting the tests or they can not be implemented with the refernces.

# Chapter Three

# Scientific research

A. **entrance**
B. **Research and Development**
C. **Scientific Research in Iraq**
D. **scientific publishing and scientific research**
E. **Science and Technology policies in Iraq**
F. **scientific methodology and scientific research**
G. **Scientific research and translation and terminology**

## A. Entrance

The scientific research represents the fundamental basis for development where countries are divided into developed and developing, and investing in scientific research in developed countries have a distinct factor, so it requires the development of plans for the attention of scientific research for the purpose of development and narrow the gap with developed countries by allocating an annual budget aspires to reach (1%) the disposal of the national income through effective management and legislation process and the adoption of a trained human cadres.

Research is contributed to the changing of economic and social features of the peoples of the world and the groups of countries vary depending on the economic, educational and scientific level.

**Group (1)**

Includes the countries that produce Science and Technology and is located in the highest degree of peace and science, including:

- United States of America.
- Japan.
- Western Europe.

**Group (2)**

It includes countries that have elements of scientific and technological progress and seeks to develop, including:

- Russian Federation
- Some East European countries

**Group (3)**

This group includes the introduction of elements of scientific progress such as :

- Some of the Arab countries.
- Southeast Asian countries.

**Group (4)**

These group include any means of scientific progress and called consuming countries for the outputs of scientific research, including:

- Most of the Arab countries.

- Rest of the world.

The universities it may participate in scientific research, there are universities have a high reputation in scientific research, in the United States about the (3500) the University and Institute and some searches and others who teaches only.

There can be a real sense if the University is not neglected scientific research and it must be careful in its mission of scientific research, the scientific quest is necessary to raise the level of university teaching. Some universities are seeking the search for the establishment of centers of education and research units independent of the units in the education within the university.

In Iraqi universities and their activities include individual initiatives or small research groups, and is worth mentioning that a researcher isolate the means of weakness and the possibility of arriving at the sources of information and the ability to communicate with other top qualifications Scholar.

Research projects in Iraq stands in the field of research and development at the door of "academic luxury" while we find that the developed world devote a lot of potential to support various research and scientific experiments for the development, scientific quest, and in this sense is the "investment" and not a luxury.

World spends about 2.1% of the total national income on the areas of scientific research, which is equivalent to about 536 billion dollars. There are3.4 million Researcher in scientific research institutions in the world and approximately, an average of 1.3 researchers per thousand of the workforce.

United States of America, Japan and the European Union are Spending of estimated on the research and development of approximately 417 billion dollars, and the United States alone spends annually on scientific research more than 168 billion dollars, or about 32% of the total spent the whole world. Japan comes after the United States:

(130) billion dollars, equivalent to more than 24% of spending Statesand then after that the order of the Advanced world: Germany, France, Britain, Italy, Canada, for a total spending of the seven countries more than 420 billion dollars. In these seven countries there are two million and 265) thousand Researcher, representing more than (66%) of the total researchers in the world.

The percentages of the Scandinavian countries allocated by those countries for research and development this way: Sweden (4.27%), Finland (3.51%), Denmark (2.6%) and Poland come by (0.59%) ranked last among European countries. EU budget reached for Scientific Research from 2002 to 2006 (17.5) billion euros, representing 3.9% of the total budget of the EU in 2001 and other sources amounted to the EU budget for scientific research during the period from 2007 to 2014 about (300) billion euros.

The South and East Asian countries are increasingly important for research and development, South Korea has raised its spending on R & D ratio of (0.6%) of the gross domestic product in 1980 to (2.89) in the year 1997. The research is directed in its attention to electronics, , the seas and oceans, environment and technology, information technologies, standardization tools, new materials, and space science and aviation.

China has planned to raise its spending on R & D ratio of 0.5% of GDP in 1995 to 1.5% in 2000. The proportion of spending on scientific research in China have risen recently to nearly 2.5% of the total national expenditure, China's budget was for scientific research and nearly 136 billion dollars at a time when the budget did not exceed 30 billion dollars Only in the year 2005.

The rest of the world does not exceed their spending on scientific research more than 116 billion dollars, and most Arab countries do not show figures and statistics on researchers and scientific research,: United Arab Emirates (0.6%),

Kuwait (0.2%) Jordan (0.3%), Tunisia (0.3%), Syria (0.2%) and Egypt (0.2%) and also the researchers in the Arab countries less than (16) thousand Researcher, and in statistics issued by the Arab League in 2006 Each million Arabs there are (318) researcher, at a time when the ratio of up in the west to the (4500) Researcher per million people, and Israel has spent on scientific research (9) billion dollars by the year data 2008, which is equal to (4.7%) of the production National in Israel.

The companies in North America, Europe and Japan constitute (95%) of the total spending on scientific research. Spending was recorded on research and development parallel to growth in 2006, as spending rose on research and development by and the first public spending in 2005 by forty billion dollars to $ (447) billion, an increase of the amount (10%). US companies raised total spending on R & B (13%), representing the largest source of growth of spending between the companies. And there is increased in total spending on research and development by 21 billion dollars in 2006, compared with China and India, which significantly increased by (400) million dollars during the same period. The spending on companies in China, India and other developing countries is estimated around(5%) the world.

Japanese show great interest in the idea of cooperation between universities and enterprises productivity. And the number of private universities in Japan exceeds public universities almost four times as of private universities (372) University compared to 96 public university.

The government of Japan support the scientific research and does not exceed (21.5%) than is spent in this area, while the industrial sector alone provides more than (68%) and about (11%). In more modern sources, the proportion of private sector participation in the financing of scientific research in Japan rose to nearly 85% of the total budget of scientific research.

Israel spends on scientific research is equal to (1%), which is spent in the whole world, and Israel spends twice as much of Arab States (combined) on scientific research and development. And Israel is the highest in the world as a whole in terms of the proportion of spending on scientific research of national output, the United States is spending (3.7 percent) and Britain (1.8%). And Germany (2.6%).

The UNESCO issued Reports by for Science and Culture in the year 2008, that the Arab countries spend (14.7) dollars on the individual in the field of scientific research, while the United States spends (1205.9) dollars for every citizen, and European countries about (531) dollars. Jordan was spending on scientific research (0.34%), and Morocco (0.64%) and Syria (0.12%) of the total national expenditure as the Arab countries, in general lacks well-defined goals and scientific and technological means policy! And it does not have the so-called information industry, and there is no information networks and devices for coordination between the institutions and research centers, and there are no specialized funds funded research and development.

**B. Research and Development**

The research and development two activities vital for them strategic importance in maintaining the quality of scientific personnel and in ensuring access to advanced science and in promoting the transfer of technology, as well as providing an early warning system in preparation for technological change which reflected the results on the technological and industrial, agricultural and health progress, it is in fact investment is guaranteed.

The efforts in the field of research and development in Iraq may vary on the progress similar efforts in other countries in terms of sum and it is still inadequate to meet the challenges posed by scientific and technological developments and the globalization process in spite of the allocation of funds in this area and was the last twenty years the best proo.

And it is supposed to take over Iraq's great attention to research and development in all fields as a base progress in all fields and allocate good proportions of their national incomes for this purpose and that is to develop the contribution of scientists and creators Iraqis and technicians as well as research centers in the state institutions and advisory offices in universities, taking into consideration that the benefits of R & D activities can not be performed without complex networks work and relationships.

• Research and Knowledge Society

• scientific research and higher education

• The importance and goals of scientific research

• policy and scientific research

Knowledge and scientific research community

There is an urgent need to invest capital to build a knowledge society and that in the operation and sending graduates to get degrees in all fields of knowledge and that the state adopt the deduction of estimated proportion of the state budget for scientific research and contribute to the private sector to allocate an estimated percentage of the profits to support the universities and scientific research centers and finding a link closely between the public and private sectors and institutions of higher education and improve the quality and efficiency of the harmonization of scientific research to the requirements of society through the development of standards and foundations for the adoption and quality control conform to international standards and keep abreast of developments in information and communications technology and employment in academic administration in terms of content and methods of scientific research, taking into account the economics. And give the role of the private sector to participate in the future of scientific research industry, and by increasing the development of research centers and the introduction of the concepts of quality and quality control in its various components and stages, and to reconsider the policies of scientific research to achieve the greatest possible harmonization between the wishes of the community and disciplines available to them, and through the provision of the mechanisms necessary to excellence and innovation.

The strategic axes for scientific research could include quality, creativity, management, legislation and executive plans and public policies in higher education institutions such as universities and other tuning. Given the importance of scientific research in higher education institutions and the needs of society, it is assumed that the Ministry of Higher Education and Scientific Research formulate the general policies aimed at the development of scientific research as these policies are based on the key elements is to connect scientific research plans and development technological scheme of development and the needs of society, and in the Cooperation with the private sector, and in the necessary human and financial resources, resource development, and in promoting the use of the Arabic language in scientific research and in the creation of a database for research and development, and to measure the extent of this research interaction with the service sector.

Public policies aimed at the development of scientific research in higher education institutions based on the key elements is to connect scientific research and development technological scheme of development and needs of the community plans, and in cooperation with the private sector, and in the integration efforts of higher education institutions and coordination in this area, and in the development of financial resources the necessary human resources in this sector, and the use of the Arabic language in the scientific and research activities by increasing the interaction between the scientific community and higher education and in promoting the use of the Arabic language in scientific research, and the adoption of scientific terms to impose mainstream Arabized terms and standardization in general use, and to encourage scientific research in the field of Arabization of Science. And the preparation of publications dealing with specialized scientific terminology in cooperation with Arab institutions and distributed to the local and Arab universities, and the creation of a database to search, as well as agreements between the Ministry of Higher Education and Scientific Research, governments and other agreements in order to activate and maximize the benefit and encourage researchers and teaching staff to do the research carried out by independent research teams or joint and reconsider the foundations of promotion in universities so that give more value to the research carried out by research teams and institutions used for preparation of research published titles in patrols court, and summaries of the research presented at scientific conferences, and books author published, and the number of researchers at the university level of doctoral and master's degree, and patents recorded. That each university to report on the envoys PhD in various fields of knowledge and inventory scientific capabilities available from the library, scientific periodicals, information networks and scientific equipment from the importance of scientific research in higher education institutions for the development plans and the needs of society, the establishment of a comprehensive electronic library are managed joint, forming a body for the purposes of regional and international conferences that each University to provide this body with its plan proposed annual conferences, seminars and workshops. And that each University to build a database of scientific production of the members of the teaching staff and researchers and master's theses and doctoral messages all. Database and be available on the electronic library network. As well as the necessary financial resources development to support scientific research of the profits of large companies and endowments local, Arab and foreign grants and the introduction of a special fund for students, as well as human resources development and development technological in higher education institutions and evaluation of cultural and scientific agreements concluded between Iraqi universities and universities.

The importance and goals of scientific research

The main goals of scientific research shall be limited to: -

1 information.

2. Knowledge configuration.

3. Change.

Of the goals of the university and duties when they are created to turn into research centers as well as education and therefore there are other reasons borne by the decision makers and political ones, double the budget of scientific research and that their responsibilities and perception of the importance of science and education, spreading the culture of scientific research through education and the media platforms, and they embrace scholars and thinkers.

Scientific research is divided into developed and developing countries, including investment in scientific research. Therefore, it requires development of plans in scientific research and development for the purpose of narrowing the gap with advanced countries by allocating appropriate annual budget of up to 1% of national income available through the effective administration and legislation.

Scientific research has contributed to changing economic and social features of the peoples of the world and the contemporary world to total countries vary according the level of economic, educational and scientific.

The university has a role in scientific research and has a high reputation in scientific research and even some of them have no good scientific research. In the United States of America there are around 3500 university and institutes, half of them carry research and teaching together and the other half mostly teaching scientific research is necessary to raise the level of university teaching.

Methodological constraints are related to education in the methods of scientific research. Those associated with legislation is a clear view of the fact that instruction of promotions, for example, encourage individualism. Those of the administrations are found in the department, colleges, universities, the ministry, the specialized research centers… etc., they all suffer from lack of coordination between different departments of research. The Obstacle of funding is one of the main channels in the country which is much lower than the proportion of GNP, representing one of the most major constraints for scientific research. The cadres' constraints refer to the wishes of the students involved to obtain certificates to improve their livelihood rather than being concerned with scientific research and increasing burden of teaching in universities and the lack of clear criteria for selecting cadres which assume responsibility for scientific research centers. Policy and scientific research

The is between political decision-makers and scientific research, what more Theses value that doomed the shelves of libraries, and more relevant messages severed from the needs of society, as if the goal is to get degrees and careers. There are other reasons borne by political decision-makers including: the weakness of the scientific research budget.

One of the direct results of this low rate in spending, the low level of university libraries and the lack of modern books and references and the lack of scientific laboratories to sophisticated equipment.

In fact, we desperately need to develop the way we deal with science and knowledge The interest paid by our universities is not so far beyond the transport and quotation for the other stage was overtaken after the real challenge phase during which the start of the human mind and liberation from various crippling restrictions, running behind the scientific truth by Research hard. The total spending on scientific research in the Arab countries does not exceed (0.15%) of the gross domestic product, which is a very low rate compared to the Israeli entity - for instance, which cut into the gross domestic product for the purposes of scientific research percentage equivalent to 17 times the percentage of the proportion of the Arab world.

That scientific research is the basis of the progress of nations, peoples, and therefore the establishment of scientific research centers of various kinds sponsor the transfer of community of inertia and backwardness to the stage movement and stage of growth. That any attempt to reform the status of scientific research should start first real conviction of all parties responsible, and this is only possible if he got a real democratic and interaction with all classes of people and educational and cultural institutions.

The Planning Research and balance between the numbers of qualified people to carry out scientific research and between laboratories and centers with all its possibilities is a matter of paramount importance must be taken into account. higher education without throwing them in the process of scientific research does not benefit the country unless accompanied by something suitable for them to secure conditions contained in the laws and regulations live up to those in place in developed countries.

The phenomenon of globalization will be imposed on us new facts can not be ignored, economy of the new global market, which exceeds the political and geographical borders existing create a system competitive severe Kan education and knowledge and imposed globalization increasing deal with advanced technology in various spheres of life, including the production and communications. So we take our place in this new world order we raise the level of intellectual and material our production even be able to compete and this can only be achieved to possess advanced technology through scientific research in our universities and our companies development and scientific research

Represents scientific Qaeda Find the basis for development where countries are divided into developed and developing, and investing in scientific research in developed countries be distinct, so make plans for the attention of scientific research for the purpose of development and narrow the gap with developed countries by allocating appropriate an annual budget of up to 1% of national income behavior through effective management and legislation process to adopt human cadres trained to turn scientific research projects.

Research contributed to the changing economic and social features of the peoples of the world and the Department of the modern world on groups of countries vary depending on the economic, educational and scientific level.

Development and economic considerations and involves scientific research on paramount importance in the present age are related, and witnessing increasing investment from the presence of the developed countries in the number of areas and fields, as well as the close connection and interaction between scientific research and its applications technological national development and reconstruction, and it seems that the industrialized countries, ingenious in establishing this link and take advantage of it to the max, and that the achievement of sustainable development is supposed to meet the needs of the present without prejudice to the capabilities and needs of the future, States that know how to apply the output of scientific research found occupies pride of place in many areas such as machinery and civil war and appliance manufacturer and this makes them superior militarily, culturally and scientifically in the provision of various services to its citizens.

According to the Arab Human Development Report 2002 that he became a clear and acceptable in general, that knowledge is the main ingredient in the production, primary productivity and schedule, and human capital. And stop the

Arab Human Development Report for 2003 on the knowledge society, the conditions which hinder the development of a knowledge society in various Arab countries.

In general, and on the critical situation faced by many Arab universities and poor coordination between the Arab scientific research institutions.

Indicate statistics released in 2005 that the Arab countries combined, allocated for scientific research is equivalent to 1.7 billion dollars, the equivalent rate (0.3%) of the gross national product while spending ratio reached on scientific research in Israel during the same year to ( 4.7%) of its gross national product and indicates that Arab scholars contributed about eight thousand scientific papers in 1996 for international magazines Court This is equivalent to (60%), which was produced in China and (50%), which was produced in India and that the scientific output of the Arab country currently stands at (72%) only from the production of Israel alone in this regard.

You need the Arab countries to the development of scientific research and betting in the development, due to a series of bottlenecks that hinder this vital sector The financial assistance offered allocated to him within a number of these countries have not yet reached (1%) of the Internal Nantj a weak proportion stray too far from the global average ( 2.3%). The low funding of scientific research by the productive and service sectors in the Arab countries explains to some extent the limited research activity qualitative in Arab countries, especially if we take into account that the government funding which up to another (89%) of the total funding, which consumes in coverage workers in this sector salaries. Also emphasizes the various reports of local and international statistics weakness allocated to the sector of modern technology in a number of Arab countries possibilities. It has also pointed to the dilemma you know education system in many Arab countries, as a result of the inability of educational institutions to keep pace with various modern scientific developments. Also, most of the R & D centers in the Arab countries are not ready to convert the output of scientific research to the investment product, because of the absence of universally adopted orientations or because of lack of knowledge, expertise and capabilities necessary to carry out activities of innovative required, in addition to the lack of a number of Arab countries to researchers qualified full-time for Scientific Research only, as is the case with the leader in this area of the States. As well as the scarcity of specialized magazines and periodicals and the absence of a sophisticated equipped science laboratories. The scientific research in the humanities today do not enjoy academic freedom where researchers prevented from addressing the many research dilemmas Unlike freedom of practice in developed countries.

Report 2005 for Development and Humanitarian proposed to adopt a set of measures and solutions to solve the funding problem, such as imposing a tax for the benefit of research and development on the industry, and set up endowments for scientific research and the allocation of a percentage of the income of customs for this purpose as well, or to issue a postage stamp the name of development back part of its revenues on scientific research, or Add tax on tickets prices, private donations from national and foreign invested enterprises. The interest in a field of scientific research in the Arab country and its development, requires exceeded outlook, which sees in science and technology just two commodities can of developed countries, and is supposed to start elaboration of a real and clear strategy to develop education and scientific research, as the two sectors vital inseparable, and monitoring necessary for the use of technology resources appropriate for economic development and social welfare, and the rehabilitation of universities and support publications, books, magazines and academic and scientific journals in a way that makes them accessible to researchers.

World spends about 2.1% of the total national income on the areas of scientific research which is equivalent to about 536 billion dollars, and works in scientific research institutions in the world approximately 3.4 million Researcher, an average of 1.3 researchers per 1000 of the workforce. The United States, Japan and the European Union has been spent on scientific research by nearly 417 billion dollars, exceeding (75%) the total global spending on scientific research and spend the United States alone each year more than 168 billion dollars on scientific research, any about 32% of the total spent the whole world and then come Japan (130) billion dollars (24%) and then dribble then the order of the developed world, Germany, France, Britain, Italy, Canada. Scandinavia has topped the list of European countries for scientific research and that in relation to the national outputs, came percentages allocated by those countries for research and development on this growth: Sweden (4%), Finland (3.5%), Denmark (2.5%) was the EU budget amounted to search Scientific from 2002 to 2006 (18) billion euros, which represents 4% of the total budget of the EU in the year 2001, South Korea lifted its spending on scientific research ratio of 0.5% of GDP in 1980 to 3% in 1997 and 4% in 2010 and directed its attention to science and electronics seas and oceans and environment technologies and information technologies.

But China has planned to raise its spending on scientific research ratio of 0.5% of GDP in 1995 to 1.5% in 2000 and to 2.5% in 2010 and were also the goals of the five-year plan during that period towards improving technology applications in the agriculture sector, and the development of national infrastructure foundation of information, and increasing development in manufacturing processes.

As for the Arab countries ratios were in the order following: United Arab Emirates (0.6%), and Kuwait (0.2%), and Jordan (0.3%) and Tunisia (0.3%) and Syria (0.2%) and Egypt (0.2%) and in the statistics of the Arab League in the public 2006 that there is (318) each researcher million Arabs, while the figure is in the west to the (4500) Researcher every million people. It is noteworthy that Israel spends on scientific research (9) billion dollars, according to the general data 2008, which is equivalent to 4.7% of the production of national Israel as the rate of disposal Israel on civil scientific research in higher education institutions (34.6%) of the allocated to education the government budget Higher entirety. The Arab countries combined spending on scientific research half of what it spends on Israel despite the fact that the Arab national product has a 11-fold national product in Israel and the space is (650) vulnerable. According to reports issued by UNESCO for Science and Technology in 2008 that the Arab countries spend 15 dollars on the individual in the field of scientific research, while the United States spends (1205) dollars for each citizen and European countries about 530 dollars.

The ranking Arab country in average in terms of spending on scientific research comes at the lowest among the regions of the whole world has twice the productivity of Arab researchers to lack conviction Arab governments usefulness of scientific research and the reluctance of the private sector for conducting scientific research and a weak back income researcher compared to other people who work in the sectors Other, non-availability of equipment and means of advanced scientific and the lack of strategies or policies in the field of scientific research and finally migration to Arab scientists outside their home countries.

The role of the private sector globally has expanded in scientific research, Fanfguet, for example, US companies (4.8%) of sales on research and development in the year 2006 while the spent Japanese companies (3.5%) of sales, and companies of the European (3.4%) note that more than two-thirds of spending the total for 2006 focuses on three sectors only: informatics and electronics (29%), health (22%), cars (17%). There is a joint cooperation between Japanese industries and internal and external universities and especially American universities, according to statistics from the Japanese Ministry of Education, the Japanese industrial enterprises have spent in the period between 1980 to 1990 amounted to 95 million dollars in exchange for expertise and research for the benefit of Japanese companies. In Germany equivalent provided by the industrial sector there for scientific research with what provided by the same sector in the United States, and according to other information, the private sector in Germany contributes a percentage of (70%) compared to (30%) of the public sector and be equally ratio between the two sectors in the United States US.

The Chinese experience restructuring own scientific research and scientific policy process has been conducted since 1980 included the revitalization of technological innovation and the development of operations and the development of high technology and its applications and modernization of manufacturing processes and raise the technological content of Chinese products. Chita program led to the development (50) technological park during the nineties and succeeded in establishing (500) technological incubator until

2002, which made China the second place in the world in the number of incubators. Chinese universities have either returned restructured and put a plan to develop one hundred leading Chinese university. Note that many Chinese universities have their own companies.

The Malaysian experience involved a number of companies and institutions introduced by the State, including the development of the Malaysian technology company in the year 1995 and aims to transfer innovative ideas issued by the Malaysian universities. And it embraces the company's small specialized vital projects and This company also activate and the results of research and local development and the development of treaties and opportunities for strategic cooperation and long-term benefit from the facilities provided by the university, such as equipment and laboratories and the development of labor relations and personnel in the field of technology and research and development and engineering consultancy and technology transfer high common international cooperation and the development of human resources and support quality development and manufacturing operations management software.

There are so-called French Association of incubators has identified the legal form of government, including incubators and incubators, which funds a number of technological incubators. And belong to incubators established within the faculties of Engineering and various scientific institutes and incubators within the plant research centers and incubators owned by big companies and the houses of international experience which incubators is encouraging the development of new projects, particularly in the economic and technological fields, such as electronics and areas of biotechnology incubators private sector, an investment incubators. Despite serious attempts by Iraq historically to set up scientific research institutions and technological canceled some of them, are still some institutions face major problems preventing them from starting and productive work, most notably the absence of a clear scientific policy, define meaningful scientific research scientific objectives, pour in the development socio- economic, and in the establishment of a scientific-technological base and a national independent, not subject to blackmail powers and international companies.

## Information technology

Iraqi educations are able in very limited ways to build information networks that can be developed to provide internet services to workers and students. The ratio of the number of students to the number of computer available is still very high and where there is no general frame work for the plans in information technology noted. A state of chaos exists in the process of developing these plans which are often duplicated. In the light of randomness, there must be a general framework that defines the requirements and competencies to be developed.

In (1964), 119 countries approved at the general conference of UNESCO, the recommendations that include the different forms of education outlying school and adult learning must be considered an internal part in the education system, and that the opportunity must be available for males and females to continue to lifelong education. Patterns of continuing education in Iraq education remain valuable and suffer in breach of the continuous weakness in performance, despite the long period exercised.

Students are able to learn through self effort, re-engaging directly by attributable steps to the year (1926), when Percy invented machine that include a series of questions and answers. Note: this education is missing in Iraq schools.

These elements are of good education to pursue promptly to search for useful work. They invest time, but this kind of study is not available n Iraq.

Students from different educational levels are received, then provided with different types of studies required by developments of the individual and society. There is weakness of Iraqi universities to accept this kind of education.

Communications satellites are able to grant broad prospects for the educational process at all stages of education, especially, in higher education. It has become possible to lecture from the university to house or transmit a lecture to another university.

University of Hawaii Islands began testing the use of satellite ATS- I to transfer voice message and printing between its various islands in the year (1971). Then established the university plans to send ground- receiving television broadcasts in educational exchanges between libraries and medical conferences, and student-teacher training joint researches. In (1971) satellite ATS-I was used to provide medical treatment and educational programs for rural schools and some guidance to some medical lectures in the college of Medicine. University of Washington has also further tests at Stanford University in conjunction with Brazil using the moon ATS-6, as well as experience of the territory Rocky Ponte Moon using the ATS-6 and the experience of using the territory Appalachian Satellite. In Ottawa, Canada, Stanford and California Universities in the United States, exchanges experiences between teachers via satellite and mutual distances are often performed. The experience of the Arab satellite ARBSAT from the Arab Satellite Communications Organization will be successful solutions to the problems of education in rural areas. Also, it will help to compensate and cope up with the serious shortfall in the preparation. There is no education this kind in Iraq.

Banks Data are information outlet to the problem of traditional storage and retrieval of information which can be stored in these centers. By using one of the elements described as document number, copyright, title research, objective data banks, we can link centers and scientific institutions and universities. These are tools of opening the doors in front of knowledge and information gathered by international experiences. In Iraq there are no techniques or terminology to this type of education.

Iraqi education have the option for the development and technological change, specifically technology knowledge, which is the most important factor in the production of knowledge and logic that provides technology and thus access to the production of more advanced technology.

The developments were strengthened in technology and communications to a pattern of Open Education, because of its reliance on techniques of knowledge and information, given the rapid development of societies and the transition to a knowledge society. Knowledge society has expanded greatly in introducing this type of university education. The justifications mentioned previously include:

- Integrated use of other media technology.
- Defeat of the many obstacles to the normal university education.
- Submission services to individuals of all ages.
- Defeating of the barrier location.
- Allowing the teacher to work and learn.
- Entrenches a culture of continuous learning.
- Development of opportunities for developing the performance of workers in state institutions.
- Increasing social demand for tertiary education.

Information Technology knowledge/ concept and relationship to the Iraqi education

There is a clear correlation between knowledge society and technological knowledge; both are two sides of one coin. The role of the knowledge society is clear and consists of dimensions as a result of spectacular progress in technology applications, where communications have evolved staggeringly. Turning to technology revolution is one of the important creations performed by rights in the late twentieth century and the beginnings of the twenty-first century. Then this session has widened in areas essential to humans, including:

- Chemistry, Medicine
- Bio- Engineering
- Other sciences

Substantial changes in the pattern of human life become a modern mind and thought that is the basis of profit and invest. Therefore, there are sufficient justification that emphasizes the importance and necessity of Iraqi to take this kind of education, namely as follows:

- Rapid technological evolution.
- Large increase in the numbers of educated and willing to education.
- Democratic education and rights of citizens to education.

To meet the challenges of community learning, it is required to work with attention to information technology, knowledge-building rules, modern communications networks, integration of technology in teaching and learning processes, and research. The success of these universities and transforming them into learning societies, depend on the extent of interest in the development professionally, and adoption of the principles of participation planning and stimulation of the use of modern technology and transforming classrooms into environments for active and effective learning.

The challenges of the knowledge society depends on the nature of the nature of the information and knowledge which will be published daily in different parts of the world and very quickly, knowledge of various types, including the globalization of knowledge and virtual knowledge and technology knowledge will be dominating the world.

**C.Scientific Research in Iraq**

The effect of the intention of officials to shift public research funds toward research programs that serve the national priorities has already affected the nature of the funding available at the funding agencies. For example in USA, at the National Science Foundation, a small increase in funding for the research is directed toward so-called strategic research initiatives that involve, for example, advanced materials and processing, biotechnology, environmental chemistry and high-performance computing.

It is likely that this trend will continue. The indications so far are quite clear that the government expects to shift publicly funded research activity into the areas that are deemed strategic There also has been in the United States, a major Industry-University cooperative research program conducted by the National

Science Foundation. The motivation of this cooperative program is to develop and transfer industrially relevant technologies from the university into practice. Major investments have been made by sponsor organizations, based on center technologies. There are also many other industry-university collaborations that are not part of the National Science Foundation program.

Interest in scientific research began to change after the concept of education. Teaching is no longer the sole function of education. However, the insight into scientific research in Iraq, shows that scientific research suffers from many problems including lack of funding, as the salaries of members of the teaching staff amount to nearly 90% of the budgets of education.

The remaining is distributed on all other aspects of spending, and thus it negatively affects the requirements for scientific research devices and other supplies, as w as other sufferings of scientific research in Iraqi education. It is merely considered functional performance by teaching, to achieve the goals of self separate from the research.

The reality of scientific research in Iraqi education before (2003) was in accordance with the following elements:

- Forces and human cadres.
- Communication and networking.
- Funding.
- Research basic and applied sciences
- Elements of research and development

Mainly, include teachers who consider scientific research, as well as community services. The correlation of promotion with scientific research is done in institutions holding the doctorate degrees. Moreover, human cadres including graduate students and new graduates play a positive role in completing advanced research.

It is clear, that reality of scientific research in Iraq does not live up to expectation of a country trying to contribute to the progress of scientific research, access to higher share of the global economy, and invest knowledge as a result of scientific research and development.

There are systematic obstacles and constraints linked to legislation and administrative. Other impediments are related to financing of scientific research, and some are related to human cadres, constraints, interface and communication.

It addresses the Human Development Report in the Arab countries, including Iraq, for the year 2003 to the difficulty of reducing the gap in the knowledge that experienced and investment in the quality of higher education and to encourage the dissemination of information as well as the strategic vision of the establishment of a knowledge society based on freedom of opinion and expression, organization and dissemination of education, development and localization of science and building the potential of scientific research and the shift towards knowledge production style, and the establishment of a knowledge model and the presence of scientific backwardness stand out at a time when a new global culture is characterized by self-confidence and social competence and the ability to think and quantitative expansion in the field of education.

Furthermore, the report points compared to the achievements of other countries and the high rate of illiteracy in the Arab countries.

The reasons for the underdevelopment of knowledge in Iraq and the Arab countries according to the report to the unilateral economy which leads to the decline in demand for the knowledge economy. As well as the low status of freedoms and human rights that lead to the killing of incentives and the flame of knowledge.

The report also indicates that there is a strategy for knowledge requirements or society in the Arab countries, including Iraq and how to see it, including:

- Arabic language support and advancement as well as linguistic and cultural diversity.

- The development of a knowledge model in Iraq and the Arab region and to stimulate diligence and honor him.

- The trend towards knowledge production graded pattern in Iraq and the Arab countries.

- The possibility of building a self in research and development in all scientific, political and social activities.

- Do sophisticated quality education.

- The launch of the freedoms of opinion, expression and regulation.

Interest in scientific research in Iraq after the change the concept of university education, so that teaching is no longer the only function of the universities. But the look in-depth scientific research in Iraqi universities in general show that scientific research suffers from several problems, including lack of funding, as the faculty members' salaries nearly 90% of the budgets of universities and the rest is distributed to all other aspects of spending, and thus reflect negatively on the requirements of scientific research equipment and other supplies, as well as the suffering that the Iraqi scientific research in universities, than as merely functional performance, doing the teaching, to achieve their own targets for separate needs.

According to the following elements can be studied and the reality of scientific research in Iraqi universities before:

- forces and human resources.
- communication.
- financing.
- Research basic and applied sciences.
- the elements of research and development.

The powers and human resources in universities, which primarily include basic scientific research faculty and community service, as well as it has played other human resources, and that graduate students and recent graduates include a positive role in the completion of advanced research.

It is clear that the reality of scientific research in Iraq does not rise to the level of expectations for a country seeking to do its part in contributing to the progress of scientific research and investment knowledge as a result of scientific research and development, there are obstacles methodology and obstacles linked to legislation and other related administration and constraints relating to the financing of scientific research and those related to human cadres and constraints related to the communion scientific universities in Iraq one of the three basic functions of the teaching staff as well as teaching and community service, and is associated with academic promotion campaign PhD, the remaining proportion of the teaching staff and most of whom have master's degree full-time to understand the process of teaching. There is a continuing increase in the teaching staff in the Iraqi universities commensurate with the continuing increase in the number of students enrolled in universities, and the cursor on the high ratio of students to teaching staff which constitutes an additional burden on the teaching staff, which in turn reflected negatively on scientific research in quantity and quality in universities. The graduate students in Iraqi universities make up an important part of the manpower involved in scientific research, it is these students formed the future researchers in universities in developed countries have an important role to these students to participate in the completion of scientific research. Despite the expansion of universities in the graduate program in Iraq, but their contribution to scientific research are still limited understanding of studying in difficult conditions. The current reality of human power in the relevant scientific research in Iraq is characterized by a lack of the number of full-time researchers in spite of the trend and communication. methodology relating to the methods of scientific research, for example, while the upgrade instructions that encourage individual and lacks legislative systems and include those relating to management dilemmas in departments and colleges and universities and the ministry and specialized research centers and the lack of coordination between different research departments. Funding constraints are one of the main channels in the country, which represents much less than the allocated GNP ratio represents one of the most major obstacles to scientific research. While the constraints of human resources refers to the wishes of the students involved get certificates to improve their standard of living more of their attention to scientific research and increased teaching loads for teachers at universities and lack of clarity for the selection of cadres who take responsibility for scientific research centers standards.

Requires proper scientific planning the development of university scientific research and the study of the current reality as it is a scientific study designed to find out the possibilities available in the universities and what has been accomplished in this area and identify the negative aspects of it and the scarcity form an integrated research teams, and the high proportion of the number of students to teaching staff for global ratio accepted and the scarcity of opportunities for research assistants and technicians for training in developed countries to deal with specialized and maintenance services in research laboratories. According to statistics available that Iraq spent a few sums on research and development after the (2003) and of which focus on supporting research projects, publishing and missions.

The productivity of the teaching staff of doctorate in Iraqi universities campaign later (2003) was (0.44) Search for each doctor each year while total (1.00) a year ago (2003), as it has made the Ministry of Higher Education in the year (2002) in collaboration with universities and strides to issue scientific journals specialized global in nature and in the court of various scientific disciplines and humanity developed in 2006 and it is hoped that these magazines are of high scientific level of classified and adopts world.

The research of investment nature after 2003 and that may lead to creative results and related industrial applications and the patents are rare and it is clear that its contribution is very modest, indicating a lack of institutions of higher education's ability to transform the results of scientific research to investment projects, and is due to several obstacles has paid Iraqi universities after a year (2003) set the legislation issued by the special attention to scientific research and issued its instructions to be encouraged and organized, especially in 2006. The included legislation issued by these universities founded to support scientific research and to support publication and instructions for scientific journals that Add to issue the patent and instructions to send faculty to attend conferences and seminars.

Separated scientific research in Iraqi universities for the problems of society, where this research aims to get a degree or upgrade either his role in the production of knowledge and problem-solving.

The graduate students in Iraqi universities make up an important part of the manpower involved in scientific research, it is these students formed researchers future and are these universities creating material and moral right conditions to attract, and despite the expansion of universities and the growing number of graduate students in, but their contribution to the development of scientific research does not still limited and are pursuing their studies in difficult circumstances.

Characterized by the current reality of human power related to scientific research in Iraqi universities, including the following:

- Not to give the opportunity to campaign doctorate from new graduates on training and direct involvement in training.

- Scarcity form an integrated research teams.

- The high proportion of the number of students to teaching staff for global proportions accepted.

- Concern for a large number of teaching staff in overtime.

- The scarcity of opportunities for research assistants and technicians for training in developed countries.

Available statistics show that Iraq spent in 2006 on research and development accounted for 0.1% of gross national product has contributed universities accounted for 31% of the volume of expenditure on research and development.

In an analytical study of the budgets of public universities in 2006 in relation to the expenses of scientific research (support research projects, publishing, magazines and scientific conferences, books and periodicals), compared with expenses of scientific research and development in 2000, indicating that these expenditures increased in 2006 and that most of the increase has been concentrated in the components of scientific research ( support research projects, publishing and conferences) and (missions and training) directs the charge of scientific journals in Iraq as containing low-level articles, which publishes Sometimes articles low-quality or questionable in value and validity and taken them as well as the lack of mental pluralism and critical tradition.

And it indicates the number of approved studies in developing countries, which can be applied in Iraq, particularly those that rely on global cited as evidence is not convincing in the evaluation of scientific production.

As for the negative aspects of the face of scientific research and the brain drain and talent can do the following steps:

(A) allocation (1%) of the budget for the benefit of scientific research, and financial and social privileges for scientists and researchers as individuals and institutions.

(B) the relative efficiencies of scientific and spammers migratory benefit, through the organization of conferences for expatriates, and share advice with them, whether in connection with access to the latest means of medical and pharmaceutical processors, or on the transfer of scientific expertise and technology, or even the purpose of the financial and economic participation in the implementation of vital projects.

C-Arab cooperation for the establishment of projects and centers of scientific and university applied research, capacity building for the purpose of an Arab expert to alleviate the negative aspects of the Arab brain drain and competencies.

(D) requirements focus on providing social, economic, scientific and political conditions necessary to create a favorable environment for linking science and human frameworks qualifying comprehensive development policies.

(E) activating the role of the private sector.

(F) the availability of good equipment and advanced scientific and means in scientific research centers and universities.

(G) the development of strategies and policies in the field of scientific research.

(H) to stop the migration of Iraqi scientists.

There is a gap between what may be scheduled, from research and the recipients of the outputs of such research, this is what we hope to be avoided through what will do the Iraqi Ministry of Higher Education, and universities of the drawing of the policies of scientific research taking into account the ways of directing scientific research to serve the current development and future needs and the development of mechanisms for linking scientific research and service sectors and productivity institutions. It is well known that the ways to strengthen and develop national capacity in scientific research and development can be done by finding appropriate and effective mechanism involved by research bodies and beneficiaries in the public and private sectors working on the research efforts of institutions and integration and to stimulate and promote economic and social demand on the activities of national institutions for scientific research and development coordination Technical all possible means and the establishment of new units for research and development in the public and private sectors, with an integrated technical capabilities in leading strategic areas of the national economy. And provide resources necessary to improve research centers in higher education institutions, and develop, to become a pillar of President for Research-oriented service development, and actively participates in the contemporary scientific and technical progress and adopt effective mechanisms to document the relationship between scientific research and technological development on the one hand and the productive sectors and service institutions on the other hand, and to encourage the exchange of researchers various categories of research and development centers, in universities and institutions production and service in the public and private sectors and work to adopt the main trends of scientific research and technical development to meet the priorities of the comprehensive national security and sustainable development requirements such as directing scientific research and technological development to achieve water security and directing scientific research and technological development to secure the strategic needs defense and national security. And directing scientific research and technological development to enhance the competitiveness of sectors of oil, gas and petrochemical industries and take care of the technical scientific research in the field of electronics, telecommunications, information and development.

The global ranking of universities gave the importance of scientific research in mind one of the main elements adopted by the organizers of an arrangement that and this raises the problematic complex, where the fledgling Iraqi universities are, its main task at the beginning of its work rehabilitating human resources for the purpose of building or establishment and the provision of competencies to ensure to achieve its mission, which is educational universities in the first place, constitute undergraduate students (90%) of the students and 10 percent of them graduate students, while the developed world described universities as research universities is a graduate students proportion (50%) of the total requested. As it does not increase the proportion allotted Iraqi universities for scientific research (0.08%) of the gross domestic product, while Japan spends only about 150 twice what is spent by all the Arab countries of institutions and universities, but the University of Berkeley US nine branches get in exchange for research services at six times What is spent on higher education in all Arab universities, given that universities in the developed world is a research universities and foremost first allocated industrial and productive institutions large budgets for the purposes of research and development, and that most of our scientists in the Arab universities research published in Arabic in court that scientific patrols of up to about 450 scientific journal, and this is what constitutes a barrier to their spread in scientific forums, due to the weakness of the capabilities available to these scientists and their production and dissemination of winning opportunities to compete with other product of other scientists.

And that most of what conducted by researchers at the Iraqi and Arab universities reflect personal and individual concerns of faculty members such as searching for promotions, and the lack of a clear policy of linking scientific research and the needs of institutions of Arab production and reliance on researchers and faculty members to provide them with studies for the development of their products, and to provide the necessary financial support to them. In light of the above in the context of the need to develop evaluation and university ranking systems, there is a need to take into account the different education systems and differences in the environments and the capabilities and different purposes and functions of the universities in the North and the South and in the various regions of the world to adopt all of the scientific research vision and intellectually and practically, and it requires a research review based on multi-founded scientific studies and administration in accordance with the terms of reference and the various perceptions Wares.

- Activating the role of specialized scientific staff working in universities and to enable the performance of these cadres and determine what is required of them teaching and scientific research.

- Conducting studies basic theory to develop a field for future research plans.

- To hold and guide the development of research.

- Carry out scientific and technical consultancy through consulting firms or in other ways.

- A review of the structure of the teaching work.

- The development of research centers in the light of future research proposed for Science foundation.

- Enhance knowledge and experience in various areas of the modern science and modern information.

- The development of competencies required to meet the needs of oil and mineral sectors to various scientific disciplines. In addition to the terms of reference to the need for future surface and underground water resources, and the fight against desertification, and the need to solve the geological and environmental problems expected from structural and industrial expansions of urban and urban growth in Iraq as well as other specialties.

Suggestions and general recommendations

- The development of intra-studies with the availability of potential in preliminary studies has major competence and a sub and sent abroad in the future to get a graduate degree in jurisdiction on stage.

- The establishment of a specialized center in mathematics professors foreigners operates and foreign expertise or send missions outside Iraq depends on the use of resources paid by Arab students.

- My pension to provide a decent standard of teaching makes him a full-time scientific element allows him enough time to meditate and see and compensation for teachers.

- Development of sectoral body computer science at the Ministry of Higher Education to develop a curriculum.

- The introduction of the Smart Graduate factors and the opening of research centers for the software industry.

- Send new missions and innovative specialties.

- The foreign committees of the developed countries for the renewal of the curriculum.

- Development of Graduate Studies in inside Iraq to compensate for the difficulties faced by our students in obtaining acceptance outside the country.

- The use of experts from developed countries to train workers in the country.

- University system and change radically rethink form.

- To do with a contract to modernize the advanced university education in Iraq universities.

- Undertake new studies bear specific figures.

- A new calendar of the curricula in force.

- Evaluation of the policy of admission to graduate studies.

- Calendar global scientific developments and our position from them.

- Calendar-level graduates graduate (master's and doctoral) and their eligibility for studies.

- The introduction of a specialized center for the attention of the Arabic language and to conduct computer research and all aspects of speech and writing, translation or languages of the search or retrieval software or Arab operational.

- The introduction of a specialized center in the field of industry support code.

- The introduction of a specialized center in computer use in design and production.

- The introduction of a national center for information.

- Focus on graduate studies for the preparation of scientific cadres of teaching in the proposed future disciplines in order to prepare the teaching staff necessary for the establishment of new scientific departments.

- Future development of the most important disciplines in the life sciences.

- Environmental Science.

- Aware of the life of the cell cell biology

- Immunology

- Toxicology

- Biological neurology neurobiology

- Bio-medical sciences biomedical sciences

- Human genetics human genetics

- Molecular Life molecular biology

- Biochemistry

- Bioethics bioemics

As well as some of the proposed sections in light of global developments:

- Department of genetic science.
- Department of Biochemistry.
- Immunology Department.
- Department of the cell life.

Tracks and disciplines each section

The main paths for each section can devise courses of study, including preliminary studies, upper and by stage through which the section in question, as determined research lines and disciplines within that section and the following are some of them: -

Department of Environment

It includes the following tracks:

- Contamination of the agricultural environment.
- Pollution prairies and deserts.
- Pollution of the cities.
- Water pollution.
- The effects of development projects on the environment.
- Biodiversity biological diversity
- Desertification.
- Biological control biological control

- Alternatives to ozone

- Biogenic alternatives radioactive isotopes.

- Climate science life environmental biology

The role of scientific personnel:

- Re-establishment of palm research.

- The establishment of (bank) to preserve and protect the Iraqi germplasm and seeds.

- The establishment of the uses of computer in the life sciences center.

- Create a new nature reserves.

- The establishment of a specialized research center to limit the health and environmental effects caused by the unjust siege to be able to identify and eliminate the effects.

- The establishment of a national center for research environment and biological diversity and the protection of national germplasm, especially those related to the strategic crops.

Investment modern trends in the life sciences: -

- Focus on the completion of efforts to identify the backup geographic areas in Iraq and various environments and biodiversity in all of them for the purpose of preparing a full scan of life of wealth in Iraq Studies.

- Focus on the life of wealth that are unique to Iraq like palm trees, camels and buffaloes, goats, reeds and truffle.

- The aquatic environment and fisheries studies.

- The study of the marsh environment.

- The reduction of desertification studies.

- The study of ecological problems in industry and manufacturing.

- Research and studies to clean up the environment from pollution.

- Studies and research on inherited diseases in Iraq, especially in captivity based on the marriage of relatives.8Policy and sPhysics

Future terms of reference in physics are determined by the following: -

- For a study that rely certain axes within the same specialization, it can expand its base by adding electives or perhaps the introduction of certain themes, such as the physics of modern technologies and physics environment and refer to it again.

- For studies interfaces that work is under way in a limited way, they In undergraduate can develop a plan to set goals so that integrated approaches mode, a joint study of the physics and life sciences or medicine can make a lot of industry and industrial parties, for example, and some of the departments of Engineering role of physics in this The area is known as the (biomedical engineering) The physics solution in the field of sports mainly involved as physics and engineering in the field of electronic physics and electronic engineering.

- Issue remains central to the learning process is a matter of teaching staff and the necessity of survival in touch with the latest developments knew physics evolve with amazing speed and multiple areas of applications as the determinants of doctoral studies to provide a shortage of staff can be cured by a physical determinants often include the upgrading of laboratories and methods of communication in the world devices, These things live up type, something always required.

Geology

Threads of study in the initial stages: -

- Modernization of the curriculum in general vocabulary of basic topics for Geosciences and the adoption of modern books in it.

- Teaching threads modern within the jurisdiction lessons like: geology environment, geophysics engineering, software, mathematical modeling, geological mining, Layering seismic, stratified relay, facies sedimentary analysis, sea geology, geological relics and afternoon drive, soil science, geochemistry membership, modern applications in fumbling remote, mineral exploration, Jesse wells, structural metals Engineering Physics.

- Specialized scientific departments proposed: Petroleum Geology, Geology and mineral resources and industrial rocks, geology and water resources, environmental and engineering geology, the geology of the sea.

- Graduate curricula promote applied in graduate and the emphasis on topics related to the mineral and oil and water resources in addition to the pure scientific disciplines.

- The opening of a modern research centers: Water Research Center, Oil Research Center, Desertification Research Center, Research Center of Engineering Geology, Quaternary Geology Research Center and effects.

- The role of scientific personnel: to take advantage of specialized scientific personnel from outside the university and take advantage of fellowships and training opportunities abroad to promote knowledge of and experience in the fields of modern science in the ground and to participate . Translation as a Source of Knowledge

**D.Scientific publishing and scientific research**

Many scientific groups have produced literature that describes, in terms of many examples, how curiosity driven research has led to important developments in the interest of society.

. A major part of this issue was devoted to the matter of basic research.. Such research is the seed corn of the technological harvest that sustains modern society." In an article on the laser, **Nicolaas Bloembergen**points out that "the first paper reporting an operating laser was rejected by*Physical Review*

*Letters* in 1960. Now lasers are a huge and growing industry, but the pioneers' chief motivation was the physics."

In an article on fiber optics,. " In an article on superconductivity, Theodore H. Geballe states that "it took half a century to understands' discovery, and another quarter-century to make it useful. Presumably we won't have to wait that long to make practical use of the new high-temperature superconductors." Other articles concerned nuclear magnetic resonance, semiconductors, nanostructures and medical cyclotrons, all subjects of great technological and medical importance that originated in basic physical research.

In a preface for a publication of the American Chemical Society, *Science and Serendipity,* the President of the ACS in 1992, Ernest L. Eliel, writes about "The Importance of Basic Research." He writes that "many people believe - having read about the life .

But the examples given in this booklet show that progress is often made in a different way. Like the princes of Serendip, researchers often find different, sometimes greater, riches than the ones they are seeking. For example, the tetrafluoroethylene cylinder that gave rise to Teflon was meant to be used in the preparation of new refrigerants.

And the anti-AIDS drug AZT was designed as a remedy for cancer." He goes on to say that "most research stories are of a different kind, however. The investigators were interested in some natural phenomenon, sometimes evident, sometimes conjectured, sometimes predicted by theory. Thus, Rosenberg's research on the potential effects of electric fields on cell division led to the discovery of an important cancer drug;.

**Kendall**'s work on the hormones of the adrenal gland led to an anti-inflammatory substance; Carothers' work on giant molecules led to the invention of Nylon; **Bloch** and **Purcell**'s fundamental work in the absorption of radio frequency by atomic nuclei in a magnetic field led to MRI. Development of gene splicing by Cohen and Boyer produced, among other products, better insulin. Haagen-Smit's work on air pollutants spawned the catalytic converter.

Reinitzer's discovery of liquid crystals is about to revolutionize computer and flat-panel television screens, and the discovery of the laser - initially a laboratory curiosity - is used in such diverse applications as the reattachment of a detached retina and the reading of barcodes in supermarkets.

Some of the other topics in the brochure on *Science and Serendipity,* that were included to document further the importance of basic research, concerned several examples of the impact of chemistry on medicine.

There are, in fact, countless such examples. The Federation of American Societies for Experimental Biology (FASEB) in their Newsletter of May, 1993 considered basic biomedical research and its benefits to society. I quote from the FASEB Public Affairs Bulletin of May, 1993.

"There have been recent suggestions that tighter linkage between basic research and national goals should become a criterion for research support. Concerns also have been raised that science is being practiced for its own sake, and that it would be better for the nation if research were oriented more toward specific industrial applications.".

"A critical factor in sustaining the competitive position of biomedical-based industries is for basic research to continue to provide a stream of ideas and discoveries that can be translated into new products. It is essential to provide adequate federal support for a broad base of fundamental research, rather than shifting to a major emphasis on directed research, because the paths to success are unpredictable and subject to rapid change.

"History has repeatedly demonstrated that it is not possible to predict which efforts in fundamental research will lead to critical insights about how to prevent and treat disease; it is therefore essential to support a sufficient number of meritorious projects in basic research so that opportunities do not go unrealized. Although its primary aim is to fill the gaps in our understanding of how life processes work, basic research has borne enormous fruit in terms of its practical applications.

FASEB continues with a discussion of economic benefits and a number of examples of basic research-driven medical breakthroughs. "Society reaps substantial benefit from basic research. Technologies derived from basic research have saved millions of lives and billions of dollars in health care costs. According to an estimate by the National Institutes of Health on the economic benefits of 26 recent advances in the diagnosis and treatment of disease, some $6 billion in medical costs are saved annually by those innovations alone.

FASEB continues with thirteen examples of contributions by basic research to the diagnosis and treatment of numerous diseases, most of them very serious. Also noted in this Public Affairs Bulletin is that "our ability to know in advance all that is relevant is very poor" (Robert Frosch) and that, in suggesting new ideas for the management of funding for science, never considered were "the serious consequences of harming the system."

Every year universities, governments and other organizations spend in excess of $10 billion dollars to buy back access to papers their researchers gave to journals for free, while most teachers, students, health care providers and members of the public are left out in the cold.

Even worse, the stranglehold existing journals have on academic publishing has stifled efforts to improve the ways scholars communicate with each other and the public. In an era when anyone can share anything with the entire world at the click of a button, the fact that it takes a typical paper nine months to be published should be a scandal.

Tonight, I will describe how we got to this ridiculous place. How twenty years of avarice from publishers, conservatism from researchers, fecklessness from universities and funders, and a basic lack of common sense from everyone has made the research community and public miss the manifest opportunities created by the Internet to transform how scholars communicate their ideas and discoveries.

I will also talk about what some of us have been doing to liberate the scholarly literature – where we have succeeded and where there is more work to be done. And finally, with these efforts gaining traction, I will describe where we are going next.

While I talk, I want you to keep in mind that this is about more than just academic publications. This is about the future of the Internet and what we are willing to do, as individuals and societies, to ensure that information that should be free IS free. One last bit of introduction. I am a scientist, and so, for the rest of this talk, I am going to focus on the scientific literature. But everything I will say holds equally true for other areas of scholarship.

.

Most people date the birth of the modern scientific journal to the middle of the 17thcentury, when the Royal Society in England took advantage of the growing printing industry to begin publishing proceedings of their meetings for the benefit of members unable to attend, as well as for posterity.

Like their predecessor, these journals were enabled by the technologies of the industrial revolution – steam powered rotary printing presses and efficient rail-based mail service. But they were also severely limited by them. Printing and shipping articles around the country and the world was expensive, and because of this, two key features of modern journals were established.

First, journals limited what they printed, choosing for publication only those works deemed to be of the greatest interest to their target audience. And second, they sold subscriptions – sending copies only to those who had paid. While intrinsically restricting, this business arrangement made sense

As science grew, so too did science publishing, with increasingly specific journals emerging to cater to new disciplines. By 1990 there were around 5,000 scientific journals in circulation, all of them printed and shipped to subscribers. And the costs were skyrocketing. If you were lucky enough to be at a major research university, you could find most of these journals in the library. But most scientists had to make do with a small subset – .

Scientific journals, serving a computer savvy audience with access to fast Internet connections through universities, were amongst the first commercial ventures to take advantage of this new technology. Within a few years – from 1995 to 1998 – virtually all major publishers put versions of their printed journals online.

, until they finally have a result they want to share with their peers.

So they sit down and write a paper describing why they were interested in the question, what they did, how they did it, what they found, and what they think it means.

And then they hopefully submit it to one of the 10,000 journals currently in operation – choosing based on scope and importance. With few exceptions, these journals work the same way. The paper is assigned to an editor – sometimes a salaried professional, but usually a practicing scientist volunteering their time. They read the paper and decide who in the field is in the best position to evaluate the authors' methods, data and conclusions.

They send the paper to these scientists – who again are volunteering their time as a service to the community – who read it and render their opinion on the paper's technical merits and suitability to the journal in question. The editor looks at all these reviews and decides whether to accept, modify or reject the work. If the paper is accepted, the journal takes the manuscript, converts it into a publishable form, and posts it on the web.

They didn't come up with the idea. They didn't provide the grant. They didn't do the research. They didn't write the paper. They didn't review it. All they did was provide the infrastructure for peer review, oversee the process, and prepare the paper for publication.

And yet, for this modest at best role in producing the finished work, publishers are rewarded with ownership of – in the form of copyright – and complete control over the finished, published work, which they turn around and lease back to the same institutions and agencies that sponsored the research in the first place..

Universities are, in essence, giving an incredibly valuable product – the end result of an investment of more than a hundred billion dollars of public funds every year – to publishers for free, and then they are paying them an additional ten billion dollars a year to lock these papers away where almost nobody can access them.

This is most obviously a problem for people facing important medical decisions who have no access to the most up-to-date research on their conditions – research their tax dollars paid for. In a world where patients are increasingly involved in health care decisions, and where all sorts of sketchy medical information is available online..

But this lack of access is not just important in the doctor's office. Scores of talented scientists across the world are blind to the latest advances that could affect their research. And in this country students and teachers at high schools and small colleges are denied access to the latest work in the fields they are studying – driving them to learn from textbooks.

Back in the 1990's several people began promoting a simple alternative model. The idea was to treat science publishing like a service, with publishers getting paid a fee for the value they provide, but once this fee is paid, the finished product would effectively enter the public domain rather than the publishers private one.

After all, universities were already forking over billions of dollars to support publishers. We were offering them a better deal – access for everyone at a lower price. But, while logic and value were on our side, and we got statements of support from within and outside the scientific community, when push came to shove, only a small group of pioneers joined us. And the reason was that publishers had one very powerful card up their sleeve.

. We hired professional editors from others in the industry, built fancy editorial boards and had a suite of Nobel Prize winners singing our praises.

But prestige is a difficult thing to engineer. Colleagues, friends and even family members would stipulate all the flaws in the current system and praise what we were doing, but, when they had a high profile paper, would turn around and send it to the same old subscription journals. It was a very frustrating experience.

Fortunately, publishing decisions are not entirely in the hands of individual investigators. In 2008, under pressure from Congress to provide taxpayers access to work they fund, the National Institutes of Health – who funds about $30 billion dollars of research every year – implemented a public access policy requiring that grantees make their work available through the National Library of Medicine.

This was an important landmark in the history of the access movement, as, for the first time, a major funding agency was making it a condition of receiving a grant that authors make their works available to the public. And the policy has been successful – 80% of NIH funded works published in 2011 are now freely available online – there's nothing like the threat of losing funding to get people to do the right thing.

Unfortunately, under heavy lobbying pressure from publishers, the NIH policy allows for up to a years delay between publication and the provision of free access. While better than nothing, delayed access to the literature no more provides the public with access to the latest advances in biomedical research than handing out year old copies of the New York Times keeps everyone up to date on the latest World events.

However, despite these failings from scientists, funders and universities, the facts on the ground are changing rapidly. In 2007, PLOS launched a new journal – PLOS ONE – that not only provided open access to all of its content, but also dispensed with the notion – central to journal publishing since the 17th century – that journals should select only papers of the highest level of interest to their readers.

Rejecting papers that are technically sound is a relic of the age of printed journals, whose costs scaled with the number of papers they published and whose table of contents served as the primary way people found articles of interest.

But the battle is by no means won. Open access collectively represents only around 10% of biomedical publishing, has less penetration in other sciences, and is almost non-existent in the humanities. And most scientists **still** send their best papers to "high impact" subscription-based journals.

But as frustratingly slow as progress has been, I believe we are close to a tipping point with most members of the scientific community believing that open access is the future, and a growing and diverse set of publishers engaged in open access businesses.

But being able to access papers is just the beginning. We can now finally start to actually take advantage of computers and the Internet to not just make scientific publishing open, but to make it better.

The multilayered, hyperlinked structure of the Web was made for scientific communication, and yet papers today are largely dispersed and read as static PDFs – another relic of the days of printed papers. We are working with the community to enable the "paper of the future", that embeds not only things like movies, but access to raw data and the tools used to analyze them.

There is also no need for papers to be static works fixed in a single form at their time of publication. Good data and good ideas in science are constantly evolving, and scientific papers should evolve over time as new data, analyses, and ideas emerge – whether they support or refute the original assertions.

But the biggest target of our efforts is peer review. Peer review is the closest thing science has to a religious doctrine. Scientists believe that peer review is essential to maintaining the integrity of the scientific literature, that it is the only way to filter through millions of papers to identify those one should read, and that we need peer reviewed journals to evaluate the contribution of individual scientists for hiring, funding and promotion.

Attempts to upend, reform or even tinker with peer review are regarded as apostasies. But the truth is that peer review as practiced in the 21st century poisons science. It is conservative, cumbersome, capricious and intrusive. It encourages group think, slows down the communication of new ideas and discoveries, and has ceded undue power to a handful of journals who stand as gatekeepers to success in the field.

Each round of reviews takes a month or more, and it is rare for papers to be accepted without demanding additional experiments, analyses and rewrites, which take months or sometimes years to accomplish.

And this time matters. The scientific enterprise is all about building on the results of others – but this can't be done if the results of others are languishing in peer review. There can be little doubt that this delay slows down scientific progress and often costs lives.

•

So, while it is a nice idea to imagine peer review as defender of scientific integrity – it isn't. Flaws in a paper are far more often uncovered after the paper is published than in peer review. And yet, because we have a system that places so much emphasis on where a paper is published, we have no effective way to annotate previously published papers that turn out to be wrong.

So what would be better? The outlines of an ideal system are simple to spell out. There should be no journal hierarchy, only broad journals like PLOS ONE. When papers are submitted to these journals, they should be immediately made available for free online – clearly marked to indicate that they have not yet been reviewed, but there to be used by people in the field capable of deciding on their own if the work is sound and important.

The journal would then organize a different type of peer review, in

review system while shedding most of its flaws. It would get papers out fast to people most able to build on them, but would provide everyone else with a way to know which papers are relevant to them and a guide to their quality and import.

There is nothing technically challenging about building such a system, and it makes so much sense that it can't help but happen. But, of course, we've been there before. Science is oddly conservative, and there is enough money and power at stake to ensure that people will try to stop this from happening. So if you care about making the scientific literature open and accessible, I urge you to do whatever you can to make it happen. If you're a scientist, get with the program – there are so many open access options around today, you no longer have any excuse. And try to stop looking at journal titles when you evaluate people and their work. It's a poisonous process that has to stop.

If you're not a scientist, but are interested in this cause, you can do all the normal things – write your members of Congress and the such. But I also encourage you to find scientists whose work you find interesting, but can not access, and send them an email. Or better yet, give them a call. Let them know you want to – but can not – read their work. And remind them that, in all likelihood, you paid for it.

If we all do this, them maybe the next time someone like Aaron Swartz comes along and tries to access every scientific paper ever written, instead of finding the FBI, they'll find a giant green button that says "Download Now".

The general features of scientific publishing in Iraq

In Iraq, the scientific journals are accused of containing low scientific- level articles. These journals adopt a committee to evaluate essays and articles, publish sometimes items of low quality or of questionable value and validity, items that lack of scientific mentality as well the tradition of pluralism.

The number of studies approved in developing countries, which can be applied in Iraq, particularly those that rely on international scientific site, is not convincing in assessing the scientific production of the third world, including Iraq. The measurement and analysis of the overall scientific production of these states scientifically is impossible, since no rules are approved to the local scientific work. Most of those who tried to estimate the scientific production and dissemination of scientific information are dependent on the rules of (International Institute of Scientific Information IST) and the adoption of the Guide to international scientific cite SCI.

Statistics indicate that significant portion of tie work of researchers published locally are distributed rarely outside national borders. Even if deployment of Iraq's scientists took place in the research journals of international scientific reputation, these articles would not be quoted as articles of western scientists and colleagues, thus it is clear that phenomenon is different and arguably broad. Other statistics indicate that more scientific production in Iraq in the field of pure science researcher at an article published annually is mostly inside Iraq.

The articles published locally read and quoted faster than those published abroad and thus 60% of local authorities cited a recent research, while the proportion of foreign modern references was declining.

Statistics of the earlier Ministry of Information show a total of that more than 44 scientific evaluated journal had granted the right to publish and issued by:

- The earlier ministry of Information.
- The Ministry of culture.
- The Ministry of Higher Education and Scientific Research and its institutions.
- Associations and trade unions.
- Other ministries.

These journals have been granted the right to survive in Iraq. Most of them have been given to the associations, trade unions and official institutions and semi-official activity in the various aspects of scientific and technological advances.

This large number of journals has not succeeded in filling the void created by the need, not quenched thirst of researchers who are looking forward to science and knowledge and the reasons go back to:

- The Ministry of Higher Education and Scientific Research which issued 32 journals, some are annual and others are quarterly on semi-annual basis.
- Some journals published sporadically in the dates which are not fixed.
- Some journals are issued without serious preparation and real efficiencies.
- Thirty two specialized scientific journals that are published by the Ministry of Higher Education and Scientific Research were not chosen clearly; clear follow- up has not been issued.

- Low percentage of publishing attendance and continuation, is finding their way with difficulty and hardship to the various impediments and obstacles.

Besides, the journals or periodicals of the Ministry of Higher Education and Scientific Research, that issued according to decision No. 4660 of 5/7/1998, by the council of ministers, 44 periodicals magazine called the evaluated, issued by colleges and universities, associations and trade unions, federations and scientific institutions scientific centers and some of the specialized services.

Evaluated journals are characterized and listed as follows:

- Mostly depend on the fields of specialized fields such as medical, legal, pure science, engineering and pharmaceutics, electronics, agricultural and veterinary
- Some of these journals rely on strict specialization fields such as endemic diseases and digestive system, chemistry, social science and embryos research and infertility treatment.
- Some of the valuated journals are dependent on areas of scientific and humanitarian, especially those issued by private colleges and some trade unions and centers.
- There is no documented information about the continuation of these magazines and the nature of their issuance (annual, quarterly, semi-annual).
- These journals do not cover all specialized fields; they lack fields of the life sciences, biotechnology, genetic engineering, computer engineering, architecture and others.

These journals suffer from:

- Financial difficulties.
- Limited distribution
- Non-programmed continuations.
- Absence of uniform methodology for the titles.
- There are no uniform regulations for publication.

- There is no clear formula for the dissemination of scientific research and documentation of a confidential nature (limited circulation) well as the methodology documented.
- There is no danger in the dissemination of scientific research inside and outside Iraq.
- Responsibility for issuing magazines published by the Ministry of Higher Education is transferred to associations and trade unions which have long experience in the field of scientific publishing.

In the light of what has been referred to, we could raise the following proposals and recommendations:

- Support for scientific journals:
  - Increase the number of scientific periodicals published in Iraq.
  - Increase budgetary allocations to cover the cost of issuance and the development of permanent secretaries.
  - Linking the information network with international.
  - Adoption of the principle of part-time to oversee the journals.
- Relationship of journals outside Iraq.

- Urging scientific journals in the country to adopt a more open policy towards foreign searchers and invitation from abroad to contribute to the publication.
- Facilitate access to the local science publication by external scientific societies.

- **Scientific Publications and Research**
- The pressures of globalization makes it necessary to allocate sufficient resources to the higher education sector, and must be reform of the sector at the enterprise level and the system together, which, of Higher Education is may be the essential its tool because it helps developing countries to reap the benefits of globalization, the latter can help countries

attract investment foreign and help any country to take advantage of higher education products. Finally, the linking of globalization in higher education provides opportunities to improve living standards, India, for example, benefited from globalization by building computer industry, some believe that universities and higher education international perception of nature and global the trends, not globalization trends, where the world is different from globalization, higher education is under Globalization is seen as a commercial commodity, governed by market forces, if the positive of globalization from the perspective of increasing access to higher education opportunities and for the globalization of the economy listed Higher Education in the agenda of business for the World Trade Organization WTO is not for the contribution of higher education in development, but as a service of trade services or commodity commercial, and became the Higher Education market generates a lot of money, as well as it has affected globalization extensively in universities as a result of the policies of globalization, low The support of the public sector in the free market economies, led to universities are managed on a commercial basis is very clear, but many European universities that rely cooperative economics of the market did not follow the Anglo-American way of making universities are managed on a commercial basis.

- Knowledge and Education
- Knowledge (characterization)
- Knowledge is the mental output operations education and thinking is the basis of power and gain, and the most important ingredients contained in the Labour and activity, especially in relation to the economy, culture and education, which is also an economic asset, as well as it is a man and take care of it and prepared supplier basic knowledge.
- And see another section that knowledge is the awareness and understanding of the facts or the acquisition of information by trial and means to acquire the unknown and self-development are linked to the information, education and communication directly linked.
- A report by the World Bank entitled "Knowledge path of development" about the conditions of the knowledge and Fjutea between the

North and the South at the end of the twentieth century, and between this report the scientists in the world are not distributed equally among the countries of the north and south as follows:

- Africa 0.7%
- Arab countries 1.5%
- North America 19.8%
- Europe 20.2%
- Asia 32.4%
- Moreover, a knowledge-based society of being the most important products or raw materials, as it used the knowledge of modern technology in this society and not be restricted to co-exist in the geographic location of the same and it's become the most important capital components in the current era and is supposed to be their applications free and free of charge for the benefit of society.
- The transfer of knowledge to the individual call it learning, which is precisely the process of receiving knowledge through study and learning, while education is a process in which the learner acquires knowledge and skills, practiced university education and student learning exercise.
-
- Knowledge society Knowledge Society
- The knowledge society is defined as a group of highly convergent interests, who are trying to take advantage of their knowledge aggregation in areas that they are interested in.
- Called knowledge society postmodern society, and linking knowledge and economy, generating a profitable commodity.
- It requires a knowledge society, potential and special skills and super abilities and well-developed infrastructure and natural resources and minds notebook capable of producing knowledge and turn it into an economic power outweigh the process progress.
- As for the challenges of the knowledge society emphasizes the importance of university education, and its responsiveness to the requirements of the knowledge society and therefore requires improvements

to university education systems in order to turn it into knowledge production operations.

- Higher Education is the foundation of the relationship with the perpetrator of the knowledge society, which facilitates the exchange, advanced stage of acquisition of knowledge and the entrance to the broader knowledge society.

- Featuring a knowledge society which is called atheist-first century society (third millennium) its ability to produce knowledge and convert them into profitable commodities, a set of values has led to the adoption of this society and power, including:

- - Intellectual flexibility.
- - Teamwork.
- - The adoption of democratic values.
- - Rumor intellectual diversity.
- To the knowledge society infrastructure includes:
- - Material elements.
- - Technological elements.
- It is intended to infrastructure facilities and basic services and equipment needed by the community, such as:
- - Transportation.
- - Means of communication

- It is worth mentioning that the knowledge society is the basis of the information society, the report of the United Nations Educational, Scientific and Cultural Organization "UNESCO" for the organization in 2005, has been issued entitled (of the information society to the knowledge society) was contained in it under the title: (can not be reduced to knowledge societies to the information society) .

- In Iraq, knowledge Iraqi university community defies easy challenge The community of advanced knowledge for being the Iraqi and the university backward Entering Iraqi University knowledge society requires them to provide a sophisticated infrastructure of ICT favorable and a climate of stability and systems teach contemporary take new technologies and emphasizes the Supreme actual operations.

-

- Knowledge and scientific publishing output

- Forms of knowledge production advanced stage to gain knowledge and can generally measure this production through scientific publications, patents and innovations, as observed a significant increase in the movement of scientific publishing in the Arab countries, the number of Arab scientists publications has risen annually from (465) in Bulletin (1967) to about (7000) Bulletin year (1995) in the fields of medicine and basic science implant. According to the Human Development Report for the year (2003) of scientific publications index, which is measured in the number of research articles published in the Court of global patrols per million people in 1995.

- 26 in all Arab countries.
- 840 in France
- 1253 in the Netherlands
- 1878 in Switzerland

- It is noted in Iraq, said the large gap between him and the countries of the world in the production of knowledge, in order to reduce this gap has to be substantial changes to the education systems in Iraqi universities, most notably the need for a modern communications network and the rules of knowledge and information new and participation in information systems and a culture of education in the spirit of building the team has taken the Ministry of Higher Education and strides to issue scientific journals specialized court and in various scientific disciplines and humanity.

- Salah al-Ahmad et al was presented (1987) a report on the period from 1980 to 1985, between that productivity in Arab countries was 0.44 lookup for each doctor and the number of scientific articles (2002) and published by researchers in some Arab countries in international journals classified: In Egypt (2500) and Saudi Arabia (1300), Lebanon (300), Oman (236) and Syria (108).

- The patents registered in the United States to some countries for the period (1976-2002) vary as the number of (104), while South Korea (27 298), Sweden (26318).

- The number of workers in the production of knowledge, it is steadily rising, but that there is a weak response to market demands of higher education systems in the Arab countries and the shortcomings in the education requirements in the areas of science and culture.

-

- Knowledge economy and electronic babysitter

- Knowledge-based economy represents a new kind of economy, different from the old economy that was dependent on the land, labor and capital as factors of production while the new economy called knowledge economy depends on the other, including technical knowledge, creativity, and intelligence agents, and information. The United Nations estimates that the knowledge economy accounts for 7% of the global total domestic Nantj. Accordingly, the knowledge economy must be a knowledge engine president for economic growth, and is characterized by knowledge-based economy by:

- - Innovation.
- - Education.
- - The infrastructure of information technology.

- And that the major powers outs of the knowledge economy is changing the trade rules and national capacity in the knowledge economy, including:

- - Globalization.
- - The information revolution.
- - The spread of communications networks.

- Knowledge-based economy plays an important role in the knowledge society, which is also associated with globalization depends on the economy and grew up in it, and that contributed to the result of developments in technology knowledge:

- - To find a new economic regulations and rules.
- - Opening up of world markets.
- - Redraw the economic map of the world.
- - The emergence of new centers rely on global trade.

-

- The emergence of knowledge-based economy led to the confirmation of the importance of education is key to economic success and the knowledge-based society, which is closely linked to the knowledge economy is assumed that each of them depends on the brains contemplative creative generated from higher education that contribute to a major role in the production of knowledge, so the relationship between the universities and the knowledge economy document too, because the knowledge economy based on knowledge production, and the production of knowledge is one of the most important functions of contemporary college.
- According to the logical perceptions of the importance of the knowledge economy and the adoption of private standards him () indicates the UNESCO Institute unesco statistics institute for statistics in 2004 to the need to adopt indicators:
- Overall spending total expenditure
- The intensity of spending intensity expenditure
- For comparison between countries of the world in research and development potentials, rises each in industrialized countries as both of them in the consuming countries of the industry decline, confirming the link between him and the arbitrator Higher Education.
- And within the horizons of knowledge economy is also the government support multiple channels of higher education, including research and development in the field of health and medicine in particular, noted an increase in the average age of the man during the last century (1900-2000) about 30 years old and thus resulted in an increase in imports of society about $ 2.4 trillion.
- In order for the Iraqi University to play a prominent role in the knowledge economy must be tailor learning processes as necessary economically active, contributes an essential role in the knowledge that represent a necessary chain in the production of a knowledge society, and that the Iraqi University also solving a lot of problems that faced, including:
- - The link between higher education and the labor market.
- - Strengthening the IT infrastructure of knowledge.

- - Trying to push the traditional features of all these universities. Knowledge methodology

- Human knowledge passed on three main stages theological Femitaveziqih then a scientific first position include the supernatural, interpretation and human evolution towards mental abstraction and abstract metaphysical and the second phase refers to the union of the Church with the power became a religious text is the reference cognitive basic In other words, the rule of the religious text in the perception of the universe and man. The third phase refers to the situation based on scientific discoveries of Copernicus which has shaken confidence religious sources and philosophical foundations of mental Descartes, which tried to bring out others with knowledge of the church to the mental visualization and the Count of trying to establish a so-called positive religion and turning religion into social phenomena.

-

- Knowledge and Education

- Believes that cognitive activities include knowledge generation research, development and dissemination of training and education means different media and using them and using them to improve services and in humans and its potential. Also it believes that these cognitive activities fall within the functions of higher education institutions. And it requires the development of a knowledge society system with integrated form the basis of a major affect motivate and accordingly, the need to build a knowledge society mainly to higher education advanced knowledge and education.

-

- Human knowledge

- Human knowledge refers to the problematic dialectic in humans Some thinkers is Almtdnyin believes that the main sources in humans sense and experience and denied the role of reason and revelation, or what they called metaphysics either including Islamists, have believed that sense and experience are two of the main sources of knowledge, but the reason .

- There is a common science goals are:

- - The collection of information description and interpretation and forecasting
- - Formation of knowledge Problems faced several of them the exact wording that exceed the quantity to quality measurement process and to understand and shape to subcontractors.
- - The introduction of change to look for the value of science and scientific fact and solving real problems.
- - To obtain the certificate.
-
- There has never been a scientific renaissance without knowledge of production through scientific publishing, which it can be found on the latest scientific achievements in the whole world. Talking about scientific publishing is very important, because in this stage of our development and scientific can lead a very effective role in enriching the scientific research, and to expand its range and providing it the achievements of researchers as well as to compensate for the lack of specific activities and creative achievements.
- Scientific publishing methods
- Scientific publishing methods include the following:
- - Scientific Publishing specialist (patrols).
- - Scientific Publishing Mass (radio, TV, press, popular magazines).
- - Scientific documentation (a confidential nature) of scientific publishing.
-
- As for the scientific publishing specialist It lends character Khasas follows a systematic and clear, scientific organization and the way in writing, exposed in detail, to the process of scientific research, involving technical complexities and difficulties, and many problematic. One of the important points that interest in them is supposed to detect the reality of scientific publishing specialist as an essential step to improve this reality and develop it, Vtchkas scientific publishing problems Introduction and the starting point to describe her treatment, so as to ensure the benefit of scientific research, which is a scientific publication, including a key part. In the light of the

existing problems, for example, point out that there are, above all, an urgent need to develop a well-developed standards and effective legislation to regulate the process of scientific publishing and editing of improvisation and randomness.

- - As it was necessary to issue bulletins patrol at the level of each country's business documents that have been published.
- - It also highlights the need for the need to provide incentives for greater moral and material.
- - As well as the need to pause and clear for the global network of the role of information (the Internet) as an important means to obtain scientific information.
- There is no doubt that improving the scientific publishing should not be limited to professionals, but must exceed that of scientific publishing to the public (press, radio and television public and scientific journals).
- Scientific Publications / globally
- Scientific Publishing concentrated globally in the advanced industrialized countries and economically. As it produces only nine countries more than (90%) of the global scientific production, which ever recorded by scientific information institutes.
- Scientific production to developing countries represents about 6% of world production, including:
- 3.74% of the Asian continent
- 1.15% Latin America
- 0.59% Middle East and North Africa
- 0.38% black Africa
- It is worth mentioning that India China People's excellence in scientific production and by (1: 5) Egypt has maintained a leading position in the list, and despite the apparent increase in the amount of scientific production to developing countries.
- If we compare the rates of local publishing and on the outer continental level, we find that East Asia and Latin America, researchers are more of the locally published either in Africa is higher than the proportion of research published abroad. But in developed countries, researchers published

what these countries in foreign periodicals does not exceed (20%) and the French (25%) and the Japanese (12%) of the overall European researchers.

- All the periodicals issued in languages other than English does not impair the right of attention.

- - In statistical included researchers from the English-speaking countries show that very low rate published in other languages.

- - While we find that (17%) of the French-speaking researchers and published in English (36%) of the researchers speaking in Spanish or Portuguese published in English.

-

- The researchers of the English-speaking countries is almost no point in mentors to research published in other languages.

-

- Scientific Publications / developing countries

- Studies that rely on banks and international information to researchers of the third division of the world into two categories tend: -

- - Those who spread their production abroad in her weight international patrols and who are taken into consideration.

- - The second category are the ones who provide locally note it is usually a great value despite the fact that modern research in the areas of modern indicate the slimness of this science despite the wide spread, and that local knowledge is not necessarily a low quality.

-

- It was clear from other studies, conducted on East Asia, said the local scientific journals which had reached an advanced level, and the researchers spread by choice and not because they are forced to because they can not external publication.

- Also, articles researchers developing countries fry great interest if published with the participation of researchers from the industrialized countries and this raises a very important point is the choice that must be done by the third researchers world between research in mainstream science globally and the trend towards solving local problems, it became apparent that the local scientific journals which reached an advanced level, and where researchers publish their choice.

- The measurement and analysis of the overall scientific production to developing countries, including Arab and Iraq impossible scientifically where supported in the scientific local business rules are not available most of those who tried to estimate the scientific production and scientific publication on international rules depends (Institute of Scientific Information ISI and scientific citations SCI guide. Where there is a scientific publication In third world countries, Iraq and one Mnha- more than 5% of the global publishing, and can cope with this kind of statistical Calendar if we know that the rules of this selective information for scientific journals and because most of the output of science in Iraq is not taken into consideration.
- Of the more than 100 thousand scientific periodical issued in the world today it is using the rules of the information listed about five thousand only patrol, which has been the practice to be called mainstream science, so the image formed incomplete and erroneous in relation to the contribution of researchers of the country in the global science.
-
- Scientific Publishing characterization in Iraq
- Statistics indicate that a significant portion of the work researchers Iraq published in the local patrols are not distributed, but rarely outside the national borders. Even if the deployment of Iraqi scientists in their research with an international scientific reputation, the essays will not be cited in the articles colleagues Western scientists magazines and so it is clear that reliance on the phenomenon of martyrdom greatly raises controversy and broad. In other statistics indicate that more scientific production in Iraq in the field of pure science at the rate of an article to researcher annually publishes mostly within publishing scientific output in the local leagues represents a reality strategically and a decision on the part of the researcher local easiest and surest of correspondence scientific journal foreign publishing the . It also allows the researcher identified himself to local readers of colleagues, students and others, and they would not have seen the publication of his research if in a foreign magazine.

- Notes that the articles published locally read and cited faster than those published abroad Thus, (60%) of local references cited recent research, while the proportion of references modern foreign dwindle, and cited local articles remains small because it was represented only 2% of the total References mentioned in the research published locally in Iraq.

- Statistics show the total of the Ministry of Information previously that more than 44 refereed journal has been granted the right to be issued at the breasts are: -

- - Formerly the Ministry of Information.
- - Ministry of Culture.
- - The Ministry of Higher Education and Scientific Research and its institutions.
- - Associations and trade unions.
- - Other ministries.
-

- It is clear that these magazines have been granted the right to chests in Iraq has been mostly given to associations, trade unions and official institutions and the semi-official in various aspects of scientific and technological activity.

- That such a large number of magazines did not succeed in filling the void created by the need not extinguish the thirst of eager researchers of science and knowledge and the reasons go back to factors such as:

- • considered the Ministry of Higher Education and Scientific Research issued each annual magazine and some quarterly and semi-annual basis.
- • Some magazines published sporadically in the non-fixed dates.
- • Some magazines are issued without serious preparation is based on a real efficiencies.
- • was selected specialized scientific magazines published by the Ministry of Higher Education and Scientific Research formula is not clear and did not pursue issued.
- • suffer from these periodicals: -
- - Physical difficulties.

- - Irregular breasts.
- - Limited distribution.
- - Subscriptions is programmed.
- • lack of a unified methodology addresses these patrols.
- • There are no uniform regulations for publication.
- • There is no clear formula for publication of scientific research and documentation of a confidential nature and limited trading, as well as in methodology documented.
-
- In the light of what was referred to can make proposals and recommendations as follows: -
- • Support scientific journals:
- - Increase the number of scientific journals published in Iraq.
- - Increase the financial allocations for the patrols to cover the cost version and the development of a permanent secretariat.
- - Linking patrols international information network and locate her.
- - The adoption of the principle of part-time to oversee the patrol.
- - The provision of other possibilities for the issuance of the patrol.
- • relationship patrols outside of Iraq.
- - Urged scientific journals in Iraq to follow the more open towards the outside, inviting researchers from abroad to contribute to the publication available at the local science published by scientific third parties and coordination with banks and international information to the attention of larger science product locally and to allow for patrols and researchers for the World Leagues and facilitate faster and easier.
- But with regard to the international information network requires: -
- - Expand and develop the use of the network and stay away from the approved methodology for the time being.
- - Allow professors, researchers and graduate students access to a network of scientific facilities.
- - Allow teachers and researchers access to positions in the network.
- - Re-consider controls blocking some foreign websites scientific beneficiaries inside Iraq.

- General features of scientific publishing in developing countries and the situation in Iraq

- First: bring charges against local scientific journals in third world countries, such as the country has a low level, with articles taken on patrols following observations:

- - It's a committee to evaluate the articles do not support it sometimes publish low-quality articles or questionable scientific findings and authenticity.

- - Lack of proper research methodology accepted for publication.

-

- Second: it indicates the number of accredited studies in developing countries, which can be applied in the country, especially those that rely on global cite evidence that it's not convincing in the production of scientific evaluation of the Third World, including the diameter. The measurement and analysis of the scientific output of these countries is almost impossible because of the lack of scientifically supported information the local rules of scientific work.

- Third, most of the scientific tried to estimate the production of scientific and publishing information on international rules, including the ISI Information Institute and the adoption of SCI cited scientific evidence supports.

- - It does not represent a scientific publication in third world countries (including diameter) more than 5% of global scientific publishing.

- - And you can cope with this kind of statistical Calendar if we know that the rules of this information selective scientific journals that are selected note that the majority of patrols did not fall within that choice yet.

- - Of the more than 100 thousand scientific journal published in the world currently using the rules mentioned information about (5000) Selected thousands patrol only, which has been the practice to be called mainstream science (Main stream science) So is the image the world formed incomplete and erroneous in relation to contributions to the country researchers in the world of science.

- Scientific Publishing characterization in Iraq

- First: Statistics show that a large part of the country spreading the work of researchers in the patrols are not distributed, but rarely outside the national borders.

- Second, even if scientists publishing their research in the country with an international scientific reputation, the magazines will not be their articles cited in the articles as much as their peers scientists in developed countries This makes it clear that the scientific citation is mainly based on research published scientific originality and depth.

- Third: Hchir other statistics that the majority of the scientific output in the country in the field of pure science published in the inside diameter.

- Fourth: The deployment of the scientific output in the local leagues represents a reality Stertejia by local researcher Valnscher easier, faster and guarantee of foreign correspondent scientific journal for publication in it. And also it allows the researcher identified himself to local readers of colleagues, students and others, and they would not have seen the publication of his research if in a foreign magazine due to unavailability locally.

- Fifth: noticed that articles published locally read and cited faster than those published abroad and so little about (60%) of local authorities in recent research while male lineage of modern foreign references diminished.

- Those responsible for the deployment in Iraq

- First: Indicates the total statistics from the Ministry of Information that more than 44 scientific journal court may grant the breasts in diameter and shall be issued the following authorities: -

- - The Ministry of Information.

- - Ministry of Culture.

- - The Ministry of Higher Education and Scientific Research and its institutions.

- - Scientific societies.

- - Other ministries.

- Second, it is clear that most of these magazines shall be issued scientific societies official and semi-official institutions and dealing with various aspects of scientific and technological activity.

- Third: that such a large number of magazines did not succeed in filling the void created by the need not extinguish eager thirst for science and knowledge of researchers and the reasons for this as follows: -

- - Considered the Ministry of Higher Education and Scientific Research to issue (32) annual scientific journal each court and others quarterly and semi-annual basis.

- - Some magazines published Sdora is spotty at fixed dates.

- - Some magazines are issued without serious preparation is based on a real efficiencies.

- - specialized scientific journal Issued of the Ministry of Higher Education and Scientific Research have been selected format is clear and has not been issued follow-up rates.

- - The proportion of part at least, which issued assiduously and he continued his way with difficulty and hardship in front of various obstacles and hurdles.

-

- As well as the magazines or the Ministry of Higher Education and Scientific Research patrols issued on according to the Council of Ministers Resolution No. 4660 on 07.05.1998, there are a number of periodicals named magazines Court and number issued by colleges and universities, associations and trade unions, scientific institutions and institutes and scientific centers and some Aldoaihr We specialized in various ministries such as the Department of Health Nineveh and follow-up committee of agricultural magazine in the province of Nineveh.

- Characterized magazines Court mentioned the following: -

- First, most of which depend on specialized medical and legal fields Pure and engineering, pharmacy, electronics and agricultural and veterinary sciences and science.

- Second, some of these journals rely on Specialization Kalamrad settlement and digestive system, chemistry, social sciences, research embryos and infertility treatment.

- Third: Some of these journals rely on scientific and humanitarian fields and public, particularly those issued by community colleges and some trade unions and centers.
- Fourth: There is no documented information on the continuation of these magazines and continuity in the breasts and the nature of issuance (annual and semi-annual, and quarterly) have not been fundamentalist her calendar.
- Fifth: This Court magazines do not cover all the minute to the terms of reference are lacking in the life sciences, and biotechnology, genetic engineering, and computer engineering, architecture and others.
-
- C-set fixed editorial boards accordance with international standards to be enjoyed by the editorial staff advanced scientific competence in the field of competence and does not require administrative position mainly in the selection and includes a choice of who are in all parts of the country.
- D-reconsider the editorial boards periodically re-especially for those periodicals that have not persevere on the issue.
- E-The Ministry of Higher Education and Scientific Research issuing registration scientific journals Court in coordination with the Ministries of Culture and Information instructions.
- And-The Ministry of Culture to issue special instructions patrols relevant mass culture.
- G-revision in rates of issuing official journals, particularly those that are issued annually and semi-annually and the adoption of the pace of issuing quarterly basis.
- H-only issue the official journals in the case of a number of research at least six and that is the version in time.
- Fifth: the confidential nature of research.
- Proposes in this area follows:
- A-activate the work of committees launch information in ministries and departments not associated with the Ministry and choosing the trading stamp on the gradient according to the following: -
- - Trading in a research that can be deployed.

- - Limited trading.
- - Secret.
- - Top secret.
- (B) shall be reviewed annually in the trading stamp from the Commission to launch information or modify the nature of trading and be under the official records of the ratification of the competent minister.
- C-development of the concept of scientific documentation, methods and reconsider the law No. 70 of 1983 (Act preserve documents) that writes research methodology as a binder for scientific research and publication itself, either studies and reports is recorded as usual.
- Sixth: sporadic cases
- A-The Ministry of Higher Education and Scientific Research, the overall supervision of the official journals through the competent department in the ministry is working to support patrols and coordination between them and work to be issued on time specified.
- B-position indexing references and articles in diameter system to be a reference for researchers inside and outside the country.
- C-support and documentation centers in universities and research institutes libraries in the country in the use of the international information network.
- D-character provide support public science education and scientific journals (such as Science magazine) is similar to the support provided for magazines and specialist periodicals.tion are two of the supScientific research and the curriculum (83)
- • Provide
- • General features of Islamic scientific approach
- • methodology in Higher Education
- • methodology models
- Submitting

**E.Science and Technology policies in Iraq Technology**

Up to this point, we have been concerned with basic science and its support by government funds in a modern society. Although there is also some support by private institutions established for that purpose and also some industrial investment in generally product-oriented basic research, the greatest amount of support by far comes from public funds. One of the ways that the public is repaid for their support is through the technology that fundamental research generates. I suspect that the economic return from technology alone more than compensates for the monies expended for the entire basic research effort. I have no estimate, however, of whether my suspicion is true or not. It should be noted that the public gains much more than the economic value of technology. It gains culture, comfort, convenience, security, recreation, health and the extension of life. What monetary value can be put on the triumphs of health over debilitating or fatal disease? The monetary value has to be higher than the purely economic savings that were noted above in the 26 examples referred to in the FASEB Bulletin.

The word "technology" means industrial science and is usually associated with major activities such as manufacturing, transportation and communication. Technology has been, in fact, closely associated with the evolution of man starting with tools, clothing, fire, shelter and various other basic survival items. The co-evolution persists and, since basic science is now very much a part of developing technologies, the term co-evolution of science and society which is used at times very much implies the co-evolution of both basic science and industrial science with society. Advances in technology are generally accompanied by social changes as a consequence of changing economies and ways of carrying out life's various activities. An important question arises concerning how basic scientific discoveries eventually lead to new technologies and what that may mean to the rational support of basic research and the future of science and technology in the developed and developing world.

There are great uncertainties in the process that starts with basic research and ends with an economically successful technology. The successful discovery of a new development in research that appears to have technological significance does not ensure the economic success of technologies that may be based on it.

Nathan Rosenberg of Stanford University, in a speech, "Uncertainty and Technological Change", before the National Academy of Sciences (April, 1994), pointed out that there are great uncertainties regarding economic success even in research that is generally directed toward a specific technological goal. He notes that uncertainties derive from many sources, for example, failure to appreciate the extent to which a market may expand from future improvement of the technology, the fact that technologies arise with characteristics that are not immediately appreciated, and failure to comprehend the significance of improvements in complementary inventions, that is inventions that enhance the potential of the original technology. Rosenberg also points out that many new technological regimes take many years before they replace an established technology and that technological revolutions are never completed overnight. They require a long gestation period. Initially it is very difficult to conceptualize the nature of entirely new systems that develop by evolving over time. Rosenberg goes on to note that major or "breakthrough" innovations induce other innovations and their "ultimate impact depends on identifying certain specific categories of human needs and catering to them in novel or more cost effective ways. New technologies need to pass an economic test, not just a technological one."

What does this mean with regard to government managed research? I quote from Rosenberg's speech.

"I become distinctly nervous when I hear it urged upon the research community that it should unfurl the flag of 'relevance' to social and economic needs. The burden of much of what I said is that we frequently simply do not know what new findings may turn out to be relevant, or to what particular realm of human activity that relevance may eventually apply. Indeed, I have been staking the broad claim that a pervasive uncertainty characterizes, not just basic research, where it is generally acknowledged, but the realm of product design and new product development as well - i.e., the D of R&D. Consequently, early precommitment to any specific, large-scale technology project, as opposed to a more limited, sequential decision-making approach, is likely to be hazardous - i.e., unnecessarily costly. Evidence for this assertion abounds in such fields as weapons procurement, the space program, research on the development of an artificial heart, and synthetic fuels.

"The pervasiveness of uncertainty suggests that the government should ordinarily resist the temptation to play the role of a champion of any one technological alternative, such as nuclear power, or any narrowly concentrated focus of research support, such as the War on Cancer. Rather, it would seem to make a great deal of sense to manage a deliberately diversified research portfolio, a portfolio that will illuminate a range of alternatives in the event of a reordering of social or economic priorities. My criticism of the federal government's postwar energy policy is not that it made a major commitment to nuclear power that subsequently turned out to be problem-ridden. Rather, the criticism is aimed at the single-mindedness of the focus on nuclear power that led to a comparative neglect of many other alternatives, including not only alternative energy sources but improvements in the efficiency of energy utilization."

To these words, I add those (noted by FASEB) of Bruce Ferguson, Executive Vice President of Orbital Sciences Corporation, a space technology firm. Ferguson said, "The federal government should focus its research and development spending on those areas for which the benefits are diffuse and likely to be realized over many years, rather than areas for which benefits are concentrated on particular products or firms over a few years. These areas are not well covered by corporate investment, yet are vital to the long-term economic strength of the country."

Some reactions to "strategic" research are recounted in an article in *Nature* of February 10, 1994 (Vol. 367, pp. 495-496) from which I quote some passages. The concept of strategic research "is not an unfamiliar cry, witness last year's debate in Britain about harnessing of research to 'wealth creation.' Nor, of course, is the objective in any way disreputable; what scientist would not be cheered to know that his or her research won practical benefits for the wider world as well as a modicum of understanding? The difficulties are those of telling in advance which particular pieces of research will lead to 'new technologies' and then to 'jobs'.

"The recent past is littered with examples of adventurous goal-directed programmes of research and development which have failed for intrinsic reasons or which, alternatively, have been technically successful, but unusable for economic or other reasons."

The article goes on to say that the affection for strategic research in the United States may prove short-lived. "In Britain, much the same seems to be happening. Having pinned its reorganization of research on the doctrine of science for wealth-creation, the government appears now to be more conscious of the problems it has undertaken to solve. Indeed, the prime minister, John Major, seemed to be suggesting in a speech last week that the British part of the research enterprise deserves respect of the kind accorded to other social institutions at the heart of his 'back to basics' rhetoric. After more than a decade of needless damage-doing, that would be only prudent."

As a final remark, the article ends with the statement: "On the grander questions, on both sides of the Atlantic, it seems likely that the first flush of enthusiasm for turning research into prosperity will be abated by the reality of the difficulties of doing so. When governments discover in the course of seeking radical reorganization that the best they can do with their parts of the research enterprise is to cherish them, the lessons are likely to be remembered. If the outcome in the research community is a more vivid awareness of how much the world at large looks to research for its improvement, so much the better."

**The Future of Science, Technology and Society**

In discussing the future of science (including industrial science) and society, it is valuable to recount some of the important points that emerged from the previous discussion.

1. As a consequence of recognizing the economic benefits that derive from the development of novel, successful technologies, governments have been attempting to direct research, supported with public funds, toward subjects that are perceived as national priorities. This contrasts with broad-based "curiosity" oriented basic research.

2. The views of scientists, a distinguished economist, some industrial leaders and an editorial comment in a distinguished science journal provide very strong indications that governmental management of goal-oriented research is replete with uncertainties and pitfalls and, although well-motivated, may cause serious damage to the scientific culture. This, of course, would defeat the original purpose, since the co-evolution of science and society is a very-well documented and irrefutable phenomenon.

3. Strong arguments are presented in this article by individuals and groups that support the current system of governmental funding of a very broad range of scientific efforts as probably being as close to optimal with regard to national priorities as is possible. No one can predict with any certainty what the most successful inventions and technologies will be in the future. The economic return on federally supported funding was the subject of a report by the Council of Economic Advisors to President Clinton. This report was released in November 1995. It documents high returns to the economy and the importance of governmental involvement. [1]

4. By any measure, basic scientific research has made monumental contributions to technology and national priorities. The bond between basic research and the development of both novel and current technologies has been and is well in place.

There is no question that science and society will continue to co-evolve. The nature of this evolution will certainly be affected by the extent to which governments set funding priorities. Societies whose governments recognize the dependence of the development of successful novel technologies on broadly supported basic research are more likely to be healthier and economically prosperous in the future than those that do not. Because of the unpredictability of the details of the new science and technology that will evolve, the details of social evolution are also unpredictable.

Science and Technology play a crucial role in shaping the challenges faced by individuals, organizations and nations constantly, (discoveries of genetic engineering, industrial human parties, mobile phone, biotechnology…etc.). The challenge that was facing by mankind at the beginning of the twentieth century is how all countries could benefit from the strength of science and technology.

However, Iraq was unable clearly to improve the use of available science and technology, like other Arab countries, despite the availability of consulting firms and construction companies, millions of university graduates and about one million Arab engineers and hundreds of industrial companies and thousands of universities teachers.

The challenges facing Iraq, lies in the two groups, first resulting from major developmental problems, food security, health, housing human rights, education, transport, and difficulty caused by the absence of the required scientific culture. The second is cultural in nature and includes a special site independently. According to this scenario it requires the creation of systems of national science and technology take upon themselves the development of science and technology policies.

Iraq made over the past decades considerable progress in several areas, and has increased the resources allocated to education, social services and infrastructure which has had a positive impact on the average per capita income and quality of life, followed by problems of concern (during the nineties). The per capita of the total GNP was decreased, the blockade halted efforts to diversify the economy and the adoption of the main sources of GNP on non-renewable mineral wealth.

There are some hypothesis for scientific and technological policies in the country which are limited in scope, effectiveness and ineffective strategies in the best of circumstances. So it was proposed to create new structures after some test on developed countries and developing countries.

The availability of financial resources in Iraq, especially before the nineties of the twentieth century has made great efforts in manufacturing, especially in the field of Military Industrialization. Iraq succeeded in building an independent military industrial base by enabling technology in modern manufacturing processes, however, strategies and manufacturing policies lacked effective action to develop local capacity and providing appropriate incentives for local people to be able in modern industrial technology.

Despite this, Iraq still needs to pursue innovative methods to meet the daunting challenges of a large number of other sectors of production and services. It also needs to develop scientific and technological capabilities of local traditional industries to modernize and to address a variety of social economic problems.

The scientific efficiency in Iraq itself has a capacity of innovative crucial role in facing challenges, in spite of standard conditions faced by the departure of many of them outside Iraq.

The limited scientific successes that have occurred in Iraq are the result of the efforts of some institutions of science and technology (Scientific Academy, the House of Alhikma, and institutions of higher education). On the other hand, important achievements were made in building the institutions mentioned, as well as in human resources development. But the institutions mentioned and others are still far from enabling their employees from playing a distinguished role in development. The linkages and synergies between the scientific and technological institutions of governmental organization, and the business world remain weak, despite the existence of some versions of contracting.

The spending on research and development in Iraq, at best, less than its counterpart in the Arab countries and more discouraging regarding the outputs of science and technology, to the degree that scientific and technological publications in specialized areas and the number of patents granted to institution and individuals are much less than the average of the corresponding figures in other developing countries.

As a result, the status of science and technology in Iraq needs a lot of attention. The input and output of scientific and technological point indicate the deficiencies in information networks, computers, advanced equipment, and scientific research.

These include, institutions (universities, research centers universities, atomic energy, etc.). Their shortcomings in Iraq are one of the reasons that led to the absence of scientific and technological policies, and the recognition of the limitations. The ineffectiveness of administrative practices and the existence of structural deficiencies are symbolic recognition of the need to develop capabilities in

Scientific Research in Iraq

Interest in scientific research began to change after the concept of university education, so that teaching is no longer the sole function of universities. However, the insight into scientific research in Iraqi universities shows that scientific research suffers from many problems including lack of funding, as the salaries of members of the teaching staff amount to nearly 90% of the budgets of universities. The remaining is distributed on all other aspects of spending, and thus negatively, it affects the requirements for scientific research devices and other supplies, as w as other sufferings of scientific research in Iraqi universities. It is considered merely functional performance by teaching, to achieve the goals of self separate from the research.

The output of scientific research in higher education institutions

The reality of scientific research in Iraqi universities before (2003) was in accordance with the following elements:

- Forces and human cadres.
- Communication and networking.

- Funding.
- Research basic and applied sciences
- Elements of research and development

Forces and human cadres

Mainly, inc1ude teachers who consider scientific research, as well as community services. The correlation of promotion with scientific research is done in institutions holding the doctorate degrees. Moreover, human cadres including graduate students and new graduates play a positive role in completing advanced research.

Impediments to scientific research

It is clear, that reality of scientific research in Iraq does not live up to expectation of a country trying to do in turn to contribute to the progress of scientific research, access to higher share of the global economy, investment of knowledge as a result of scientific research and development. There are systematic obstacles and constraints linked to legislation and administrative, other impediments related to financing of scientific research and those relating to human cadres and constraints related to the interface and communication.

Methodological constraints is related to education in the methods of scientific research, while those associated with legislation is a clear view of the fact that instruction of promotions, for example, encourage individualism while those of the administrations found in the department, colleges and universities, the ministry and specialized research centers suffer from lack of coordination between different departments of research. The Obstacle of funding is one of the main channels in the country which is much lower than the proportion of GNP, representing one of the most major constraints for scientific research. While the cadres constraints refer to the wishes of the students involved to obtain certificates to improve their livelihood rather than being concerned with scientific research and increasing burden of teaching to teach in universities and the lack of clear criteria for selecting cadres which assume responsibility for scientific research centers.

## Evaluation of university scientific research in Iraq

A sound scientific planning for the development of scientific research of the university requires first the study of the current reality as it is a scientific study designed to learn the possibilities available in the universities and was done in this area to identify the negative aspects of it.

The scientific research at universities in Iraq, is one of three core functions of teachers as well as teaching and community service is linked with the promotion of academic holders of doctorates, while the remaining percentage of teachers holding a master degree of the operation full-time faculty.

There is a constant increase in the number of teachers in Iraqi universities that commensurate with the continued increase in the numbers of students enrolled in universities, and the indicator on the high ratio of students to teacher, an additional burden on teaching, which in turn reflected negatively on scientific research quantity and quality in universities.

The postgraduate students in Iraqi universities are an important part of the manpower involved in scientific research. Those students shape the future researchers, at universities in developed countries and have an important role in the completion of scientific research. Despite the expansion of universities in postgraduate programmes in Iraq, their contribution to scientific research is still pursuing their studies and specification in difficult circumstances. The current reality of manpower in the relevant scientific research in Iraq is characterized by low numbers of full-time researchers in spite of individualism to conduct research and the scarcity of complementary research teams.

The high proportion of the number of students compared to the teachers of acknowledged international figures, and the scarcity of opportunities for research assistants and technicians for training in developed countries to deal with the specialized instruments and maintenance of research laboratory, is a problem that should be considered seriously. The available statistics indicate that Iraq had spent a few sums of money on research and development after (2003). In an analysis of the budgets of universities refer to focus on supporting research projects, publishing and missions.

It is well known that the funding of scientific research in a university in developed countries comes mostly from the industrial sector and has reached the proportion of the funding year (1996) in countries like Japan to 67% and the United States to 63%, while the contribution of the industrial sector in supporting scientific research in Arab countries is very modest, almost Non - existent in many universities. Statistics indicate that the total expenditure on research and development at universities and colleges in America (2002) to 36,333 billion dollars when the federal government contributed.

The productivity of doctorate-holding teachers in Iraqi universities in (2003) was 0.44 researches for each holder of a doctorate each year, while it was 1.00 a year ago (2003). Ministry of Higher Education in (2002) in collaboration with universities strides for the issuance of specialized scientific journals in nature and in various scientific disciplines and it is hoped that these magazines with a scientific level are classified and adopted globally.

The nature of research and investment that may lead to creative results after the year (2003) and industrial applications and patents are rare. It is clear, that contribution in this area is very modest, because of the inability of the institutions of higher education to transform the results of scientific research to investment project because of several obstacles. Iraqi universities after the year (2003) have received a package of legislations that have given special attention to scientific research. Instruction were issued own encouragement, regulations adopted by all universities and research achievements as a condition for promotion of distinguished professor status. The legislations included those issued by the university foundations to support scientific research, publishing support and instruction to published journals by the addition to patents and instruction: for the dispatch of teaching attending conferences and symposia.

Iraqi universities and scientific research

Scientific research in universities is inseparable about the problems of Iraqi society, where research aims to obtain a degree or promotion, its role in the production of knowledge and solving problems is very limited, and worth noting that universities operate on the developed world.

- Scientific research and its role in the production of knowledge and solving problems.
- Development a clear policy for scientific research.
- Availability of information to help scientific research.
- Service sectors of society productivity through scientific research.

The graduate students in Iraqi universities are an important part of the manpower involved in scientific research, those students shape the future researchers, therefore these universities, create conditions of physical, moral, appropriate to attract. Despite the expansion of universities and increasing umbers of graduate students, the development of scientific research is still limited and attending school is in difficult circumstances.

Human manpower related to scientific research in Iraqi universities is characterized by a strong current reality with the following:

- The opportunity for holders of doctorates from the newly graduated in training and direct involvement in training is very limited.
- Scarcity of complementary research teams.
- The proportion of the number of students to the teachers of acknowledged international proportions is high.
- Preoccupation with a large number of teaching staff to work overtime.
- Opportunities for research assistants and technicians for training in developed countries are very limited.

Available statistics indicate that Iraq spent in 2003 on research and development rate of 0.4 % of GNP, and the universities have contributed hitting 31% of the volume of expenditure on research and development.

The analytical study on the budgets of university official in 2003, indicates that the expenses of scientific research (to support research projects, publications, journal and scientific conferences, books and periodicals) had reached in universities, compared with expenses of scientific research and development, in the year (2000), increased by 1.4% as it in (2000) and that most the increases were concentrated in the creators of scientific research (supporting , research projects, conferences and publishing missions and training).

**F.Scientific Methodology and Scientific research Methodology in Higher Education**

Strategic planning for higher education faces challenges through

- Future directions of higher education and future goals and objectives.
- The participation of diverse segments of society in the formulation of the strategy.
- Invent ways and new mechanisms work.
- Identify areas of change facing the educational system.
- The external environment assessment of the institutions of higher education.
- Evaluating the higher education system.

Described science policy and technology as a range of sectors development deals with some direct scientific activities and technology and their relationship in the establishment of infrastructure national science and technology, and account for a large range of investment and financial policies and laws relating to intellectual property rights and trade policies, export and transfer of technology and research and development.

And involve policies adopted by government departments in general and institutions to contribute to the national innovation, both based on local technological capacity or to the imported technologies. However, these policies may be unrelated, but may be sometimes contradictory Therefore, the main task of the policy of science and technology and create a framework for policy coordination and increase the collective impact.

After reviewing a number of scientific policies and technology in the countries of the world it is clear it is found to operate by institutions and centers and independently on the assumption that many developing countries, including Iraq does not have the actual capabilities of the use of science and technology in the development process and this will help in assessing the effectiveness of scientific institutions and technology as well as create and activate National capacity to coordinate the contribution of these policies in the development and rely according to the following criteria when evaluating science policy:

- The existence of scientific policy.

- The existence of a scientific policy is effective (the inability to use science in development).

The scientific research in theoretical and applied higher education curricula constitute the foundations of scientific research total spending on scientific research in Iraq and the Arab countries does not exceed (0.15%) of the gross domestic product, which is low note that the foundation in the progress of peoples.

The scientific research in developed societies finds the generous support of formal and informal institutions due to the fact that scientific research turn into an investment product and not a luxury academically was spending the United States and Japan and the European Union as much on scientific research (417) billion dollars has been South Korea raised the proportion of its spending on research and development (0.6%) of GDP in 1980 to (2.089%) in 1997, focusing on electronics and marine science and space science and information technology, new materials and planned China to raise the proportion of its spending on research and development and the goals of the five-year plan.

According to reports issued by UNESCO for Science and Culture in the year 2008 that the Arab countries spend $ 14.7 on the individual in the field of scientific research while the United States spend $ 1205 for every citizen and that the proportion of spending-rich Arab countries of their national income on scientific research much of expenditures by the countries least poor Arabic. And it requires the establishment of centers for scientific research of various kinds to move the community to advanced stages. And that the government will not be the only financier for scientific research, but must contribute to public and private companies to donate as well as it requires the programmer planning for scientific research and careful and balance his addition, flexible legislation and issue laws live up to those that are in force in developed countries also requires the development of a full mature plan to develop infrastructure for scientific research, including the establishment of specialized laboratories. The scientific research methods are closer to art to science, including a term that expresses the compound include semantics way in which the scientists in the development of the rules of science.

Methodology models

The Research Methodology multiple rely on the conversion of science to the benefit and the establishment of science on the basis of extrapolation to nature and Muslim scholars deliberated rules curriculum and emphasized that the induction is only a stage of the construction of systematic The al-Hasan ibn al-Haytham dish measurement approach in the field of light.

Lacks much of the world to the methodology and the policy of specific scientific and technological milestones and goals and means there is no information networks and devices for coordination between research institutions and centers, and there are also fiscal policy to fund research and is considered the Arab region, the lowest in terms of spending on scientific research, said Dr. Ahmed Zewail, in his book "The Age science "that the proportion of research provided by research universities ranging from (0.0003) of the total research Court offered by universities of the world and back twice the productivity of Arab researchers to several reasons, including:

• reluctance of the private sector for conducting scientific research.

• migration to scientists outside their own countries.

• the lack of a realistic and clear scientific policies for scientific research.

• lack of cutting-edge research supplies.

• Weak researchers income.

• Do not convinced governments and research the feasibility of scientific research.

One of the important phenomena lack of partnership between the university and companies in Iraq and the Arab nation and when a comparison with international companies, we find the following: -

• US companies maintained her lead for spending on scientific research and innovation has spent 4.8% of sales on research and development.

• In Japan budget for scientific research budgets are one of the most generous in the world and show the Japanese as well as the idea of cooperation between universities and private institutions productivity and innovations in modern technology.

• In Germany notes that the industrial sector sits as provided in the United States and that the private sector contributes to (70%) compared to (30%) of the public sector.

The Chinese experience systematic process of transformation and substantial restructuring of policies for scientific research policy and scientific research has been conducted there is a gap between the political decision and scientific research in terms of makers of political decision-makers bear weakness budget scientific research, especially in the Arab countries.

## G.Scientific research and translation and terminology

The translation of philosophical texts reached an apogee during the Caliphate of Abdullah al-Ma'mun (813-33 CE) and his successors. The translation movement declined and ended during the Buwayhid period (945-1055CE). These men took a personal interest in the progress of theology, philosophy, science and literature. Some families associated with the Abbasids became patrons of scholars and translators.

Most notable among the early translators were Banu Musa Bin Shakir, Abu Ishaq al-Kindi, Masarjawaih, Yuhanna ibn Masawaih, Hunayn ibn Ishaq al-'Ibadi, Thabit ibn Qurrah and Qusta ibn Luqa. Some of these should be examined more closely.

The astronomer Musa bin Shakir was associated with prince Abdullah al-Ma'mun before his rise to power. When Ibn Shakir died prematurely his three sons, Ahmad, Muhammad and al-Hasan (who became celebrated as mathematicians) were the wards of al-Mamun, and each achieved success as a patron of translators. Muhammad, the eldest of Ibn Shakir's sons, employed Thabit ibn Qurrah in his house (library) and other translators worked for him at *Bayt al-Hikmah.*

One outstanding translator of this period was Hunayn ibn Ishaq who worked under Harun al-Rashid, al-Ma'mun, al-Mu'tasim and al-Muwakkil 'ala-Allah. He was familiar with Syriac, spoke Arabic and late in his career mastered Greek at Alexandria or Byzantium. He travelled from Baghdad through Syria, Palestine and Egypt in search of Syriac and Greek manuscripts. the Hippocratic Oath as a genuine work, which he translated into Arabic. His translations from Syriac and Greek inspired his son, Ishaq ib. Hunain and his nephew Hubaish, whose works he supervised. According to Strohmaier, he was 'the most important mediator of ancient Greek science to the Arabs .

Thabit ibn Qurrah (d.288/901) of Harran, a Syriac speaking person who wrote and translated into Arabic, was associated with Banu Musa ibn Shakir by whom he was inspired to learn mathematics, astronomy and philosophy. Other celebrated translators included Qusta bin Luqa, a Syrian Christian from the Ba'labakk region who was well versed in the Syriac,Greek and Arabic languages and collected Greek manuscripts from Byzantium, which he carried to Baghdad to translate. According to Ibn al-Qifti, he was a contemporary of the first notable Arab philosopher, Ya'qub ibn Ishaq al-Kindi. He was known to be a versatile scholar, knowledgeable in contemporary astronomy, geometry, mathematics, natural science and medicine , and like many of his c

**Features historical methodology for translation**

The Assyrian King Sargon the use of many languages

and was talking to the people during the reign of Hammurabi of Babylon, about a year (2100 BC), multiple languages and translators employed by the ancients to the creation of dictionaries, preserved on clay cuneiform patches.

In about a year (2400 BC) translated to Ivuos Andronicus Odyssey of Homer's poetry to the Latin language, as Tivuos and Aionnbos transfer of a number of plays, Greek and Latin languages. Thus, the methodology focused on poetic terms.

Follow Jerome in 284 AD in translation principles include transfer of meaning in the sense and not word by word transfer methodology and this stage it is called without translation methodology (Figure 1).

Taatler has explained in 1790 methodology for translation of other principles, including:

- Writing for similar pattern continued.
- Develop translator original strain.

It means Translation Transportation words from one language to language, and in the San Arabs indicates meaning Translation Transportation speech from one language to the language and came in dictionary is the interpreter is divided translator generally into two types the first interpretation or instant and the second biblical translator specializes Translators usually in several fields of science technology and the social, economic, political, creative, legal and business issues and philosophical and technical, linguistic and literary. scientific research that contributes to the development of society.

So there is a need for coordination between linguistic and scientific academies and look at the business as a localized scientific efforts and publish a magazine belonging to the translation and localization.

**language translation**

The relationship between translation and language require full Tdilaa in two languages (source and receiver) understanding of the structures and meanings of linguistic structures syntax.

In addition believed that the method in which the expression of the meaning process conducted by the language component of the group icons are arranged in a certain order by logicians and scientists semantics wordy.

There are linguistic applications in translation based on grammatical rules, for example, rely on to avoid the passive voice.

Example:

The cell was destroyed by the enzyme

Cell destroyed by the enzyme

As long as the actor information translated into Arabic language it is understood format and on this basis the correct translation is the phrase mentioned above (destroyed cell enzyme).

If participated in added to the last one shall be limited to between.

Example:

The ovaries and of the woman

Do not translate them into the ovaries and the uterus to the ovaries of the woman, but the woman and her womb

And see goal of the translation is the expression of the content (sense) and the drafting of using the means of a new language (target) for the types of translations:

- literal translation.
- translation meaning.
- literal translation-moral.
- explanatory translation.
- redemptive translation.

Also, the mechanics translator can include (Figure 5) a number of factors including:

- Starter target language.
- standards in mind and the language of the target.
- surroundings and traditions of the language and the target vein.
- translator.

There are a number of difficulties for the translation and the young ones:

- translation examples which bear more than one meaning.
- translation of new words or new developments and metaphors and words not found in dictionaries.
- lack of technical terms the interview in the Arabic language.
- risk some linguistic terminology.
- choice between foreign and used word in the Arab and Arabized word.
- linguistic duality and sometimes triple and quadruple A good example is the translation of the Koran, where the translator is located in a grave mistake, or in poetry translation where beauty loses.

There are a number of difficulties for the translation and the young ones:

• translation examples which bear more than one meaning.

• translation of new words or new developments and metaphors and words not found in dictionaries.

• lack of technical terms the interview in the Arabic language.

• risk some linguistic terminology.

• choice between foreign and used word in the Arab and Arabized word.

• linguistic duality and sometimes triple and quadruple A good example is the translation of the Koran, where the translator is located in a grave mistake, or in poetry translation where beauty loses.

**Translation theory**

The translation process its tool language movement (message) between the two Parties, the sender and the receiver and witness of our century, in particular, the trend to theorizing the translation process, do.

• transfer and convert a linguistic process (one to one).

• the translator to discuss in mind the language components.

And paints the translation process of planning just two steps and the translator that the two operations consecutive timetable:

• diagnosis of the text in the language of sense.

• re-composition of the text in the target language.

Kochmidr that emphasizes the translation process sometimes require modifications in a ferry focused on the composition of wholesale or Tag features.

**Modern translation movement**

The translation movement in the country in the early nineteenth century, was limited initially on religious and literary books, either in Egypt has been adopted by Muhammad Ali Pasha, and a means of modernizing the Egyptian emerging state. He founded the School of the Age in 1835, and took Sheikh Tahtawi supervision. He gave Muhammad Ali translation of important French, Italian, Turkish and Persian languages scientific and literary books. A number of writers and writers who came to Egypt from the Levant effectively in the movement translator scientific, literary and renaissance experienced by Egypt in the final of the nineteenth century quarter also contributed, and took over the newspapers and magazines like "extract" and "Crescent" and other translation research and articles and publish them.

The translator has historically prevailed in the methods of teaching foreign languages in Europe on essentially between 1840 and 1940, and continues in various forms in language teaching curricula in different parts of the world. With the development of linguistics and multiple applications abounded views and ideas since the sixties of this century about learning and teaching foreign languages, and made extensive and unremitting efforts in order to find the best way to get the best results in this field. And it became a translator creates panic impedes the functioning of the educational process and prevent the education of new generations.

The newly translation is a vital area moving in airspace Vigtna creativity and produces better, and that the process of translation and localization corner of the scientific work that can contribute effectively to the development of the Arab community, to enrich human knowledge. In order to achieve this clarity, the link has to be reference to the following conditions: -

- the need to link the processes of translation and localization scientific research.
- the establishment of centers for the translation of the Arabic language and to promote their productions in the Arab countries and across the world.
- Support institutes practiced and taught translation and localization and encouraged to play their role better, and the establishment of departments for translation to and from international languages necessary to produce qualified translators enjoy thoughtful scientific approach.

• the need for coordination between linguistic and scientific groups with a view to cooperation and to agree on formats and terminology one to be adopted and disseminated.

• translated works regarded as scientific efforts are taken into account in university promotions like research, investigation and authoring.

• Issue magazine specializes in translation and localization, especially in the university community, each field of knowledge fields.

• raise the financial rewards for translators commensurate with the efforts. The distinction between the translation of scientific and literary text position translator of each scientific text must be objective and should be committed to accuracy and transmits the text you translate it as accurately as possible the secretariat, taking into account the order of the elements of sentence the same way that arranged in the original text. It is worth mentioning that little effects that may result from the erroneous translation in a way the use of a drug or run electrical device. On the contrary, the literary text translator enjoy a degree of freedom creative scientific text and creativity means to a great extent the ability to imagine.

And it became the scientific translation of the most important demands of the creation of the Arab to enter into a knowledge society. It seeks Arab Organization for Education, Culture and Science at present to put an Arab strategy for the translation and publication of the mothers wrote scientific culture, including the foreign scientific sources which plays both the scientific and the Arab publisher of scientific Arab translator.

This translation requires a set of basic infrastructure such as dictionaries, dictionaries and terminology banks and institutions for the rehabilitation of translation. And it requires a translator of scientific institutions of the media system centered on the role of media in spreading scientific culture as well as information that rely on machine translation between different languages as well as automatic indexing, automatic summary and building electronic dictionaries technology. It is problematic scientific translator own individual aspects of copyright and ownership of the authors of the original texts.

- Translation and Globalization (Acceptable for unity)

Perhaps the obviousness of people say that is culturally identical, and all the people of specificity that differentiation from others. But the cultural differentiation is not a privilege, and difference does not eliminate the existence of the bonds of common humanity.

If there was today directed to a global culture, Van such a call may pose a risk under the age of globalization, which, though characterized by tremendous speed at the level of the flow of information and the flow of knowledge, it nevertheless contributes to devote unequal technological and media, and directed in the direction of reducing the gap between the diverse cultures, and thus attempt melted into a single world culture, are unipolar culture, the other Western culture, which began its shares rise on the shares of other cultures account, including the Arab culture.

In this context, according to a study of the United Nations Environment Program published in 2001 that half of local languages in the world on its way to demise, the study warned that ninety percent of the local languages will disappear in the atheist and the twentieth century.

If the language in itself represents another face each manifestation of culture and identity and see the world, reducing its role in civilized coexistence process or marginalized or even is tantamount to marginalize of culture and identity and to see the world, which meant about reducing the role of the translator, as synonymous pluralism, and diversity, pluralism and the various different aspects: cultural diversity, multilingualism, the multiplicity of meanings and connotations, the multiplicity of interpretations and readings, multiple translations ... etc. Accordingly, the translator as the other side of the found at odds with the logic of globalization aimed at synthesis of culture after one.

Based on the foregoing, it is evident that if the translator and Longhorn, has launched a series of dialogues civilization, their role at the moment began to gradually shrink with shrinking influence and the presence of the languages and cultures of the multiple in the global scene by the wave of globalization, which confiscated the right to live and the right to difference and diversity, in the sense that translator which aims to create a culture of trying to achieve pluralism. So if there is a trend today to do a global culture, like the call may pose a risk in the absence of parity, media and technological knowledge, and cultural interaction between the various and diverse to enrich world culture of civilizations only come to accept the cultural parity and guaranteed and sponsorship. **Translation in the Arab-Islamic times**

Methodology adopted in these covenants on the developments that occurred in other countries. The translation contributed to the promoting Arab-Islamic entity covenants, which began in the first century of migration and ended in mid-fourth century.

The translation passed stages started from extrusion specifically in the Umayyad era to its end around the middle of the fourth century. It has appeared to us that the Abbasid period alone was characterized by the emergence of more systematic methodology in the translation movement, including the systematic phase of Mansur and methodology. The next stage was the era of Al-Mamoon that lasted until the middle of the Fourth century of the Islamic Calendar (started with the Prophet Mohammed migration).

Therefore, the development of methodology for the translation movement may represent the four main trends and Al- Mamoon in the Umayyad era in the age of Mansour, Rashid and translation in the era of Mamoon. The methodology of the especially the first Abbasid era depends on:

- Fluency in Arabic.
- Conservation of the stock, verbal foreign language.
- Definition of foreign terms on the structure of the Arab tongue.

## Literal translation

Characterized by translating the original text word for word, which made it necessary to put the word translated as foreign, causing in the receipt of many of the terms of non-Arab, which remained in use until the present.

**Translation sense**

The translation was carried of the sentences together but not separate word. Han Ben Yitzhak Abadi was considered as the most famous who followed this method of translation.

**Assessment of the two methods**

The first method as a suitable as a language school for students word without meaning, either way, but the second way draws the meaning and to transfer the language-borne sound. The translations which were made by the Siryac Greek science, which were transferred accurately and honestly such as, medicine mathematics with the occurrence of the circumstances from the Greek into Syriac and Syriac into Arabic. Despite the fact that translators Hanin Ben Yitzhak Abadi and his son, Yitzhak Ben- Hanin and a nephew and Habeechebn bin Thabit bin Kasna and Luqa, and Jacob Ibn Ishaq Kindi, concentrate on those of the famous capacity of knowledge and understanding of the subject.

The prosperity of translation was due to the care of the caliphs to translation for some families to the translation and the desire of some ministers and doctors access to the ancient civilizations.

The features of the methodology of this era with the following:

- The need to key factors that contributed in the translation into Arabic.
- A literal translation which was led by John the son of Al- Batrig.
- Sense of translation, mainly contributed by Hanin Ibn lshaq.
- The existence of stages in the translation such as an implicit way of translation from Greek into Arabic language through mode of mediation is Syriac.

## Translation

Translation in most cases is considered using translated art forms individually and others tended to convert the translation into science. What supports the process of translating the language itself is characterized by key features of grammar and morphological settings and the fact that translation is central to deal with two languages that have the advantages of each. Translation accompanied by the growth of human groups and collected them throughout different history links, and about the differences of these groups in the language that had to be a bilingual to secure the understanding among them. Assyrian King Sargon has published in many languages, as were the people of Babylon during the reign of Hammurabi, 2100 BC. using multiple languages. Rashid Stone gave the key to the mysteries of Egypt and save the text of a treaty between the Egyptians and the Hittites in the Egyptian and Hittite language.

Schools for the training used over time, the translators and many of the researchers in many languages ideas regarding the process of translation. But these ideas did not come out of the theory in translation studies, or clear the issues raised by this theory. Therefore, the difficulties facing the formation of the clear theory in the translation mainly by the multiplicity of kinds, including for example, word for word translation and literal translation, faithful translation and translation of the moral, free translation, translation of conventional interpretation, translation service, prose translation of information and knowledge and thus academy translation. Then it is difficult to develop the theory and a clear methodology and parameters to deal with all types of translations.

## Features of the old methodology for translation

The Assyrian King Sargon of used the many spoken languages, the people of Babylon during the reign of Hammurabi, about the year 2100 BC. Multiple languages and translators employed by the ancients to the creation of dictionaries, preserved on cuneiform clay patches.

At approximately 2400 BC., Gaius Andronicus translated Homer's Odyssey, a poem to the Latin language, and also quoted Tivuos Aionnbos a number of plays to the Greco- Latin languages. Thus, the methodology focused on the terminology of poetry. The Rashid stone the second century DC, has been known to write the key to Egyptian Hieroglyphics and wrote two forms based on a systematic basis Aldematip language and Greco- Egyptian. Jerome follow in 284 AD, the principles of translation that involves the transfer of meaning and sense, but not systematic transfer of word and by this stage, it was so- called translation with no methodology.

(The clear role in the field of translation during the sixteenth century AD). This research has formulated the systematic translation through the followings:

- The use of terms to help pottery connectivity.
- Use vocabulary and expressions neglecting the specialized vocabulary.
- Use special phrases when necessary.

Played a clear role in the translation during sixteenth century and he establishes the theory of special courses and special rules included in the recipes below:

- Knowledge of the language that translates it and, which translates to it.
- Knowledge of the contents of the writer.
- A way from the translation word for word.
- Use formulas to speak in circulation.

Dryden said at the name time not to carry out the translation of characters where they are as described by dancing on the ropes does not lead to clarity of vision.

It has been shown in 1790 by Taitler methodology for translation from the other principles, including:

- Writing style with similar to the original.
- Translation mode through strain of originally.

The methodology for the translation as art has dealt with many researchers, including Dalembert and Aboboto and Dr. Campbell Hulusi that believes Dalembert's remarks can not be considered rules or even principles of the art of translation, while the Abu Bhutto developed basis for the methodology of grammar, Campbell also presented the criteria for good translation methodology included the following:

- The clear language of the original text.
- Transfer the spirit of copyright and its ways to the text of the interpreter.

The benefits that can be obtained from the study of translation as a methodology art is represented:

- Comparism of the language with another language.
- Strengthening the capacity of writing in two languages (the sender n1lreceiver).
- Auditing the terminology and the development of rules.

And can be summarized in the subsequent stages that we have mentioned:

- Jerome and principles of translation.
- Luther and concepts of thc new translation.
- Dolly which developed the first theory for translation.
- Titlei with his extra counterpart.
- Dalembert and Aboboto and Campbell Translation as an art.

**Methodology of translation in Arab- Islamic times**

Methodology adopted in these covenants on developments in other countries, contributed to this translation entity in the promotion of Arab and Islamic eras, which began in the first century of Islamic migration and ended in mid- fourth century. Translation passed through stages of translation that began its emanate from the specific of the Umayyad period and even eclipse around the middle of the fourth century. All then has appeared to us that the unity of the Abbasid era was characterized by the emergence of more systematic translation movement and there is a systematic phase- Mansour and the rational and methodology for the safe and the next stage for the era of safe and lasting until the mid- fourth century AH.

Therefore, the development of methodology for the translation movement may represent the four main directions, the translation in the Umayyad era and then in the age of Mansour, Rashid and translation m the era of Mamoon and translation in time after Mamoon. The methodology of the Abbasid era, especially the first depends on:

- Fluency in Arabic.
- Conservation of the stock, verbal foreign language.
- Definition of forcign terms language on the structure of the Arab tongue.

In the light of the general features of this methodology resulted in it:

- Literal translation characterized by translating the original text word for word, which made it necessary for the translator to put foreign word as it is causing in the receipt of many of the non- Arab terms, which remained in use even at present.
- Translation in the sense: The translation of the sentence together and not a separate, established by Hanin Ben Yitzhak Abadi of the most famous that was followed by this method of translation.
- Evaluation of the two methods: The first method is a suitable school method for students without the meaning, either way; the second

method draws the meaning and transfers it by language- borne sound. Take, for example, a piece of the novel of time machine (writer HG. Wells) translated by as described by the D. Safaa Khulusi.

The darkness grew a pace a cold wind began to blow in freshening gusts from the east, and the showering of white flakes in the air increased in number.

**Literal- translation of Ibn- Batrik**

Soon the darkness increased and took a cold wind blowing from the east refreshing and the number of snow flakes in the air.

The translations which were made by Syrial, who transferred accurately and honestly Greek science, such as mature, medicine and mathematics, characterized by mistakes in the translation from Greek into Syriac and Syriac into Arabic, despite the fact that translators Hanin Ben Yitzhak Abadi and his son and nephew and Arabic bin Alasam and Thabit bin Qura and Kusta bin Luka Al- Baalbaki Yaqub Ibn Ishaq al- Kindi, were characterized by the famous capacity of knowledge and understanding of the subject.

The prosperity of translation was a result of the care of the caliphs for translation and care for some families to the translation and the desire of some ministers and doctors access to the ancient civilizations. And families and ministers, each of Masawaiyh in the year 543 AH, Gabriel Ben Ben Bukhtishu Bin Georges and Gabriel Ben Ben Ben Slamueha Bukhtishu bin Gabriel Lebanon (who dies in 225 AH / 838 CE). The Minister Mohammad bin Abd al-Malik died 233 AH, Yahya ibn free Barmaki (d. 190 AH).

Then summarize the features of the methodology of this era with the following:

- The need to key factors that did in the translation into Arabic.
- A literal translation was led by Yuhanna son of Al- Batrik.
- Sense of translation, mainly contributed by the Hanin Ibn Ishaq.

- The existence of stages in the translation where there is an implicit way of translation from Greek into Arabic, by the language mode is Syriac.

But during the past sixty years has obtained a qualitative and quantitative in translation, including what comes.

**Translation concepts and dimensions**

The meaning of the translation into another language by the way that the author wanted the text and thought Mounin ideas of the inability of the translation on the reproduction of the original. Grant says (that the translation is something malicious on the creation and inspiration) and that translated material if it is to meet the key requirements that it must contribute to:

- Performing the meaning.
- The emphasis on the style of the original text.

According Cocheraiwan purpose of the translation is the expression of substance (mind) and the formulation used by means of a new language (target) to get the kinds of translations:

- Literal translation.
- Translation o the meaning.
- Literal translation- morale.
- Explanatory translation.
- Redemptive translation.

The mechanisms could include the translation a number of factors including:

- The author target language.
- Criteria of the language and the target point.
- Environment and traditions of language and the target point.

**Translation and language**

The relationship between translation and language requires full knowledge in both languages (source and receiver) understanding of the linguistic structures and meanings syntax structures. In addition it is believed that the way the operation is conducted, the expression of meaning by the language component of the set of icons is arranged in a particular system by logicians and scientists semantics.

There are applications for language translation based on grammatical rules that depend on, for example to avoid the passive voice.

Example: The cell was destroyed by the enzyme

As long as the actor known to be translated into Arabic form is well known and on this basis the correct phrase mentioned above (enzyme destroyed the cell). If two additions participated in the genitive, then we must be limited to the last one of the two additions.

Example: The ovaries and uterus of the woman

**Scientific Translation**

**Characteristics of scientific language**

The scientific text characterized by special features including:

- Descriptors feature: It provides features by writer and not self-descriptive of the scientific truth. The author writes of scientific text directly depend on the coherence of its impressions of self.
- Formal feature: The language of the scientific text and way to deliver content but it is not an end in itself and the form of scientific work is not an integral part of the content.
- The Oneness of meaning in the scientific text. Scientific meaning of the text is likely of one explanation.

- Lack of scientific text of linking with defined time. Skip business major scientific barrier of time and place.

The distinctive features may be invested of the scientific text, reliable and scientific texts translator seeks to communicate a message with aesthetic though the aesthetic quality the desire for greater clarity and precision of expression.

The translator of the scientific text adapt what he meant with the nature of the message that it is delivered, the text of scientific education in the first place and where some of the aesthetic values.

A translator of scientific text needs to acquire daily a huge amount of new terminology, while the writer is moving in the area much narrower than the field of science, whatever the language, whatever the imagination because its development is much less rapid. The scientific text in some cases carrying an aesthetic adds to the content. The translation of the text distinguishes between scientific and literary translator and determines the position of each.

The translator of scientific text must be objective and should be committed to accuracy and to convey the text as evidenced as much as possible from the secretariat, taking into account the order of the elements of the sentence the same way as arranged in the original text. It is worth mentioning that to remember that the effects that may result from the wrong translation by way of using a drug or operation of an electric device. On the contrary the literary translation characterized by scientific innovation and creative means to a large extent the ability to imagine.

The importance of scientific translation increase as a result of the explosion of knowledge, and vast technological progress in all areas of life and is worth mentioning new to the Arab Human Development Report for 2000 that the total has been translated since the establishment of Dar al- Hamah in the era of Mamoon so far, an estimated ten thousand books.

The scientific translation became the most important demands of the scientific creation of Arab societies to engage in a knowledge society of the Arab Organization for Education, Culture and Science is seeking to develop an

Arab strategy for translation and publication of major books of scientific culture, including scientific foreign sources involves both the translator and publisher.

This translation requires a set of infrastructure including dictionaries, terminology banks, and qualified institutions for the translation, the translation of scientific institutions requires of the media system centered on the role of media in spreading scientific culture, as well as information technology, which rely on machine translation between different languages as well as the indexing mechanism, and the summary and automatic building of electronic dictionaries. It is problematic translation of scientific aspects of copyright and intellectual property of the authors of the original texts.

**The theory and scientific of translation and concepts**

The translation is process with a language as a top transfer (message) between the two Parties, the sender, recipient and witness of the twentieth century in particular, the tendency to theorize the process of translation, on the basis of rules had to be a researcher of the theory of translation and thus prevent the translation into the science of translation is aware of recent origin, it is natural that different views of the workers.

Aravin Kochmidr says in the details of the scientific translation:

- It is not a transfer of substitution and the transfer of language (one to ones).
- The translator to discuss the components of the language accordingly.

Kochmidr illustrate planning translation process steps of the two consecutive operations in time:

- Diagnosis of the text in the language of spirit.
- Re-configures the text in the language of the target.

Kochmide confirms that the translation process occasionally requires a few modifications in the statement focused on the composition of the sentence or semantic features.

**Translation approaches**

These approaches are divided into:

- Technical curriculum, they are basically the following stages:
  - Analysis of the language of the source and the receiver.
  - Examination of the text in source.
- Regulatory approaches is t represented by using the regulatory framework for translation as follows:
  - Liberalization of the first draft.
  - Review the first draft.
  - Read the translation to the achievement of the creative style.
  - Studying the reactions of the recipients.

Nobirt divided generally the texts to be translated to:

- Texts related to language of point of view.
- The language of texts related vein target and roll together.
- Texts closely fully inked to the target language.

The division of Boser on a tripartite basis:

- Texts that distinguishes content.
- Texts which distinguish the shape.
- Texts which distinguishes impact.

**Science and the theory of translation**

Phyllis Volgeram believed that the task of science is to describe and explain the focus of a domain in which this science in a media analysis by using methods appropriate to the purpose. It consists on the translation of the three areas:

- Analysis of the source of the language accordingly.
- Transport between the two languages.
- The translation process and their product.

The task of the science of translation is an innovation approaches to these three areas and determine the conditions of conformity on a scientific investigation of safety.

Eugenio Cocherio finds that the difficulties encountered by researchers in establishing the theory, of the science of translation due to ambiguity of the relationship between the theory of translation and text linguistics, where the transport process of reaching on the basis of conversion from one language to another and the translation process of converting a linguistic and translation theory in its current form based on the concept of the just.

Antoine Popovic believes that the theory of translation is part of the following concepts:

- The general theory of translation.
- The special theory of translation.
- The practice and teaching of translation.

Literary translation organized by the specific theory of translation, which takes into account the various scientific fields

The translation may reflect the face of the writing linguistics as perceived by many intellectuals, where it was:

- Knowledge or can be art.
- Does not express the thought of the needs of the owner.
- The translator to know the languages of the transferee and the transferee to it.

- The reader considers the text after its stability in the new language, and then decides on the translator.

As stated in the above, are considered the question of the identity of the text in the languages, constitute the main difficulty in the way of the theory of translation.

The Mounin thinks that the translation is friction between languages, but an extreme case of friction between the cases and suggests that the teaching of contemporary linguistics and should be concerned with questions of translation for the following reasons:

- It is not permissible for to continue to ignore the translation movement.
- The translation on the behalf of contemporary linguistics.
- Mounin confirms that the art of translation is not limited within the boundaries of linguistics, but on the different faces.
- Mounin displays at the same time theories of linguistics that may deny the legitimacy of translation.
- Mounin concludes opinion on the concept of translation (all we can be connected from one person to another in a language that can be connected from one person to another, from one language to language).

There are other criteria that are referred to by many researchers in the translation of standard Marur, which states:

- Transfer of meaning in the translation.
- Transfer of structural appearance.
- Transfer stylistic appearance.

This criterion has indicated two questions the terms and composition should reflect and be a definition of the word is something that Bulslev. The terminology raises the question of taste or irrelevant.

## Translation assistants and their means

There are two types of aids (printed reference) the first is relating to the understanding and other by expression. Examples of such assistance are: monolingual dictionaries (verbal dictionaries), spelling, grammar and cultural references, as well as geographic, historical and social knowledge references to help the reader, or the interpreter as well. The Auditor Professional is also used in the translation of scientific, cultural and doctrinal, including dictionary, definitions and keys to Jarjani Sciences and the Al-Khwarizmi may be monolingual dictionaries or bilingual dictionary, such as technical or scientific terms.

The expression assistants are used in the field of expression in the target language or language standpoint include language dictionaries and references that include bilingual dictionaries mismatch of public and private and the last under the terms of economic, sociology, medicine and engineering.

The methods of translation are intended to machine tools (EDP) equipment, which invests them dictionaries; translation machine automatic banks terms database data in the translation and e-mail. In this area it is believed that Frepts Bepkp out of focus on the value of a human translator, it is believed that the work of electronic machine translation can not be called a translation of it is the task of the transfer of meaning without understanding and groping in the text of the relations of language inside. Nayda, believes that the aid in the translation is not that simple, but transfers itself directly on the surface of the text, but the level of analysis into three phases that include the full texts related the target point and target language texts related closely in full.

- Coordination of the texts.
- E-mail.

Frepts Bepkp believed also that the work of electronic translation machine can not be called a translation but it is not based on understanding where that:

- Machine is the task of transfer of meaning only.
- The translator of human has the capacity to understand what the relationships in the text.

Nayda confirm as we stated the translation processes, is not transfer directly on the surface level of the text, but it is analytical processes based on synthetic human capacity in the first place.

**Translator concepts and dimensions**

The translator can know the meaning of the transfer of text into another language the way he wanted the author of the text is believed Mounin translator inability to reproduce the original. Grant says (something that the translation snooped on the creation and inspiration) and Article translated if they are to meet the key requirements they must contribute to: -

- performed for meaning.
- emphasis on the style of the original text.

Science and the theory of translation

Phyllis believes that the task of science is to describe and explain the area in which the focus of this science in a media analysis using appropriate approaches to its subject.

It consists on the translation of three areas:

• analysis standpoint language.

• transport between the two languages.

• translation process and its product.

And that the task of science translation is innovation approaches to these three areas and determine the corresponding conditions on about a scientific investigation of its safety can.

Eugenio to see the difficulties encountered researchers in establishing the science of translation due to the ambiguity of the relationship between translation theory and linguistics text, where the transport to reach isotopes process on the basis of conversion from one language to another and translation process of converting language and the theory of translation in its current form based on theory just a concept.

Antoine Bopovic believe that the translation theory is a framework for the following concepts:

• General translation theory.

• own translation theory.

• practice and teach translation.

For example, literary translation is organized by the special theory of translation, which takes into account the various scientific fields.

May cross translator aspect of writing linguistics according to the perceptions of many intellectuals where they are:

• aware of or can be an art.

• Do not express the thought of the owner.

• Translator does not need to know two languages, including movable and immovable them.

• reader looks at the text after stability in the new language and then governs the translator weak and attributed to him are found in the text of defects.

And as stated in the above text are considered the issue of identity in both languages, that constitutes the main difficulty in the way of translation theory.

translator that little friction between languages, but an extreme case of friction between the cases and suggests that the teaching of contemporary linguistics must take care of translation issues for the following reasons:

• may not continue in the Sunni omission of the translation movement, which means Sunni operations.

• put the translation movement on behalf of contemporary linguistics linguistics.

• confirms that the translation the art is not confined within the borders of linguistics but it faces is Sunni.

• displays at the same time linguistics theories that may denies the legitimacy of the translation.

• concludes that the opinion follows on the concept of translation (all that can be connected from one person to another in a language can be connected from one person to another from one language to language

The translator methodology shroud many researchers have taken, including is not correct to mind the rules or even the principles of the art of translation, while Abu Bhutto laid the groundwork for the methodology of linguistic rules also introduced Campbell criteria for methodology good translator included the following:

• clear the original version of the text.

• transfer the spirit of the author and his style to text translator.

The benefits that can be obtained from the study of translation as a methodology of art is represented:

• compared to the language of another language.

• Strengthening the ability of writing in two languages (sender and receiver).

• scrutiny of the terminology and the development of its rules.

And it can be summarized in the subsequent stages that we have mentioned Palate (Figure 2):

• Jerome and principles for translation.

• Luther and new concepts for translation.

• International, which put the first theory of translation.

• and Campbell and translate Methodology of translation in the Arab and Islamic eras

The methodology adopted in these covenants on the developments that have taken place in other countries, contributed to this translation in promoting entity Arab and Islamic eras, which began in the first century of migration and ended in the middle of the fourth century.

Therefore, the development of the translation movement methodology may represent the four major trends is the translation in the Umayyad period in time and Mansour Rashid and translation in the era of safe and translation in a time of post-Maamoun. The Abbasid first methodology especially depend on:

• proficiency in the Arabic language.

• Save verbal balance in the foreign language.

• definition Alaagamah terms on the Arab tongue structure.

• literal translation.

Characterized by the original text word translation word which necessitated the translator to put foreign word as it is, which caused in Rhode many terms and non-Arab, which remained in use until the present day.

• translator sense

The translator on the sentence together on the floor does not separate nostalgia and is considered the most famous who followed this method in their translations.

• Assessment methods

The first way serve as a school for student word without meaning, either way, they paint a second meanings and language-borne sound. Let us take for example a piece of machine time novel writer Ige. Ji. Wales translated both ways, as spelled :

And soon intensified the darkness and took a cold wind blowing from the east and the number of refreshing tease snow in the air.

The translations are made by who transport Greek science accurately and honestly nature, medicine, mathematics occurrence of errors in the translation from Greek into Syriac and from Syriac into Arabic in spite of the fact that translators nostalgia Ben Yitzhak Abadi and his son and his nephew Ajabih bin Alaasm and Thabit ibn Ksrh and share Ben Luke Baalbaki and James the son of Isaac Canadian, well-known capacity of knowledge and understanding of the subject.

And it can then summarize methodological features of this era, including the following:

• the need to key factors that do the translation into Arabic movement.

• A pony led by John the son of Penguin.

• A sense of translation contributed mainly by nostalgia Ben Yitzhak.

• The existence of implicit stages in the translation there is the style of translation from Greek into Arabic by the language mode is in Syriac.

## Translator aid and means

The expression aid in the field of expression in the target language or vein include language dictionaries and language references that include bilingual dictionaries public, private and recent mismatch being own terms, economic, social, medicine and engineering.

The means of translator is intended to machine tools (Computer), which invests including machine translation and dictionaries mechanism and the banks terms and the base of the data in the translation and e-mail in this field thought in terms of a focus on human translator value, it is believed to be the work of electronic machine translation .

## Translator

Translator describes being an artist is informed linguist is reading the text before it offers to translate it and some believe that he may change the original in shape and improve the translator and then keep the idea of even mistakes appeared originally advised modified.

May translator to add to the original meaning of strengthening the addition of synthetic and to use wisdom to increase and deletion and to clarify the basic properties of the style of the writer, plain or delightfully or coordinator or ornate or simple have talked about the translator said, should have know people in movable and immovable him.

It requires the interpreter to understand the source and expression of the language of the target language in the resolution of phonetic symbols or written in the translation and understand the meanings of words and idiomatic expressions in various contexts.

Translator requires the use of monolingual dictionaries, such as the mediator glossary or dictionary Ocean if the translator translates from Arabic. In addition, there is a language reference, including the express lexicon, spelling, grammar, lexicon in exchange. If the scientific and technical translator or interpreter specialized must resort to specialized dictionaries of scientific terms.

Scientific translator

Features scientific language

Featuring a private scientific text features recall include:

• descriptive feature

It provides descriptive features, not self-scientific fact of the private knew that. Author writes scientific text directly dependent manner the cohesion of self-impressions.

• morphological feature

The language of scientific text and a way to connect the content of what, but they are not an end in itself and figure in the scientific work is not an integral part of the content.

• Oneness meaning in scientific text

Potential scientific text and the meaning of one and one explanation because all the expressive meaning of a word or group of words.

• Do not given a time of scientific text link

Skip the major scientific works time and space barrier.

You can invest above distinctive features of scientific text and reliable scientific texts and seeks an interpreter to connect the aesthetic message in spite of the decline aesthetic character in front of the desire for more clarity and precision in expression ().

Fmtorgom scientific text adapts saying what he wants with the nature of which are connected to the message, the text of scientific education in the first place and where some of the aesthetic values.

Scientific texts interpreter needs to gain a tremendous amount of daily new terms, while the writer moves in the field is narrower, much of the field of science, whatever the language and whatever the imagination because its development is much less speed. The scientific text in some cases carrying a shipment aesthetic added to its content.

Translator curricula

These approaches are divided into:

• Technical curriculum is mainly of the following stages:

- Analysis of the source and the receiver language.

- The study of the language of the source text.

• regulatory approaches is represented using translation regulatory framework in accordance with the following:

- Edit a first draft.

- A review of the first draft.

- Read the translation of the investigation of the creative style.

- Studying the reactions of the recipients.

divides generally localizable text to:

• texts associated with the language of sense.

• texts associated with the language of spirit and the language of the target together.

• texts associated with the target language fully linked.

The division on a tripartite basis:

• texts which distinguishes it secured.

• texts which distinguishes shape.

• texts which distinguishes the effect.

Translator accompanied the growth of human groups and collected them over the different links of history, and about the different groups in this language had to be there to provide bilingual understanding among them. He has published Assyrian King Sargon Gnamh in many languages, as the people of Babylon during the reign of Hammurabi (2100 BC) speak multiple languages. Stone and gave the key to the secrets of good and save Egypt pinned between the Egyptians and the Hittites treaty text in Egyptian and Hittite language.

Accordingly, the difficulties faced by the composition of a well-defined theory in translation Home degree to which multiple kinds, where include, for example, the translation word by word literal translation, translation, faithful translation of the moral and translator free translation, idiomatic translation and translation service-and translator prose and translation of information and translation of cognitive and academic translator so It is difficult development of theory and methodology and clear cut deals with all types of translations.

Translator and globalization

That some aspects raised by the translator in relation to globalization includes:

- Translator communication between multiple languages and different cultures and civilizations distinct and different mechanisms bridge.

Become a translator of the most important tools used in the creation of cultural cross-fertilization among the nations and peoples through the logic of give and take, quote, creativity, comprehension and production. The inevitable result of this cosmic communication, became the interaction between national cultures and different civilizations it depends on the translation not as an intellectual luxury, but a humanitarian necessity dictated the terms of the existing differences and diversity among the nations.

In this context, the translator can be regarded as a form of communication between different cultures and to influence the vulnerability and import and dialogue and rejection and assimilation and so on, leading to the emergence of new elements in the way of thinking and style of addressing the issues and dilemmas analysis.

And us in the dialogue between the Arab Islamic civilization and Western civilization throughout history bear witness to the role of translation in this cross-fertilization and communication in the light of the recognition of cultural diversity.

This fundamental role played by the -kqamh translator Amadavh- we see clear through the theory that civilization phases and stages, and each stage of the process and the people of peoples carries torch of civilization ... as any people can not be carrying this torch forever.

Translators contributes trends of globalization in three fields: science, technology, social, economic and political issues. Two branches translator oral and written competence has become a stand-alone studying in universities in ways that private foreign, Arab and programs overseen by the departments and faculties granted her scientific degrees and at all levels.

That the issue of technology transfer occupies developed and developing countries, including Arab alike, and no one can deny the vital role that could be played by translators specialists in certain fields and modern and important scientific work in order to move the process of incorporation of technological advancement in their countries. The role of translation in the economic and trade development more evident and already the translation is always a factor of development, and I think some people for a long time and it's just a mediator in trade worker. The translation is now a vital role in investment and savings and manpower training, consumption, import and export operations and other activities. The translator an important role in the bilateral communities language or multiple languages and is gaining translator written both types and oral, especially in the communities that receive large numbers of immigrants the importance of pushing to reconsider educational programs and language curricula, bringing to refer to the national language necessary and projects of practical and scientific point of view .

Translator contributed to the growth of languages and survival process. Thc translation is terminological methods are helping to tighten and general and specialized vocabulary for all languages regardless of the extent of spread. Require translator necessarily be looking for formats and terminology suited to modern reality they are the backbone of modern life and a tool of international and cultural liaison. Ahmad Lutfi al-Sayyid and some of his disciples believe that he was our age of translation is not the era of authorship. In spite of the current of translation attempts in a lot of the country still rate the translator to the author does not exceed one percent, while this figure is in a country like Britain to 10% in a country like the United States of America 15% and if we take these two percentages as evidence of the importance of translation in there is progress creativity .

**Translator**

There are a number of technical difficulties of translation, including:

- Translation of the examples those are likely to more than one meaning.
- Translation of the new words or new developments and metaphors and words that are not in the lexicon.
- Lack of technical terms, the interview in Arabic.
- Risk of some linguistic terms.
- A choice between foreign and used the word in Arabic and Arabized word.
- Duplication of language and sometimes triples, quads and a good example of the translation of the Koran where the translator is a serious mistake in the translation of poetry or losing beauty.

According to the standard of NIDA, in describing the process of translation and understanding of the source language (language sense) it is necessary for the translator to be able to unscramble voice or written, and to understand the meanings of words and expressions in the context of various reforms. The requirements of expression in the target language is represented by the translator by using voice tags and use of knowledge and words and conventional expressions used truly and the use of grammatical structures used morphological true.

**Translator**

Nayda describes translator being an artist is informed of the language, read the text before it can be translated and some believe that he may change the original and improves in its translator and then appeared on the idea to keep mistakes to advise the original amendment. The translator can add to the original meaning in order to strengthen the addition of synthetic and to use wisdom, deletions to increase and clarify the basic characteristics of the style of the writer, easy or pleasant or coordinator, or decorated with simple or may occur.

The interpreter requires understanding the language of the source and expression in the target language and resolving voice tags or written translation, and understanding the meanings of words and idiomatic expressions in different contexts. The interpreter requires using monolingual dictionary such as the interpreter translated from Arabic. In addition, there is a language reference, including dictionary to express, spelling, grammar and lexicon in the exchange. If the translation of scientific and technical is required, a specialized translator must resort to specialized dictionaries of scientific terms.

Arabization of Science

Localization source of the act Arabs language of any knowledge of Arabic and Arabization of a foreigner refers to the name that utter the Arabs on their curricula and localization idiomatically find an Arab to teach Arabic interviews as currently intended Arabization use of the Arabic language a national language to express concepts and their use in education at all levels and localizations official language of administration. Arabization targets, including the creation of an Arab creative personality possesses the ability to produce self-knowledge and the ability to participate and interact Some believe that the invitation for the Arabization of political significance and national unity. The issue of Arabization of Sciences of the most important issues and problems facing the people of the Arabic language in this age of rapid scientific and technical progress that is born in the Arab countries and not written in languages including Arabic and published in all the country, including all Arab countries.

Arabization that teaching in Arabic as a national aims to achieve strengthen the cohesion of national unity and Collect between tradition and modernity, where the Arab Science and Civilization language as well as strengthening the bonds of the link between the university and society and the unification of culture and scientific and intellectual effort and achieve democracy Education and continuity of teaching foreign languages. The objectors Arabization referring to those that the language of science in our time but language is English and the small availability of books and references in Arabic and a few professors preparers to teach in Arabic and surprising some Arab terminology and localization slogan without substance with a problem of scientific symbols and numbers.

A number of Arab writers have published writings, claiming that:

- Waterless Arabic language and they do not contain and suitability for the development of new scientific nomenclature.

- The language of science and technology in our time is English.

- The number of the book and a few references in Arabic and what some of them badly invalid.

- A professor who can afford to lack of teaching Arabic.

- The problem of scientific terms and surprising some Arabic terms issued by some language academies.

- The problem of scientific symbols, letters and numbers.

- Publication and distribution problem.

The issue of Arabization of Sciences of the most important issues facing the people of the Arabic language in this age of rapid scientific and technical progress. If that progress is born in non-Arab countries and written in languages not including Arabic and published in every country all, including the Arab countries, this progress holds with the value of cultural invasion linguistically to Arabic and other languages that could not these languages absorb this progress attained and tongue and in their communities, and perhaps the most important facts of this era of linguistic terms, and that the English language has become accredited international language in many areas, including the areas of science and technology, is that all this is not an excuse to ignore the native language of the students which they composed, or familiarity throughout the years of their lives of study by the University and read and write as they hear it all the time outside the university, which gives them greater opportunities for expression and understanding which allows them to study very foreign.

**The term scientific research**

The term practiced a key and effective role in the formation of knowledge, brought his concept and identifies significant.

So that the concept implicit in the form of multiple term, depending on the multiplicity of fields of knowledge on the one hand and which develops in the light of that field.

Colorful terms even within the same diameter as a result of issuance of multiple sources - specialists or individuals acting on different tastes and different methodologies - analogy and etymology and metaphor and sculpture and complex and pony or or verbally even terms issued by scientific associations and bodies standardization, coordination and synagogues, not without This contrast, which operates in various Arab countries participated unified methodology all references and terminology bodies in the Arab approval, so it was said that our problem with foreign scientific term is no longer in the Arabization as it is in unification.

First-term (and the concept of localization)

The word "term" did not respond to the ancients did not use it, but they used the word "convention" instead of the term and that the word of the common mistakes that are not properly used.

And that the prevalence of the use of the term 3 sucking the term was in sync and a close with two of the biggest heritage Dictionaries and the emergence of the most famous dictionary Ocean 0 / Necmettin turquoise Abadi) and 0 for San Arabs to the son of perspective). The bottom line is that our ancestors were "term" and "convention" Akunhua together not only on any term they knew three words together.

Shukri said the late D.gabr member of the former Iraqi Academy of Sciences that the term in the language of derivatives act Magistrate Vastalh and its source terminology. And that the long term is what the scientists in the science of science, or art of the arts of the term and must overcome the linguistic meaning of the Special Rapporteur on to become a term otherwise remained linguistically years.

Walker, the late Dr. Abdul Razzaq Mohiuddin former member of the Academy of conditions that must be met for the epithet as a single term, such as: -

• The use of competent jurisdiction wrote in the category, but lost and idiomatic significance and become significant linguistic or lost significance in general.

• be placed special meaning differs from the linguistic meaning is common among people.

• that a lot of use for the purpose of installing and accepted by specialists.

The term transfer process varies (translation or localization) At the same time it highlights the need for the unification of the Arab level were numerous seminars and a variety of basic principles formulated in test terminology.

Terminology, translation and publication in the scientific circle / Baghdad compound has recognized these principles and then the complex body agreed to these principles and became a public policy of the complex in Iraq.

And as stated in the definition and principles of the reformer illustrated as follows:

- The formation process of the term affinity between language and science.
- The composition of the term long-term process, requiring high flexibility and accept a broad and rational policy.
- The process of formation of the term is not an individual but a collective boxed other opinion a significant role.
- The application of the term Creator does not require an individual decision or even permanently Mjmaaa but a dialogue between the perpetrators and the term-makers.

Many researchers have a say in the issue of the term as follows: -

- Create complex "scientific" one for the purpose of the use of standard scientific terms.
- Issuance of Arab scientific lexicon, which depends on a clear methodology in the choice of terms.
- The supreme body at the national level with excellent efficiency and expertise of a specialist in the field of translation and terminology.

Also taken into account when setting the terms for some public foundations and approved the following rule:

- use the word Arab Steel.
- revive the old Arab term.
- Ethar the right of the Arab word.
- expansion of the derivation of scientific terms.

• Arabization of international scientific formula Titles.

The key features of the methodology developed by Dr. Ahmed Shafiq al-Khatib is susceptibility to the application of rules and principles distinctive accurately and logic and in some of them:

• Audit should understand the meaning of the term before attempting to translate it for fear of entry terms violate their meanings.

• methodology assumes knowledge of the language which conveys them to or experience in the subject of research.

• is supposed to be synonymous with one of the Arab word attributed to it or add to it, as is credited or added to the foreign word.

• rule applies to what is authentic in the language translated and transmitted to him after that investigates the Arabic word meaning lead or formulated his Arabic word by other means in derivation or metaphor or sculpture.

There are a number of methodologies have been submitted by researchers in the term, including the following: -

**First methodologies:**

These methodologies are focused on individual efforts before and after the regularity of work in laying the foundations of these methodologies. It has been active in this area scholars, including Peter Gardener and Faris Ahmed Khalil Chidiac and happiness, and Jacob, and the vicissitudes of the Carmelite Anastas and Mustafa al-Shihabi and Hassan Fahmi. They have put up with the rules and methods of setting parameters and measuring and carving and included the installation of translation and localization or verbal quotation methodology.

**Subsequent methodologies:**

Several methodologies have been achieved in the development of scientific terminology in Arabic is acceptable in its infancy, based on the ideas mentioned in the First, the Arabic language academies contributed in Cairo, Damascus, Baghdad, coming on the contributions of the Office of Arabization coordination in Rabat, that the first methodology put forward before the founding of the synagogues mentioned may It is related to what has been reached and later published in stages in the records and publications of Arabic language academies, including: -

• proposals Jordanian Arabic language, a complex document to include 18 of the principles and recommendations and references to the problem of accidental precedents and suffixes.

• Special Status of scientific terms and adopted by the Conference 45 recommendations in Damascus a document that included four principles and 12 recommendations relating to all methods of setting the Arab term.

• methods to choose the term of the Arab professor Ahmed Ghazal green and include measures relating to linguistic choice social and linguistic standards.

• methods of selecting scientific term and requirements of Dr. beautiful angels a document that includes several methodologies collected in six basic principles for the development of the Arab term.

• the basic principles in the development of scientific terms and selected approved at a seminar unify methodologies put new terms in 1981.

• precedents and suffixes 5 Dr. Mahmoud Mukhtar include 120 former and 50 later.

• Develop new terminology of Professor Ahmed Shafiq al-Khatib and methodology of 600 and 150 prior to subsequent related to the field of medicine.

• Dr. Thami Alrajaa Hashemi Arabization precedents and suffixes.

• it has been approved by the Egyptian complex board a private modern recommendations approach the status of Arab scientific terms specialized at its sixtieth (1994) and one session (1995) (see Appendix 2) and also issued recommendations for the Arabic Language Academy Conference at its sixty-third session (1997).

**Third-basic principles in the development of the term and generated**

Arabic language in constant movement, as they did not know the recession in her career in ignorance that reflect the views of their owners, and when Islam appeared new his concepts are able to express these concepts and Emma expression.

In the Abbasid period Arabic language capacities that have volunteered to retain its origins and rules ancient cultures. This, if anything it shows that the Arabic language is not the language in which reading the deadlock, but is the language of genuine flexible, expressed need new era and trends, thereby increasing vocabulary sang situation sometimes, and at other times, and sculpture three times.

And it comes with trying to know the principles adopted by some linguistic groups in the development of terminology and generated by:

• The Arab Academy of Damascus.

The biography of Arabization in the range:

- Faith teaches at Syrian universities in Arabic.

- The basis of all the work is beginning.

- The adoption of the translation of the scientific languages.

- Comprehensive Atmad gradual localization within a comprehensive plan for Arabization.

- Treat all wrong training, guidance and homeliness and Zathasin foreign language.

- Arabization of Medicine comes in the first Protozoa.

• Arabic Language Academy in Cairo.

Article II of the regulations that the compound that replaces the slang words and Alaagamah that did not express other Arab wordy, so that looking first for an Arab words in scope, if not yet her search finds Arabic names put new names in ways that search known of derivation or metaphor or not So, if it was unsuccessful in localization resorted to to preserve the character of language and weights as much energy.

And authorized the use of some complex vocalizations Alaagamah when necessary and work terms in the field of complex situation and Toleda on: -

- Maintaining the Arab heritage and the preference term translation with localization vacation.

- The fulfillment of the purposes of higher education and the requirements of Translation, Authorship high scientific and culture.

- Cope with high scientific approach in the style selection term and rounding him in Arabic and in between his world languages.

- Arabization of scientific definition of each term Magamaa.

- To keep the old Arab term and preference to the new only if the new common.

- Accept what used by generators, which have been on the scales of metaphor or derive a vacation with the derivation of the names of the Senate in the language of science.

- Vacation use some wordy of necessity.

And work in the complex methodology also: -

- The old leave on foot as long as valid.

- The adoption of derivation or metaphor.

- Develop an easy and affordable terms.

- Regarded as people use the argument.

- Legalization of the outlawed for scientific argument of necessity and download status necessary.

• Iraqi Academy of Sciences.

The methodology followed by the Iraqi Academy of Sciences does not differ from the methodology for each of the campuses in Cairo and Damascus.

And the work of the compound through its specialized committees to follow the following: -

- Preference Arab word on the generator and the generator to talk only if famous Arab inherent if the term foreign obsessed with him such utter alcohol .

- Avoid Arabization term foreign only in the following cases:

• if its significance has become commonplace significantly makes it difficult to change.

• If derived from the names of the media.

• In the case of scientific names for some chemical elements and compounds.

• If the names of foreign standards or units.

• If used in the written heritage such as the astrolabe.

- And failing Arabic term in the derivation and generation, measurement and metaphor plenty of room.

• Jordanian Language Academy.

Jordan launched Language Academy in the process of developing and generated the term of his vision that term is derived from the authentic heritage or so doubtful means available to the language of the measure or derivation or metaphor must be the ultimate goal to put the Arabic term. From here the pool was keen on the other hand, the Arab translation accuracy. But the complex at the same time we see that we appreciate the advancement of science and Arabization catch up with new where if we make a priority for Arabization not for translation.

• Office for the Coordination of Arabization.

Work of the Office for the Coordination of Arabization in Rabat, one of the Arab Organization for Education, Science and Culture offices, to develop a plan for the development of terminology and coordination Mstonsa language academies decisions. It came in this plan: -

- The use of the word Arab and foreign exchange for one expression, and synonyms used only rarely and, when necessary, thereby generating the terminology is achieved.

- Develop Arabic term for every indication whether foreign term of more than one indication.

- Study term foreign study and thorough scientific and understood its significance and meaning of your exact terminological user competence in the field before setting foot on Arab interview.

- Not only the adoption of one foreign language as the only source of foreign terms.

- The use of Arabic, which traded wordy already used by ancient Arab scientists and diligence in developing a new word index taking into account the terms set by the Councils and the competent committees.

- The existence of sufficient suitable or similar between the meaning of the term and its significance idiomatic language is not required in the term that accommodates all scientific meaning.

- Stay away from heavily several meanings of words.

- Commitment statements as much as possible semantic and precedents and suffixes standard formulas prepared by the unified complex.

- Passport resort to sculpture or installation, provided that Almzja carved word would be acceptable or common.

- The use of extraneous words or undercover when necessary.

- Choose the word easier among foreign languages to take him to the Arab lightest what can the Arab tongue without the obligation of one foreign language.

- Careful in the Arabization of the word and put it in the form of easy to collect and share them and derivation of them.

Fourth-accuracy Arabic term

What exactly do we mean two things:

- But tile verbal indication scientific concept, which we pass it (the scientific accuracy).

- But significant tile idiomatic language is significant, which we pass it (the scientific accuracy).

Any lead term scientific concept intended, and that this term be sound of language and meaning of the building.

That he must be aware of some facts on the subject of accuracy are: -

A-The precision of the most important conditions for scientific term, one of clarity and brevity, the most important scientific language requirements.

(B) that the phenomenon of lack of precision is limited to words and modern vocabulary, but which is in the old Issadvina Maajmna, even those that got out of linguists investigators.

C-problematic that the lack of precision is not only limited to the Arabic language, but is a general among all languages.

Fifth-the role of terminology in the Arabization of Science

Terminology is not just a set of lexical units in one system within the specialty, but organized terminology per specialization in a specific level for use and there Mtaadh Research tried classification of these levels that move between several axes, the language of scientific specialization, the language of the workplace, consumer language. Transmission to the terminology associated with the common language of cultural norms Languages Specialization and the role of the mass media was decisive in many terms this means described as the biggest "distributor" of terms.

While modern terminology reflects the concepts composed on a global level, and for that converge modern languages from this side increasing convergence poses this rapprochement new questions about translation and accuracy, terminology and conformity and on the importance of uniformity of terms in one language into an integrated system to express these concepts.

The chemical, physical and life sciences of the first areas that searched for terminology in the Western world with the aim codified On the Arab level, standardized terminology has Zdat in the early work of the Office of Arabization coordination and settled with the approval implicitly with the rules of theory specifying the modalities of setting terms for each partial field: chemicals the names of theories, and units of measurement, the names of the agencies, the names of the operations.

The attention of medical science presented effort in the modern Arab world to put the terms in Egypt began high medicine in 1811 year began Translation, Authorship movement in this area, adding the Arab experiences Levantine important terms, including linguistic groups efforts, and with the outstanding work being done by the Regional Office of WHO Alexandria multitude of these terms and coordination have become possible to utilize them in the fields of industry there is a great diversity in levels terms, researchers have system terms and participants in the production processes they system again, and the terms of distribution and customer services have a third system, is no doubt that there is a joint deal between all these systems or between each two of them.

Terms of the role in the preparation of the cognitive and cultural books and public references can not make a real development with continued deficiencies in this aspect. But the quantitative shortage in the Arab cultural production of publications and information programs of cognitive and cultural goals and shows deficiencies in particular the lack of what is written in or translated for new concepts in science and technology in a manner suitable for the Arab reader.

The intellectual production of books the size of an investigation and to soften and translation until today without the level of ambition, where the total Arab production of books a little more than 1% of world production in the past years excelled in this area countries were in the asymptotic level such as Turkey and Spain.

**Arabization of science**

The issue of science Arabization is one of the most important issues facing the people of the Arabic language in this era characterized by accelerated scientific and technical progress, using different languages but not including Arabic. Exhibit the progress of invasion of the Arab language in some features. Note that the English language as well as the French became the languages adopted in many fields, including the areas of science and technology.

Public education and higher education in Iraq are arabized in most fields such as law, education, economics, sociology, political science, science except medicine. The teaching was in science colleges in Iraq, in lecturing discussing, testing the English language was professor's study being tested. The Arabization of science had many grounds, including:

- The Native language is the title loyalty to the homeland.
- There is a feasible terminology enough for the needs of Arabization.
- Arabic language is manageable for example:
  - Many of the weights derivatives (the names of machine).
  - Standard pronunciation of one in particular time and place.
  - Provides weights for the interaction process (attraction and interaction).
  - Industrial source terminology (sensitivity).
  - Many of the voiced are not available in English.

There are some who oppose the Arabization of pure science, which based on the perceptions of this particular language, including:

- Disqualification of the Arabic language in teaching science.
- The language of science and technology in the present era is English.
- The number of books and references in Arabic are a few.
- The teachers who are able to teach in Arabic are a few.
- Wondering of some of the terms issued by some Arab language academies.
- The problem of scientific symbols and letters and numbers.
- The problem of publishing and distribution.
- The mastery of a foreign language is essential for those engaged in science.
- The use of English in higher education raise the student's ability is most widely spoken language.
- The teaching of any foreign living language is aimed at mastering basic skills.

The permanent office of the Arabization carried out the referendum in 1966 on the validity of the Arabic language for university education. The following problems and obstacles that have emerged then and which continue to raise concerns:

- Failure of the Arab scientific research.
- The difficulty of the Arabic language in terms of the rules and writing.
- Lack of adequate references in the Arab areas of science.
- The scientific innovation and writing in Arabic are not encouraged.

There is no country in the world that did not adopt the national language in education. For example:

- A Chinese experience: was able to unite the Chinese language as it was divided to the 300 language, contains 44444 sections then canceled the dictionary of the English language in various kinds of science and technology.

- The French experience: France got rid of the teaching science in English; then has taken its Francing science until the French language has become a universal language to learn medicine.

**Notions of Arabization**

Arabization means creating opposite to Arab foreign words and adaptation of foreign word somewhat similar to the Arabic words with alphabet sounds. The Arabization of linguistical means according to a Lisan Al- Arab by Ibn Manthoor the "Arabization for the Arabic Word".

The Arabization is currently intended to use Arabic education for all levels and scientific research in its various branches, specialties. In additions Arabization actually is intended by the dominance rule of the Arabic language on its territory and the language of science and education. The old Arabization intended to keep the format of foreign word to Arab tone, note that the Arabization is language of source the act Arabs and Arabs means expounded in the sense and Arabs, if a man spoke cast his argument. The meaning of Arabization is intended to utter a foreign word language on a platform of Arab speech and weight (antidote and oxide). The Arabic language academies of assert the term, when agreed Arab word with the votes and weights. As well as the Arabization of the txt where the text is transferred from one foreign language into Arabic and Altjeem transfers the next form Arabic into foreign language and also make the Arabization of the domain where the language, such as Arabization in education. The translation is defined according to the Lisan Al-Arabs (translated speech is transmitted from one language to another language) and dictionary Al- Waseet (translated words between him and been underscored) and should differentiate between the diverse types of translation:

- Interpretation.
- Translation.

There are many examples of Arabization, including:

- Mashreqi form: has been characterized by this form:
  - Verbal attribute.
  - As specialized technical precision.
- Moroccan model: it means the process of Arabization comprehensiveness and universality of all activities within the community.

The elements of the Arabization process consist of:

- Qualified prof. to teach Arabic.
- Scientific expression and the term.
- Book.
- Foreign language.
- Lecture.

**Arab efforts to Arabization**

There are varied experiences of Arabization in the Arab countries depending on individual circumstances of each country; including in Iraq is a university education in the Arabization of laws and decisions that led to the Arabization into force. Proponents of Arabization, the national language teaching aims to achieve the following:

- Combination of originality and contemporary.
- Strengthening the binds of the link between educational institutions and society.
- Standardization of culture and scientific and intellectual effort.
- Helping learners to understand and absorption.

The result of the delay in Arabization is the use of distorted English language in most cases and a lack of understanding of what is given to the students such as the information, embarrassment and the impasse in which it resides when the student is graduated from the non- English university.

**Arabization conferences and symposiums**

The most important decisions made at conferences or responsible official bodies for the following:

- Resolutions of the Conference of Ministers of Education and Higher Education Arab/ Morocco 1970 and demands that all Arab countries take the initiative as soon as possible to take measures and means to use the Arabic language teaching in all phases of general, vocational education and university levels.
- A decision by the Third Conference of Arab Health Ministers/ Cairo 1974 No. (4), which includes the use of Arabic in medical education.
- Resolution of the Fourth Conference of Ministers Responsible for Culture in the Arab World / Algeria 1983/ held under the slogan about the Arab cultural security and contained the decision to ensure that the classical Arabic language should be used for education at all levels and to be the safety of the language.
- Resolution of the Second Conference of Ministers Responsible for higher Education and Scientific Research in the Arab/ Tunisia 1983 / and includes:
    - To speed up the national and regional Arabization.
    - The integrated policy of Arabization of the Arab countries.
- Resolution of the Supreme Council of the GCC countries of the Gulf at its sixth session/ Muscat 1985 / and provides (the obligation to Arabization of higher education and university education in all its branches and specialties whenever possible). It includes:

- Adopting the recommendation of the meeting of heads and managers of universities and institutions of higher education in the Arab Gulf States about the commitment of Arabization, and a timetable and detailed plan for the Arabization of higher education.
- Formation of a special working group for the Arabization of higher education consists of representative of universities in the Gulf Cooperation Council and representatives of other relevant parties.

- Council of Ministers of Arab Health/ Khartoum 1987 numbered 10/ these include an operational plan and realistic for the Arabization of medical education in the Arab world.
- Resolution of the Council of Arab Ministers of Health- the thirteenth session/ Oman 1988 numbered 13 / these include:
  - Raising the medical Education in Arabic to their respective governments for approval.
- Resolution of the Executive Office of the council of Arab Health Ministers of the fifty 1988 Geneva, for study of the experience of the Syrian Arab Republic in the field of Arabization of health education.
- Resolution of the Executive Office of the council of the Arab Ministers of Health in the session of Tripoli 1988 / completion of course a policy of Arabization health education.
- General Conference resolution XII of education ministers of Arab Gulf states, Kuwait 1993 and urged the universities to provide the resources and administrative structures necessary for the implementation of Arabization and supervision.
- Decision of the Conference of Ministers of education in the GCC/ Riyadh 1986.

**Conferences of Arabization (1961- 1994)**

Regular conferences called by the Arab Organization for Education, Culture and Sciences of Arab States League and are held every three years, so far seven congresses.

• Symposium of Arabization and Arabic language issues in higher education Khartoum 1979 was discussed and dealt with aspects of the issue of Arabization, such as the term translation, linguistic studies and the role of heritage in science history.

• Symposium of Arabization and its role in supporting the Arab presence and Arab unity, Tunisia 1981. Notions of Arabization, experiences, future and the symposium concluded that the issue relating to the future of the Arabization of the Arab nation as one nation.

• Conference on Arabization of higher education Damascus 1982 the conference resulted in a number of recommendations calling for a commitment to Arabization and pursues it steadily. Perhaps it is the most important seminars, this seminar held on the Arabization and resulted in the recommendations, namely:

- Adoption of view of the Arabization of education in colleges and university institutes in the Arab countries.
- Considering the next ten years, beginning in 1989 for the Arabization of the Arab medical sciences and health sciences in the Arab world.

• Regional conference for the Arabization of medical education in the Arab world, Cairo 1990. The conference concluded with a set of recommendations and implementation plan for the completion of Arabization in stages in a period of ten years by the year 2000.

• Conference on Arabization of medicine and medical science in the Arab world- Bahrain 1993 where the Action was to obtain a political decision in each country to support the teaching of medicine and other health sciences in Arabic language.

• Symposium monitoring the educational needs of Arabization in the Arab world/ Damascus 1996. To prepare a national plan of Arabization submitted to the competent authorities and to the following recommendations, namely:

- The need to publicize the issue of permanent Arabization.
- Activating the constitutional provisions that Arabic is the official language of the Arab States.

- Conference on Arabization of medical education, Kuwait 1996:
  - Inviting the council of Arab ministers of health to urge the ministers of education and higher education for Arab Arabization of medical education.
  - Urging medical schools and teachers to start teaching medicine in Arabic, according to what comes in the operational to what comes in the operational plan.
  - Development a time plan for implementation for the completion of the process of Arabization.

**Futures terminology**

It became necessary to reconsider the forms of availability terminology, the time actually between the adoption of the terms set and printed on paper in volumes and circulation in the Arab countries can be shortened significantly if it has to benefit from new technologies, where the completed office Arabization coordination in conferences Arabization many printed over about Thirty years and it became the obtained combined demand impossible and all this has become necessary to introduce all these on CD-ROM, according to a specific system gives the term with the corresponding and easy to invoke the term Tbakas Arab entrance or English or French with a statement of specialization or disciplines in which the term is used and this production CD in the framework of achieving traded on a larger scale in all Arab countries and not have to bear Office for the Coordination of Arabization production and distribution expenses.

It assumed that these terms be in the workout for a limited time period and utilized in the writing, translation and follows the Office of localization of these publications coordination selected in specialized areas through a number of scientists and linguists experts to know how to accept these terms with the registration of the proposed amendments and carried out under the terminology system.

Then it is consulted to insert these terms in the international information network in a manner accessible all over the world and to develop a clear plan for the invitation to be used in the following areas: basic scientific books translated and author, own contemporary trends in science publications, catalogs programs in the sciences, information materials prepared for broadcast by TV networks and satellites and here the role of mass media in the decisive use of terminology. It is essential to the continuation of basic research and applied research in the areas of terms with not be confused with the linguistic work documentary to the words of traditional crafts and historical research of the terms in the Arab heritage of the applied research about the reality of the terms and requirements of terminology for the present and the future on the other.

**Computing terminology**

I seemed clear that inventions mechanism in the computer field know the evolution of Ascending, offering daily results in utmost precision, making the utilization of an order imperatively demanded by the necessity of scientific progress and with everything achieved from a scientific achievement in the field of computers, still linguistics, specifically in the field of industry Arab known slower, if not delayed, if not yet able to take advantage of the Information Storage and adapted according to the needs of research and lexical .

If the language institutes, institutes of Arabization in the Arab world have capacities store a huge number of terms and glossaries in an attempt to address them, it is not yet a world achievement in the field of lexicography industry interference, and still many researchers unable traded in their research and projects and if what has been coordination between researchers and specialists on the one hand and language institutes the other hand, many of the projects in the field of terminology and dictionaries can be known through the completion of language to strengthen trading between scientific and educational circles.

**Scientific term**

The concept of scientific term

Scientific term for each word or word intervention in the scope of scientific knowledge and formulated or invented by researchers and scholars to express the results of their work.

Launches various scientific terms either on natural phenomena or elements that make up these components and either on the means used by the researcher to carry out scientific activism. When you talk about gravity gravitation or for photosynthesis photosynthesis and other illustrates all scientific ideas describe what happens phenomena in nature while when we talk about bowel intestine or mercury mercury or for microscope microscope and these terms refer to the names of calling either the live components or non-living.

The formulation of scientific terms

• scientific terms as a think tank.

When you reach the researcher to interpret the phenomenon of phenomena, it demands that he calls the designation makes the reader aware of what he wanted to give him a ride from him ideas. In this case, the researcher resorted to several ways to bring out the term came into existence, and among these ways:

- Resorting to balance the vocabulary traded for example, spontaneous generation means the emergence of a generation while single spontaneous automatic means was Louis Pasteur chose these two to express the idea of "self-generated" expressed by the term spontaneous generation.

- Resorting to the vocabulary or other scientific terms, for example, when the researchers that could explain the installation of organic matter from carbon dioxide, water and photovoltaic energy, the term referred to her by photosynthesis .

- Resorting to the names of the researchers for the formulation of a simple new terms or vehicle, as the term is derived from the entire name of the researcher for the term pasteurization was drafted out of the French researcher Louis Pasteur name. As for the term krebs cycle is complex in the sense of the term cycle krebs cycle and which is accompanied by a researcher who named the phenomenon he refers to this term.

• **scientific terms Calfaz technique.**

This type of terminology living and non-living components and stuff with which a researcher in the course of his intellectual and scientific.

- Naming bind, size and color Titles, for example, the term globule adopted in drafting the shape and the size as the term is made up of two to Feztin glob which refers to the spherical shape and ule which reLinguistic assets scientific term

Common to all languages of the world in three pictures when developing scientific term is innovation and translation and localization as in the Arabic language and as in the Vietnamese experience, but the Arabic language of the most compliant languages and amenable to build and derive scientific term.

Innovation is the development of sound Arabic phrases scientific facts have been discovered and there is no such her in other languages. The translator represents the use of the meanings of the term scientific Arab foreign words sound almost sense to mind while expressing localization of transfer to the Arab foreign word in accordance with the rulings of its assets.

It is believed D.maha religion Saber that localization is not an issue, but the language is a key cultural issue currently facing us, language is not a scream but thought and therefore has to be the development of the Arab community and absorb the culture of the times and it is only through language as a means and a tool.

In addition to the Arabic language, there are other world as Russian experiments where adoption after philosophical revolution create a machine translation huge for the transfer of science to the Russian language, as the experience of Vietnamization College of Medicine, along with the French experience important as the Hebrew language have been used in the teaching of medicine, engineering and science. D.fadil Tai also pointed to the Japanese experience in the transfer of care Science deserves despite the letters that ten thousand symbols of their number. The Chinese experience has been accompanied by the writer machine that can accommodate those letters and was able to unite the Chinese language and the French experience was useful in getting rid of the teaching of science in English knot.

Linguistic features of scientific term

Common to all languages of the world in three pictures when developing scientific term is innovation and translation and localization as in the Arabic language and Vietnamization as in the Vietnamese experience, but the Arabic language of the most compliant languages and amenable to build and derive scientific term.

Innovation is the development of sound Arabic phrases scientific facts have been discovered and there is no such her in other languages. The translator represents the use of the meanings of the term scientific Arab foreign words sound almost sense to mind while expressing localization of transfer to the Arab foreign word in accordance with the rulings of its assets.

It is believed D.maha religion Saber that localization is not an issue, but the language is a key cultural issue currently facing us, language is not a scream but thought and therefore has to be the development of the Arab community and absorb the culture of the times and it is only through language as a means and a tool.

In addition to the Arabic language, there are other world as Russian experiments where adoption after philosophical revolution create a machine translation huge for the transfer of science to the Russian language, as the experience of Vietnamization College of Medicine, along with the French experience important as the Hebrew language have been used in the teaching of medicine, engineering and science. D.fadil Tai also pointed to the Japanese experience in the transfer of care Science deserves despite the letters that ten thousand symbols of their number. The Chinese experience has been accompanied by the writer machine that can accommodate those letters and was able to unite the Chinese language and the French experience was useful in getting rid of the teaching of science in English knot.

## Scientific Terminology

It is believed that Sumerians were the first who put the foundations in building a systematic record of scientific terminology. Then the Arab Islamic Renaissance used a lot of chemical terminology during their translation of science and philosophy from Greece. The names of initial elements of chemical used the Arabic language terminology, including escalation and distillation. By the age of industrial revolution in Europe entered the terms chemotherapy clearly formulated under the founded descriptions and characteristics of chemical substances.

In the mid seventeenth century the movement began in Europe's new chemicals (chemistry construction and installation), urea was synthesized from inorganic materials by Kohler the number of materials processed before 1828 even before the first World War, materials rose 200 thousand to 750 thousand at the end of 1983, then five million chemical at present according to AUPAC. The organization of AUPAC developed rules for designing a systematic manner. Reflecting on the chemical composition of Arab term, Arabs themselves have disagreed in Interchanger Arab. For example, if we take $H_2SO_4$, the German name it "Schwcfcl Saure" and the Syrian formula addendum sulphuric acid "Kibrit" and Iraqi sulphuric acid has been used by stokes system instead of accessories (ul, ur, ic, ou).

## Assets of the scientific term

All of the world's languages participate in the three images to create scientific term, innovation and translation and Arabization, as in the Arabic language and ffatnama as in Vietnamese experience, but the Arabic language compliance and more driven to build scientific terminology.

Beside the Arabic language there are other languages, as well as further tests, where the Russian built after their revolution philosophical interpretation center of the establishment to transfer to the Russian language.

The experience of Vitamins approached in the college of medicine together with French experience as well as the Hebrew language which has been used in teaching of medicine, engineering. The Japanese experience in the transfer of science in spite of the alphabet amounts to ten thousand characters.

**Term and terminology**

The late Dr. Abdul Razak Mohyiddeen mentioned previous conditions that must be met to qualify as a single term, including:

According to the statement of the definition and principles of the term The composition of the term affinity between language and science.

- The composition of the term long- term process, requiring high flexibility and accept a broad and rational policy.
- The process of configuring the term is not individual but collective opinion the other plays a significant role.
- The application of the term introduced, does not require a decision individually or even permanently, but a dialogue between the makers of the term and its perpetrators.

The main features of the methodology developed by Dr. Ahmad Shafiq Al- Khatib are to apply rules and principles characteristic precision and logic in some of them:

- Must be checked in understanding the meaning of the term before attempting to translate it for fear of violation of terms implications.
- The methodology assumes knowledge of them or is transferred to and experience in the subject matter.
- Supposed to be the Arab equivalent of a single word attributed to them or added to, also credited or added to the foreign pronunciation.
- The rule that what is authentic in the language translated and transmitted after checking that the word Arab means or what causes to be elaborated by other means.
- Expresses wordy world derived from the Greek or Latin (for example- electron), or wordy composite characters and internationally

recognized abbreviations, or names are placed in memory of the world or the names of chemical elements (plutonium and uranium).

- Revival of the classic word to perform a new meaning provided that the chosen equivalent of a term closes to it in meaning (use of the term of the meaning of metal).
- Write a foreign flag by counties in the Arabic language and foreign word lists with letters in parentheses Latin word that was unfamiliar.
- Adjust the scientific taken into account the issue of Getting Started and the question of convergence of residents where to add a word, and proceed to move the character starting with the resident, graphite and proton.
- Attention to weight to determine the meaning of the word if the name of God or the name of the place or the time or the name of the body or the name of the actor or effect or the name of the preference.

**Methodologies for the development of the term**

There are a number of methodologies have been submitted by researchers in the term, including the following:

**The first methodologies:**

These methodologies focus the individual efforts before and after the Second Vatican Council in the regular work to establish the rules of these methodologies was active in this area scholars, including Peter Al- Bustani and Ahmad Fans Al- Shidyag and Jacob Sarroof of Anastas and Mustafa Al-Shihabi, Hassan Fabmi. These had put forward a methodology of these parameters the rules, methods to develop, measure, include sculpture, installation and Arabization of translation or verbal quotation.

Methodologies that follow many of the methodologies achieved in the development of scientific terminology in Arabic that is acceptable in its infancy, based on the ideas mentioned in the first of which contributed to by Arabic Language Centers in Cairo, Damascus and Baghdad, seeking the contributions of Office for the Coordination of Arabization in Rabat.

The methodology first introduced before the founding of the synagogues mentioned may be related to what has been reached and later published in stages in the records and publications of Arabic Language Centers, including:

- Proposals Jordanian Arabic Language Academy, a document containing 18 principles, recommendations and occasional references to the problem of precedents and suffixes.
- Recommendations for the development of scientific terms and approved by the Conference 45 in Damascus a document that included the 4 principles and 12 recommendations were related to curriculum development of the term Arab.
- Methods for selecting the Arab term by Prof. Ahmed Ghazal and include measures relating to choice of language and measures of social and linguistic minorities.
- Methods for selecting the scientific term and requirements. Dr. Jamil Al- Malaiks a document that includes several methodologies collected in the 6 basic principles for the development of the term Arab.
- The basic principles in the development of scientific terms and select endorsed in a symposium methodologies for the development of new terminology 1981.
- Suffixes by Dr. Mahmoud Mukhtar include two types 120 and 50.
- Methodology for establishing the new terminology of Prof. Ahmed Shafik Khatib, a 600 prior and subsequent to the 150 related to the field of medicine.
- Dr. Thami Alrajai Hashemi powered precedents and suffixes.
- Has been approved by the council of the recommendations of the Egyptian modern approach particularly developing specialized Arab scientific terms at its sixtieth session (1994), and sixty one (1995) (see Annex 2) as well as the recommendations were issued by the Conference

of the Arabic Language Academy at its sixty- third session (1997) (see Annex 3).

In the methods for selecting the term and practical requirements and pose d beautiful angels, proceedings of the Arabization of higher education, new methodology for the development of terminology, office for the Coordination of Arabization. Arabic tongue magazine targets has acquired specific purpose of coordination and tracking of movements and prepare for the Arabization conferences and follow- up activity and cooperation with synagogues and dictionaries have been published for this office and is still a specific methodology are:

- The term made two foreign languages together are English and French.
- All the terms that come back because of all of them attributed to its owner (Science Park, a professor of language, a lexicon known).
- Publish the form of a dictionary order is placed under the eyes of researchers for a period of not less than six months.
- Call the office to a conference of scientists held under the supervision of specialists (Arab Organization for Education, Science and Technology) to select the appropriate term.

According to the statement of the methodologies still need to use standardized terminology of the executive authorities involved in the various Arab countries, and still problems in the Arabization of the term (the multiplicity of terms, confusion and lack of precision and the lack of the term and not use formulas of commonality and foreign) as it is.

**Basic principles in the development of the generated term**

Arabic language on the move Fasttaat in ignorance that reflect the views of their owners and when it emerged that Islam is able to these concepts and to express the need for the new age trends and growing at a vocabulary richer and sculpture as well.

The principles adopted by some council in the development of terminology were generated as follows:

- Arab Academy of Damascus was within the scope of:
  - Faith in Syrian universities teaching the Arabic language.
  - Adoption of the translation of the languages of science.
  - The adoption of Arabization progressive destruction within a comprehensive plan for Arabization.
- Arabic Language Academy in Cairo: Authorized the use of certain complex vocalizations foreign language when necessary and complex work in the field of terminology and the status and regenerating on:
  - Maintaining the Arab heritage.
  - Meeting requirements of translation and authoring and scientific culture.
  - Keeping pace with the scientific approach in the method for selecting the higher term and bringing in Arabic and the counterpart in the languages of the world.
  - To accept what the scales or derived from the metaphor of a vacation with the derivation of the names of Senators in the language of science.
  - Leaving the use of some terms of foreign language is a necessity.
- Iraqi Academy of Science: The methodology used by the Iraqi Academy of Science does not differ from the methodology of each of the campuses in Damascus and Cairo and the work of the complex through its specialized committees to follow the following:

- Preference to the term Arab-born and born to speak only if known inherent.
- Avoiding foreign term in the case of scientific names of some of the elements and chemical compounds. If the names of foreign standards or units. If using in the books of heritage.
- That term could not be an Arab in the derivation and generation and measuring, and plenty of room metaphor.

• Language Academy of Jordan Started from the vision that the term derived from the authentic heritage of the means available to the language of the measure or derive or metaphor goal should be to put an Arabic term.

• Office of Coordination of Arabization: Coordination Office of Arabization in Rabat introduced a plan for terminology and coordinate resolutions of the, statement in this plan:

- The use of the word Arab and one for foreign expression.
- Develop an Arab term for each indication.
- Study of the term foreign study and the definition of adequate and understandable scientific denoted flour.
- Not only the adoption of a foreign language and one single source of foreign terms
- Avoiding the words several meanings.
- Commitment to the extent possible cases and accessories and the standard format.

**The formulation of scientific terms**

Scientific terms as a picture of thought, once a researcher to explain the phenomena, it demands that it be designated the reader aware of what he wanted to give him a ride the ideas. In this case, the seeker to a number of ways to get the term came into existence, and among these methods:

• Recourse to the balance of the vocabulary for example, asked "Spontaneous Generation" means a single generation, while the emergence of "Spontaneous" means automatically has chosen to Louis

Pasteur to express the idea of "Self-regeneration" expressed by the term "Spontaneous Generation".

- Recourse to the vocabulary or other scientific terms, example, when researchers were able to explain the installation of organic matter from carbon dioxide, water and light energy, referred to them by the term photosynthesis, "Photosynthesis".

**Arabic language and the scientific term**

The confrontation between science and the Arabic language as being sharp, primarily due to the specifications of science, not to support the capacity of the Arabic language to contain the latest developments of science and there are millions of chemical names which up till one in terms of synthetic to, more than twenty passage connected, as the term has become a modern chemical part of the fabric of the Court and precise terminology require many rules and principles. And researchers have played a role in the Arabic language in this confrontation, many of whom not to expand in the derivation and metaphor and analogy and sustained, however, the Arabic language and sophisticated old clear- cut capable of containing the changes, including modern scientific chemistry, note that there are oriental languages, including Japanese, Chinese, Korean and written symbols and signs was able to defy the English, French, German, and stand in dealing rival it, and accordingly it is necessary to expand investment in the Arabic language to the face of renewed term chemotherapy and the buildings and the Arabic versions and formats that reward ration and description.

The Arabic language of coordination between scientists and chemists and the development of foundations and international norms agreed through a comprehensive survey of the heritage of the chemical terminology available in the synagogues as well as Arab and Arab organizations and individuals, as well as the rules could be devised a methodology for chemical terms require the concerted efforts of the parties and in that there are many linguistic needs as well as requirements that rely on standardization, which is the language of the age and science- based Fbaltakiis check to its flexibility and capacity of language and accuracy as they are as old as the exchange of language and also as the basis of the Arabic language to be taken and applied on the following principles in chemistry and related subjects:

- To indicate the profession and reflects by a source on the weight of "effective" like a bottle.
- To refer to the weight function "effective" such as glass paint.
- To indicate the residual weight of the thing or things effectively.
- To indicate the machine of the act on the weight of three disabled and disabled and enabled as tripod Tracer-graph.

Activated spectrograph (intended for drawing and planning) spectrometer detector with the possibility of using the following formulas:

The name of the machine:

Effective dropper

Effective centrifuge

Effective payment

**Derivation**

- The derivation limited to acts and sources of words on the rigid foreign vocalizations attributed to the media or the names of famous built closer to the linguistic and Arabic voice compositions to Hydrogenation Ionized.
- Derivation of words is complex foreign assets according to the rules of Arabic composition add carboxyl Carboxylation.
- The act is derived from the name expressed the rigid three- act exceeds the weight required, do such as: iodine: iodine, verb is derived

from the name expressed non- rigid triangular weight and necessary such as: carbon: carburet.

**Industrial Source**

Source of novelty, in line with the weights Arabization, build rate Bia- ta-bound express on the status of the description and a rich source of industrial markers and meanings used in the formulation of which scientific term, while the source refers to the normal situation in which it shall act.

The industrial source is derived from the usual sources (grants stability) and attributes (acidic) and the names of media.

**Sculpture**

It is recommended by a lot of academies and individual researchers that sculpture is necessary in scientific texts so as to guarantee simplicity instead of using complex combinations, while these carved terms irritated other researchers; example: Electric light = photoelectric. It is believed that many wordy carved terminologies must remain within the scope of clarity and easiness, and not restricted to sculpture.

Some was signed when it is permissible for sculpture and installation of the term of the chemical words were foreign Arab and other big mistake, where the term came out of these hybrid composite, such as amygladin

However, the preference is to avoid the sculpture of the names or actions, such as extraneous Alkylation. It is worth mentioning that the term has gone through chemotherapy historical and some have called the names represent that stage, The term Sumerian simple could be the beginning of the term and systematic term Arab-Islamic and post-term EU old who grew up on the heritage of Greek and Latin languages. The term chemical talk has emerged in the first half of this century by the International Union of Pure and Applied Chemistry (IUPAC International union of Applied and pure Chemistry).

**The term scientific cooperation between the meat and retail**

It is well known that the Arabic language is characterized by weights, the names of God and their derivatives mismatch that are based on the weights, a three-act and disabled, including different weights come on the researcher produces a wide range of derivation and measurement. The lack of clarity of label machinery and scientific instruments back to researchers disagree on the "methodology" consistent with the derivation and measurement in the "Arabization of the natural sciences", held in Tripoli/ Libya 1975, that a translation of each part of a compound word is not accurate because of the prolonged, it gave some examples of attempts to rename the Syrian universities in the physical and chemical devices recalled word (gauge) for (Meter), name of the device (Voltmeter) measure voltage.

It is believed tha

Conclusions and concepts in scientific term

For example, the scientific term, including the following, according to the information came from the public and provide definitions:

- difficult linguistic structures in the face of scientific term.
- abundance of media and objects Alaagamah in general terms.
- diverse vocabulary capacity to signify when the calendar term.
- symbols and references in science capacity.
- diversity of diligence in the development of the term scientific term.

**Naming enzymes**

Systematic naming of enzymes is necessary, given the turmoil that is located in the diversity of jurisprudence in the label and from the following:

- Transfer of the enzyme into Arabic as a whole with its – ase, Urease, Phosphatase, Lipase, Amylase, Maltase, Lactase, Diastase.
- Transfer of the enzyme partially into Arabic, where the enzyme is divided into two parts where the first half gave the Arabic and the subsequent survival - ase enzymatic significance as it is where writer in Arabic, Amylase, Maltase, Catalase.
- Transfer of the enzyme by translation each section to the Arabic term and engineering assets according to the latest Arabic, Diastase, Maltase, Lipase, Lactase, Catalase, Carbonic anhydease, Alcohol de hydrogenase, Glycosidase.

Enzymes are divided in 6 groups according to the nature of the interaction of these factors motivated assistance:

- Oxidation and reduction enzymes "Oxidoreductases".
- Transferase enzymes.
- Enzymes Lyases.
- Enzyme molecular similarity Isomerases.
- Enzymes association Ligases.

The following rules have been adopted by the Commission on the current chemistry of the Academy as well as the above:

- The number enzymatic remains as it is in foreign text, (2. 1. 1. 1. …etc).
- The Group first "Oxidoreductase" and become oxidant enzymes and reduction in the case of one enzyme of the Group of enzyme oxidation and reduction.

- Expresses chemical names as contained in the original text.
- Expresses symbols, as stated when they are spoken.

**Precedents and accessories**

**Preface**

include accessories that add extra meaning to the term and what follows is some of the illustrative examples:

- Hypercholesteremia: This term consists of the root "cholesterol", which indicates the name and previous "hyper", which indicates the meaning of overgrowth (increase), while the subsequent "emia" of significance, which means there the blood.
- Cholesterol: This term consists of three sections, the first means bile "chole" and second "-ster-", which reflects the fact that the solid material.
- A false case of failure in the side of the thyroid gland "Pseudohypo parathyroidism" the term is composed of five sections:
  - Pseudo and indicates (false).
  - Hypo means a shortage or lack of.
  - Para means side.
  - Thyroid The root word means the thyroid.
- Alcohol dehydrogenase: Consists of the term, which represents the name of the enzyme (oxidize the alcohol) of the following sections:
  - ase enzyme (subsequent that represent and express the word enzyme).
  - Hydrogen the previous de-sign on hydrogen.
  - Alcohol (root name).
- Lactate: NAD oxido reductase: This is called a term (the name of the enzyme oxidation and reduction, according to the following sections:
  - ase enzyme (subsequent).
  - Oxido reductase oxidation and reduction.

- Hypercalcaemia "Hper calcemia": Term consists of the following sections:
    - Hyper the previous name of significant increase.
    - Calcium (root name).
    - Emia blood
- Sulfate (sulfur – s) "sulphate": It consists of two sections:
    - Sulph the name of the root.
- Hypoglycemia: It consists of "hypo" which reflects the lack of "glyc" sugar (the word in mind- root) and "emia" blood
- Hyper- glyc- emia glyc- sugar (industry source) – emia blood
- Derivation and composition characterized by synthetic

Languages such as English feature the words derived and composed at the same time, the word is "contrast" origins "contra" meaning against and that the term "laevo – rotatory".

Case precedent is an element to change the meaning of the word (root) that is added to it and can get different meanings of the terms in English change the previous pre- root, for example, if used the Latin word "spirare" was added to get the various precedents of various terms after you change.

Precedents have different origins of Greco- Latin, Saxon, etc., can be placed in schedules accordingly to investment in the science of chemistry, life and medicine involved with the suffixes in the composition of the term, and one of the sources of the different language; for example, the term "spector - photo-meter".

**The scientific term**

The scientific term is a pillar of science history has been followed, researchers and many ways to elect a scientific term, including:

- Using of the act to derive the outgoing term, for example went from the act went.

- The term derives from the name of an example of rigid copper.
- Terminology inherited from the civilization of Mesopotamia or the Nile Valley Civilization or others, such as saffron Term of the Greco-Arab or other, for example camphor is derived from the Sanskrit language

The scientific term for each word or a word within the scope of scientific knowledge and formulated or invented by researchers and scholars to reflect on the results of their work. Launch various scientific terms to either natural phenomena or elements that make up these components and either on the means used by the researcher to carry out scientific activism. When talking about gravity "Gravitation" of photosynthesis "Photosynthesis" and the other illustrates the ethos of science describe what happening phenomena in nature is, while the gut when we talk about "Intestine" or mercury or on the microscope. These terms refer to release the names of the components of either living or non- living.

# Chapter Four

# Science and Education

A. Preface
B.. Globalization
C. Knowledge and education
D.Quality and assessment in education
F. Brain drain and education
F.Education and the economy
G. Philosophy and Education
H. Scientific Aspects and education

## A.Preface

These include, support Of education process, which serve Iraq and its students and people, provide education for each Iraqi individual, upgrade education to imply levels of advance and enter to new century with fixed and clear steps. Furthermore, the education aims to prepare and create scientist and specialists and to achieve the justice and equity among all individuals and society regardless their ethnic or nationality.

The strategy for the future should aim to establish a stable and sustainable financial base, ensure effective leadership and management at all levels, maximation the staff potential, and promotion of collaborative research.

Science is introduced in our lives at an early stage, through our education. We learn a basic set of skills that we expand with specific knowledge in a certain direction at a later stage, but we never really appreciate just how important science is in all this equation. Education is our first encounter with science, in its various forms, so it's crucial to get it right.

Once our education cycle is complete, it's time to choose whether to pursue a career that will push science forward, and thus contribute to science directly, or opt for a career that will just allow you to get by in life.

Furthermore, to ensure the growth of a healthy and productive university research environment. The ministry of Education should undertake a critical review of the universities, and should be encouraged to seek international funding for research.

Over the last few decades of war and sanctions, Iraq has suffered great losses in death and damages. Harm has inflicted persons, communities, communications, institutions, and land. An equal and perhaps greater damage has been the outcome of the lost development opportunities: stagnated communities, uneducated children, adolescents and youth, untilled land, declining growth rates, and the construction that did not take place.

The nation has also witnessed an unprecedented devastation and a near-to-complete break-down of the education system. The prevailing state- to-affairs of higher education disheartening and presents a virtual collapse; academic staff continues to leave for fear of mounting threats and insurgency, teaching being stopped or erratic and most buildings being damaged or destroyed.

The education system is currently in a stage of profound rebuilding and restructuring particularly at a time when the demand for higher education system to address the national needs and priorities comprehensively seems to be a source of much pressure. This mounting pressure is equally evident on the Iraqi institutions and centre of knowledge to produce relevant and appropriate knowledge at an unparalled rate. The diverse demands for qualified human resources of the complex national economy with evolving global links require knowledge for decision-making and for production.

The current efforts of the Ministry of Education and institutions of learning are in the midst of accelerating change, yet they have to grapple with profound processes of decision-making addressing issues such as: relevance (curricula reforms and accreditation); quality; scientific research; teachers' and students' welfare; technical and teachers' education; academic freedom; decentralization and institutional autonomy; long-term financial sustainability; and internationalization.

Simultaneously, institutions of higher learning will also have as a goal to provide learners with academic and ethical values, as well as having the commitment to educate critical citizens prepared to engage in democratic participation. Change of the Iraqi higher education structure and of research is thus seated in an extremely complex reality, in which no self-evident choices are available and where actions have multiple effects in a dynamically interdependent environment.

Yet the challenges Iraq faces cannot be solved with piecemeal changes here and there – challenges made more trying by the strong forces that will affect Iraqi society in the future; economic reconstruction, population change, and the rapid growth in knowledge. The situation must be met by a fundamental reorientation of higher education activities under a unified perspective (new version). Providing this "new vision" and a "strategic plan" for specific and realistic short – and medium – term choices is the aim of this part.

**Past and Current Status**

Modern universities in Iraq were established in the second half of the last century, beginning with the University of Baghdad in 1957 uniting several constituent colleges in the process, during the 1960's five more universities were established: The University of Technology and Al-Mustansriya University in Baghdad as well as universities in Basrah, Mousal, and Sulaymania.

Thus, there is wide range in the size of universities as well as a lack of geographic equity, in their distribution across the country. Rehabilitation of universities, colleges, according to international levels is required.

The major fields of study offered by the universities are education, arts, law, social science, administration, economics, natural science, engineering and technology, medical sciences, veterinary medicine and agriculture.

*University education, to which students are admitted after the secondary stage, ranges from three to six years. Applicants are registered in colleges to pursue their studies in various specializations such as arts, science, medicine, engineering, etc… Bachelor programs leading to a Bachelor's degree are awarded in Architecture Engineering, dental surgery, pharmacy and veterinary Medicine and surgery.

Further developments was characterized by establishment of technical institutes reflecting the demand for qualified technicians, comprises of technical institutes and technical colleges.

Technical education in Iraq comprises technical institutes and technical colleges 22% of whom are female.

In addition to colleges and universities there are two-year post secondary institute, which train students for various technical professions. Upon completion of these two-year programmes students are awarded Technical Diploma.

Scientific research in Iraq universities and university research centers needs radical rethinking, improvement and rehabilation research centers, to take their pioneer role in serving Iraqi society.

There are private colleges and university colleges, offering programs in computer science, business administration, economics and management and contains many departments. The private sector has the potential to deliver quality education free from financial constraints affecting the state institutions.

To ensure the adherence of private institutions to higher educational standards on assessment of the existing private universities should be undertaken by an independent commission of highly qualified academics. A new law should be governing the private higher education sector and to encourage the expansion of the private sector.

**Curriculum development**

Curriculum development requires the desirability of a change to a system of modularization, learning outcomes, skills development, assessments, student-centered learning, and the increase in intake numbers.

The Commission for Computers and Informatics (CCI) and the Commission for Medical Specialization played an important role in training of research staff and promoting research activities and offer specialized courses for postgraduate students.

Of the academic universities teaching staff 56% are male and 44% female; 43% of the teaching force is concentrated in Baghdad. The average student teaching ratio is 1:13 being much more favorable than neighboring countries such as Jordan (1:30) and Saudi Arabia (1:20). There are however extreme variations among Iraqi Universities from 1:43 to 1:4.

The teaching overload of academic staff was serious obstacle to the development of high quality research. There was also limited international cooperation in research.

Iraqi scientists were publishing widely in international and regional journals; very few articles were published in the last decade. The larger universities like Baghdad, Basrah and Mousal have between 5 to 8 specialized research centers. In addition, there are other specialized research centers, the polymer research centre, date Palm research and the marine research centre. Socio-politic research was conducted at the Gulf study centre at Basrah University, International study centre at Mustansiriya University. Iranian studies centre in Basrah and Turkish studies centre in Mousal. Archeological research was undertaken in Iraqi national Museum in Baghdad.

**Collaboration and Twinning**

Iraqi universities are called upon to seek collaborative relations with high-standing foreign universities and research institutions and Collaboration with foreign universities can take various forms.

joint curriculum development, training courses, short programmed visits, schemes for visiting professors, joint research and research supervision, networking and exploring the potentialities of e-learning, online contacts by staff students and libraries and joint exploration of sources of financing of funding.

The Iraqi universities and the Ministry of Higher Education should cooperate in working out model agreements for collaboration with foreign universities, through for example, twinning which is specific act of co-operation calling for a mutual relationship.

**Higher education Policy and strategies**

The main issues needs and priorities that emerged from the UNESCO Roundtable were the following:

- The wide spread destruction of the infrastructure of the higher education system.
- The unstable and dangerous environment for normal academic activity.
- The quality of higher education has been steadily deteriorating since the imposition of authorarition rule in Iraq.
- There is a need to equip more than 2000 scientific laboratories and for 30000 computers, libraries are in a poor condition and are in urgent need for restocking with new books and journals in both Arabic and English ..
- The student population has been rapidly increasing due both to a high birthrate and an admission policy that allows all students who have completed secondary school to enter higher education.

To reconstruct the higher education, it is required to plan spanning (5-10 years), maximization higher education graduation, learning students opportunity according to the needs, development a team to inspire student, teachers to work in team, curricula should be based on appropriate outcomes and capabilities, development of teaching materials, class size should be designed to suit the student learning process, and the enrollment polices need to aim at recruiting enthusiastic students.

Consideration needs to be given to the relationships between higher education and industry and commerce. The need to up – to date information in higher education, information and learning centers should be established. A creative policy is needed for the provision of learning resources.

New challenges for education are witnessed by the technological revolution and the production of knowledge as well as the economic challenges and globalization. Technological revolution helped to create a new reality in the fields of science, knowledge, information and communications, resulted in forcing many educational institutions to reconsider their curricula and economic content. Then, the impact on the future philosophy and objectives of various educational institutions, as well as management and control has been noticed.

The major global challenges have merged, clearly defined distinguished and experienced on education institutions. Policies, led to the introduction of significant shifts in scientific developments, social and economic. These challenges are as follows:

- Globalization
- Futurism and Science
- Knowledge
- Quality and Evaluation in Education
- Brain Drain
- Reform of education
- Science and Technology

Under these changes of the third millennium, some additional challenges for education and its institutions can be mentioned including:

- Global foreign education in developing countries.
- Diversity patterns of education.
- New types of education, such as open schools, distance education and virtual school.
- Private sector investment in education.
- The limited role of government in formulating a strategy education.

In effect, countries are naturally divided into two categories: those active in education or advanced to a precedent, and those that are passive.

**B. Globalization**

Globalization is a group of political, cultural and economical operations that penetrate the borders of a single country, leading to the convergence of parts of the world, and leading to differing views of intellectuals on the concept and objectives, and societies differ on the extent of assimilation, acceptance or rejection. A proponent of globalization refers to a form of streamlining relations between the nations of the world. As opposed to globalization, it is believed as a balance between actors, the strongest is at the level of global capital that dominates large companies, and also it rejects the view that globalization is a kind of domination by powerful countries on the vulnerable, where consumers have been the last of knowledge and thus globalization have negative impacts on economy, education and politics.

In the face of globalization it requires developing countries to revert, then to revitalize the relationship between education and economy, development, culture. Then education and economics of future generations are the basis for developmental spirits of the knowledge society and culture.

Education instructions produce economic public good, where academics enjoy a level of economic and working conditions of eminent there is a difficulty in governing the university by the market, but the reality of education in developing countries, including Iraq's education structures in its content and methods, is traditional and suffers from problems of administration, organization and funding. Therefore, the first step of confrontation to operate is to understand the times and challenges. However, the positive effects of globalization on higher education have emerged by increasing numbers of education types linked to the internet that offer its services globally.

Globalization is vulnerable to education making reform of this education and the need to make it more important than ever, and that any neglect of this sector threatens the entire development process. The pressure of globalization makes it necessary to allocate adequate resources for education section therefore there should be reform of this sector at the enterprise level and the system as a whole.

## C..Knowledge and education

### Knowledge / description

Knowledge is the product of mental process of education and thinking. It is the basis of force and earnings constitute the most important components including the work and activity, particularly in the relation to economy, culture and education, which is also an economic asset. Furthermore, it includes the human being and his care and preparation the main sources of knowledge.

Other, believe that knowledge is the awareness and understanding of facts or acquire information through experience and means to acquire the unknown and self-development and is directly related to information, education and communication. The knowledge is also the most important component included in any work or activity, especially with regard to economy, society and culture.

Moreover, the knowledge society is characterized as one of the most important product or raw material. The knowledge of modern technology is used in their community and not to complain to the presence in the same geographical location and it became one of the most important components of capital in the current era which is supposed to be free; free application for the benefit of community and should remain free.

### Knowledge Society

Knowledge society is defined as a range of convergent interests, trying to take advantage of pooling their knowledge in areas that are interested. The knowledge society is of post-modernism linking the knowledge economy, generates profitable commodity. Knowledge society requires the potential and special skills and abilities, super sophisticated infrastructure, natural resources and minds capable of producing knowledge and converting it into super economic progress. The challenges of knowledge society emphasize the importance of education, and this requires improvements in education systems in order to transform them into the production process of knowledge.

It is worth mentioning that the knowledge society is the basis of the information society, according to the report that was issued by the United Nations, educational, scientific and cultural organization "UNESCO" in 2005 under the title "From the Information Society to a Society of Knowledge".

In Iraq, the knowledge society is defying Iraqi education for being easy to be challenged, since knowledge society is advanced, and Iraqi education is backward. Entering into knowledge society requires providing sophisticated infrastructure of communication technology, favorable climates of stability and contemporary education system with new techniques that emphasize the supreme actual operations.

## D. Quality and Assessment in education

The new pattern in higher education is to ensure the quality and its management, as well as that there is quality assessment and quality evaluation and quality assurance. The term assessment will receive many meanings and connotations, while the term "performance standard" refers to the level of achievements. There is confusion between standards and criteria, while using quality assessment and quality review as synonymous to evaluation.

The terms accreditation and quality assurance in education differ from one country to another. The standards are used in the USA on the same meaning of criteria, but in Europe synonymous of quality assurance; whereas the quality assurance is apart of the quality management in education.

Quality education means an estimated total characteristics and advantages of the product to meet the educational requirements of students and labor markets and society and all internal and external benefit. The achievements of quality of higher education require directions of all of human resources, systems and methods processes and infrastructure to create favorable conditions for innovation and creativity.

For quality there are several concepts such as:

- Value for money.
- Added value.
- Transformation methods.
- Fitness of purpose.
- As threshold (beginning as threshold).
- Consumer satisfaction.
- Enhancement
- Improvement.

There are five areas of quality of higher education:

- A whole system of higher education.
- Foundation.
- Programs.
- Education operations.
- Outputs.

Some propose a distinction between three sets of specifications for quality:

- Scrutiny subjects.
- Measurements.
- Inferred tools.

The traditional concept of quality of education is associated with the screening operations, focus on the test, then the great importance for the application of total quality management in the education to ensure the survival and continuity of higher education institutions.

Quality of education includes indicators for measuring the level of achievement in teaching and measuring of adequacy of the needs of labor markets, the number of students to each teacher, spending per students and repetition rate. It is the concept of multidimensional and dynamic levels.

Quality is important to improve the outputs of the educational process and develop the participation of the state and society, where the country has the responsibility to adopt quality standards that must be available for the department, may be unable to achieve on its own in certain situations.

Multiple concepts of quality as pointed by both Green and Harvey, including:

- Precision.
- Perfection.
- Constant change.

And Lim in 2001 put on mechanism for quality assurance in higher education that includes:

- Setting targets for education institutions.
- Compatibility with the general goals of the society.
- Test the effectiveness of quality management system to achieve the goals of education.

Quality depends on the context, which is called quality system and the message of the education and its goals. The quality varies in meaning depending on the:

- Components of education, students, the labor market.
- Frames of reference for quality, inputs, processes, outputs, and goals.
- Historical periods.

**Total Quality**

Total quality is the tool and process for practical application, which aims at achieving a culture of continues improvement in order to gratification, appease and please consumers and costumers, have many perceptions and concepts, including:

- Philosophy concept.
- Tools and processes.
- Continuous improvements.
- All employees.
- Gratification, appease and please of customers.

Dealing with the comprehensive quality of whole educational system requires encourage individuals to participate in decision making.

**Quality Assurance and Relevance in Developing Countries**

Multiple terminologies are used for classification of countries according to economic conditions, social, educational, cultural. The backward demonstrates the lack of hope in economic reform in the developing countries it refers to the poverty of performance. [The third world means the (non-developed countries), the first world (the industrialized countries of Europe and America), and the world II means (former socialists)].

The use of the entrance to the execution of quality assurance in education is an effective if certain requirements are available, such as qualified faculty members that devote themselves to work all the time at the education, available administrative services, and university leaders realize the importance and necessity of quality assurance.

The quality assurance is a process which focuses on quality measurement procedure for the institution or program and refers to a range of activities, methods, and procedures. The term is used alternately with quality management, quality assurance and quality control aimed at controlling the process and removing the causes of unacceptable performance. The overall quality assurance indicates:

- Policies and directed processes.
- Description of all the systems sources, and information of education.
- Planned and organized reconsideration of the institution or program.
- Planned and organized activities that apply the quality.
- Evaluating the ongoing process (assess, control, security, maintenance, improving the quality of education system).

## Quality Management in Education

Quality management in education requires a system in several sequential steps serialized as follows:-

- The task of identifying the goal of the education.
- Determining the functions of the education.
- Defining the objectives of each post.
- Establishing a management system and quality assurance.
- Placing a system of quality inspection.

Defining the mission of the education must be conducted through cooperation and understanding the vast majority of the education community in accordance with laws and aspiration of state, the education of modern concepts and futurology realism of the education. The functions of the education are supposed to be modern (teaching, scientific research, community service) with clear objectives for each these posts. Teaching has goals, such as curriculum, innovative and effective with modern methods of teaching and distinguished health teaching climate. The scientific research has objectives that deal with increase of the productivity of research, publishing with marketing and modern skills. The goals of community service function are represented by accommodation of the needs of society and cooperation with other organizations of the society.

## Parameters of Quality Management in Education

These parameters are: developments of integrated system for the education and institutions management plans, the application of the structure of modern management plans for teaching and scientific research and community service, within the clear objectives and specific strategies on several levels. These parameters require financial resources and follow, sequence in the preparation of plans at different levels of the university.

Quality management system requires several consecutive steps, including:

- Identification of the message of the education.
- Determining the functions of the education and their relative importance.
- Defining the objectives of each function of the education or college or department.
- Identification the system of quality management.
- Preparation of quality inspection system (Evaluation of the performance of the education).

The conditions for the quality assurance system include the following:

- Eligibility of faculty members with appropriate degree.
- Providing administrative services and good electronic tools.
- Capacity of education current factors in the quality of education.
- Using and adopting of scientific promotion according to academic potentials.

**Quality Inspection**

The inspection and screening process of quality carried out according to evaluation performance of the education and depend on the quality of the new file that contains:

- Message of the education.
- Specific objectives of teaching and research.
- Community services.
- Quality management system.
- Performance standards.

**The Evaluation of education**

The evaluation process is necessary as an integral part of the process of developing and colleges affiliated with. It is also necessary to identify and to achieve the objectives of the education in raising the level of scientific achievement and scientific sobriety.

Tools used to measure the advancement of the education could be summarized according to the followings:

- Measurement of the efficiency of output (graduates).
- Scientific level.
- Scientific achievements.
- Research.

The mechanisms used for evaluations depend on description and governance. Description is carried out by gathering information, whereas governance is done through combining certain characteristics that are needed often the evaluation, in addition to the identification of certain principles to be implemented, such as the purpose of the evaluation and selection tools for evaluation.

## E. Brain Drain and education

The phenomenon of brain drain has motives of social, political, and personal nature. The social is characterized by difficulties that faced the developing countries in strengthening the shaken scientific planning in developed countries. One way to keep scientists from migration is to treat the fundamental faults by working to link with national policy, to introduce the idea of scientific planning, providing possibilities for scientific work and atmosphere.

. The phenomenon of brain drain is the most important global problems which recorded at the international level and regional level as stated.

Recent studies indicated that the organization for economic cooperation and development which includes 30 industrialized states, that the immigrant enjoyed a degree of education.

It is worth mentioning that the phrase "Brain Drain" derived by British was used to describe the loss of scientist, engineers and doctors and that UNESCO defines immigration as kind of abnormal types of scientific exchange between the states, as a reverse transfer of technology.

The "World Organization for Migration" estimates those developing countries supporting the United States, Europe and South Asia at 500 million dollars annually. On the other hand, the World Bank estimates that one hundred thousand foreigners from industrial nations are working in Africa at a cost of four billion dollars annually.

Brain drain that began after World War II included developing and developed countries spearheaded by Britain, France, Germany, Sweden, Switzerland and Japan. The United States of America is not included, but limited to become a terminal brain drain, from other countries. From a historical stand point, this kind of migration was due to Phoenician and Golden ages of Greek, Roman and Arab civilization.

It is estimated that Egypt had provided about 60% of immigrates to USA, and Iraq has increased its share significantly after the nineties, followed by Syria, Jordon, Palestine. The UNESCO has chosen Egypt from among those most affected by the brain drain, but did not contribute dramatically to solve this problem.

Egypt has used students returning from study abroad in the appropriate places, but Iraq was the first to issue legislation to participate in solving this problem.

### F. Education and the Economy

On the other hand, there is a set of economic challenges related to higher education that can be considered a package of economic and educational problems, including:

- The role of capital.
- Preparation and qualification for work.
- The emergence of new disciplines.
- Education funding.
- Diversity of sources of higher education.

Capital representatives of the growing role of market regulations at the university and hit some because "the end of the university" and thus the loss of independence, leading to replace the rule of the rise of economy standards, and the rise of administrators in the face of academics, the dominance of large companies and capital to work and education becomes a captive market system.

. Some believe that ignorance of education stands behind it the economists who determine a simplified method for assessing the return on investment in education. The fundamental problem is the measure of the return on education during the distinctions of other.

The investment in education represents the budget spent on scientific research. The UNESCO institute for statistics repot in 2004, indicate two things. The first is Total expenditure and then Expenditure intensity. Both are increased in developed countries, and decease in developing countries.

## G. Philosophy and Education

It believes that the term philosophy was a Greek origin, and consists of two words (Philo-Velia) and its meaning and loving (Sophie-Sofia) their meaning wisdom, either in the Arab-Muslim heritage has found her several expressions, including science ethics. Philosophy of the most important specifications that fields wide and several kinds of sequential steps leading towards the formation of thought essential to simplify the perceptions and understanding of the humanitarian dilemmas.

But they are all looking for the nature of things by using the mind. Has spread in Greece a lot of contradictory intellectual movements philosophy idealism and realism is better see the values fixed and virtues do not change, while the second sees Care senses more important than focusing on the imagination and are consistent with the ideal in the fact that the virtues fixed, educational philosophy required that the education part of the fully human existence Calvin, science and language.

The concept of education is different from, for example, according to the philosophy that deals with the type, points out that education is the formulation of the same human being announced and goodness is believed Plato that education is consistency between the soul and body, while Al-Ghazali believes that the education her priority is the spiritual and humanitarian atmosphere.

And lead exemplary education to the highest degree of maturity of the children, either natural philosophy refers to the mental preparation of the child. While education at existential philosophy that man is free and subject to the inevitable and philosophy of pragmatism as vision Dewey suggest that continuous education organization of experience and adapt to the social reality.

And John Dewey, one of the philosophers who pointed out the principles on which the modern concept of education and the education of them small community which comes to life and forever continuing education and curriculum must keep pace with life and the task of education is to prepare the individual for life.

Among the Arab philosophers who were interested in the problems of education such as Ibn Sina and al-Ghazali and Ibn Khaldun. Son Khaldun (1332-1406m) has been limited educational principles should be in the gradient and the transition from the known to the unknown and from easy to difficult. It particles to colleges.

The European Education, which Pflasvetha including Jean-Jacques Rousseau (1712-1778m), who called for equality and return to normal life away from corruption and meet the needs of childhood. And Herbert Spencer (1820-1903m), which focused on the psychological concept of education and private psychology Tql.

The subject of philosophy is knowledge () and knowledge of natural facts (relative), facts standard (values and ethics) is the philosophy according to perceptions of Dewey, for example, the general theory of education as some mention of it humanitarian features of human beings turned into intellectual trends Others believe that celebrities philosophy are the leaders of educational thought Ksagrat, Confucius and Plato and Jean-Jacques Rousseau, Fichte, where reflect the philosophy about the conditions of education and working on.

In other words, are the educational outlook emanating from the theories and ideas of philosophical as part of a cultural and educational philosophy way not only consider abstract ideas, but consider how to use the ideas in a better way, and lead educational philosophy to the development of channels and the principles and foundations required for our daily work in the educational teaching and practice of education and how We are learning and human growth.

**Philosophy of Idealism**

And it linked to this philosophy pioneered by Plato, which is the perception of the existence of two worlds, the world of ideals (hard) and the real world (variable). The community consists of two layers, one of them thinking and other works In other words, the first class is linked to the educational framework for the purpose of access to knowledge and that requires the mind, and thus

Knowledge is that link the mind be true and unchanging. This philosophy and believe in basic principles of strong belief in the existence of an independent in a perfect world the real absolute ideas.

This philosophical school follow a curriculum N constant evolution. The idealism and teaching methods are based on the basis of mental training staffs. According to Applied features of this philosophy in the educational field accumulation of knowledge approach is clear and unchanged and adoption of tools learning as ways constant teaching and exams without the use of traditional means and focus on the mind and the lack of school trips adoption and the adoption of corporal punishment and finally that learning is the focus of the education process, as well as it sees Plato and teacher Socrates (470 BC -399 BC) to fixed values and virtues do not change and that education must take care of reflection and imagination and that human nature is made up of two-spirit and body and must take account of these bilateral upbringing of the individual in society as the True.

According to these perceptions education focuses on realism must prevail without neglecting the spirit of the body. And it provided the ideal philosophy for education and training of the Socratic idea generation which is based on the mind and stir to pay for self-search, addition, Vehtm Education at idealists experimental spiritual philosophical issues with abstract thinking and keeping information practiced by high school.

**Realism Philosophy**

According to this philosophy of Braidha Aristotle (384 BC -322 BC) that there is only one world is the world of reality is characterized by fixed principles and the most important of the senses Care Turkao the imagination and are consistent with the ideal in the fact that constant virtues.

Follow the realists narrators approach stability depends on continuing this philosophy to discover the universe and the world and work to understand the laws in force, which includes all of the facts within a stable and constant attempts world. Accordingly Valoajpat placed on realism philosophy for Educational clear according to the following:

• follow the approach which accumulate natural, social and cultural knowledge.

• acquire the knowledge, skills and habits to prepare students for life.

• Prepare specifications teacher distinct process.

• extracurricular activities necessary realism of the school.

• the learner is the focus of the educational process.

• extracurricular activities is an important part of the curriculum.

• curriculum of a set of facts discovered by scientists consists.

• The use of programmed instruction machines.

And the fact that the realist school Qdj filed down the senses and inflicted philosophy and ideals of meditation and fantasy to reality and the senses so that scientific knowledge and curricula astronomical and mathematical Kalaom Square widened.

And John Locke (1632-1704m), one of the pioneers of realism philosophy and endorsed it by his conviction gave critical thought ample room and the experience and reality and senses no example and abstraction based on knowledge and science he sees that the child a blank page draw Storha of fact, as well as it has added Luc need to study natural phenomena, along with other math and science.

The Komenus (1592-1670m) realist sensory Making of the image of the most important methods of child education in schools as well as facts enthusiasm and believed to be the founder of the first special education child of freedom. And the development of sensory approach to teaching and took care of the physical and moral education to both.

Finally, the real reached a multiple convictions, including:

• established curriculum experimentation and exchange of scientific and methodological doubt (the road to see the existence of God, the real life).

• encouraged the learner to observe natural phenomena in an orderly manner.

• called for acts of mind in the analysis and the independence of the senses.

• This school did not distinguish between the world of ideals and spirit and the world seriously and reality.

• did not elaborate on the mental meditation.

• focused on professional education.

• invited to link educational curricula Balhajiat life.

Pragmatic philosophy Pragmatism

The roots of this philosophy go back to ancient times and the writings of Hrakulais (535-475 BC) and Undtlaan (30-95m) and Harzubayrs and William James (1842-1910).

While contemporary pragmatism linked to the New World (the second half of the nineteenth century and early twentieth century) called pragmatism generally Baladhatih and pragmatism, development and operation.

Of the principles of this philosophy that education is linked to the life and education to community service philosophy with Alentalm give a measure of freedom. And the human organism adapts to the environment and that the relative biological world is in constant flux and the truth is absolute and variable society and democracy to decision-making. John Dewey has established the features of this philosophy.

Pragmatic philosophy role in the development of teaching methods and by improving the traditional way and in a manner trial and error, or experimental way. John Dewey School-based Albergmteh philosophy has historical intellectual revolution against traditional schools, which focused on the information nor firmly believes in values, but the relative morality is renewed according to the convictions of the community.

The teacher in the philosophy of pragmatism he does not teach traditional materials in a systematic way and the teacher moves from idea to another in a sequential manner and deal with all the idea on the grounds that it in itself and suggests future problems for his students. And that the values vary depending on what they put time and society and the convictions of the individual.

Education at the pragmatic philosophy is not broadcast knowledge to the student in order to process knowledge, but to help him cope with the social needs of the environment and was able to stir up the student as required by the forces of social attitudes.

And rely perceptions pragmatism philosophy on building curriculum, including confirmed by the student on what he wants from him, and using the student Alqrah, writing and arithmetic as a means rather than targets, and the curriculum is interested in the facts relating to the nature of the child, also emphasizes the pragmatic philosophy to develop vocational education, natural sciences, while human studies and languages have secondary importance.

This philosophy has paved the Educational Progression some progressive educational and considers her application and was the most prominent pioneers John Dewey at the beginning of the twentieth century (1920-1945m), which came a revolution on a massive and cons of the traditional system produced a key concepts including:

• Coordination with the environment as the best environment for education.

• reliance on textbook and curriculum need to the principle of the freedom and flexibility and experience.

• teacher prompt administrative.

• take into account individual differences for each learner.

• requires the student to ask about the surrounding environment.

• the school has to cut off from the outside environment and that depends on the style of problem solving.

• variable values and science.

**Islamic philosophy**

There are major Islamic philosophy perceptions reflected on education, and these perceptions that the Quranic verses explain the nature of the universe and man, knowledge and values, and everything in the creature universe to God and the changes that occur in the universe is governed by paths and rights in Islamic philosophy.

Education is happiness in this philosophy and that the child has the bright page where there is no hindered Education In practice Education is starting science Quran and Hadith beyond are going to learn science has refuted Ghazali arguments philosophers in his book, The Incoherence of the Philosophers while defeated Ibn Rushd philosophy wrote The Incoherence of the Incoherence Canadian and longer of philosophers who emerged in Islamic history.

**Natural philosophy Naturalism**

This philosophy believes that man is inherently good and that is what human spoil the society and its institutions, education, her goal is to provide an opportunity for the growth of the natural child, as the jam Ptsourath negative and positive learner childlike nature.

Natural philosophy appeared to calls by reference to the educational activities that are aligned with the natural laws of indigenous cared Rousseau (1712-1778) these ideas and Crystal natural philosophy and Allamadrsah cemented the role of nature in the development of children in terms of:

- frees the child from school and classroom activities.
- meet the child's needs and the need to remove obstacles facing it in accordance with environmental requirements.

Rousseau was one of the first to call for self-learning and the nature of the most important educational principles that goes with it, and that women found the man to please, and this philosophy advocated the need to respect the individual and the protection of children from societal pressure differences.

**Existential philosophy Existentialism**

Kdeckart existentialists and Sartre believed seriousness and absolute freedom and human a philosophical vision of human existence, responsibility and human nature and the world, knowledge and values existential emerged in Europe after the First World War (1914-1918) in Germany and then in France.

It is believed the existential education as perceived by them to human existential indoctrination and does not accept the promotion of education for the existence of rights and Arts, music, philosophy and the arts of the general requirements, dialogue and debate and the basics of teaching methods and programs stable unacceptable.

**philosophy Deschooling**

Each of John Holt (1992) and appeared to Cash (1994), the pioneers of this philosophy, which depends entirely on educational institutions Some believe the philosophical foundation of this school is Gandhi's theory of education, especially in environmental education and the foundations of this school:

• learner be close to its environment and interact with it.

• learner learns from his peers.

• To achieve the same learner.

**The philosophical concept of the mystery of life**

There are three philosophical concepts of the world and developed as a result of human intellectual effort concept of spiritual and physical concept realistic and unrealistic concept of the divine. It can be evaluated and try to rush to one of them or the formulation of the concept of the center between them.

The conflict between the divine and the physical manifestation of the conflict between idealism and realism and that the philosophical concept of the world one of two things ideal concept of the physical and the concept does not correspond to reality at all, realism is not according to the materialist conception as the ideal is not the only thing that is opposed to the materialist conception, but no concept last realism Divine is a realistic concept.

The concept of the divine world does not mean cutting out natural causes, or to rebel against everything from facts and sound science but it is a concept that God is a deeper reason. Physical back door is in an ideal spiritual claim to the area and either spiritual concept in the divine way of looking at it is the reality.

As for the scientific field, there is no God and material Valfelsov whether divinely or materially believes in the positive side of the flag, there is no issue in the scientific philosopher, my God another material, but there are two Filsvetan and in conflict when it was a matter of existence beyond nature.

thought to the fact that just about art, any outside experience and physical deny it is believed to be natural causes revealed by experience and extended her hand science is the primary reason for existence. The nature of evidence that can be presented by my God is mind, not unlike the experience of direct material traditionally regarded as evidence of the experiment on your concept, it believes that the concept of divine or metaphysical issues in general can not be proved experimentally.

Materialism need a guide on the negative side, which distinguishes itself from the divine and it's the direction of philosophical, because science alone does not prove that the material of the concept for the world to be a material process but whatever undisclosed knowledge of facts and secrets in the world of nature leaves room for assuming the highest reason above article

Mental theory.

Put forward this view by (Descartes) and (was) pointed to the existence of exporters perceptions firstly sense (heat, light, taste, etc.) and the second instinct (the human mind has the meanings and perceptions did not emanate from the common but are fixed at the heart of instinct). sensory theory

Sensory theory is based on experience and common sense in this theory is the infrastructure that is based on the base of the human perception.

Therefore, the theory of knowledge can be used to follow up the mystery of life and access to him, in the perception of objective value of life as expressed in the presence of thing Mdarkina and certification is an objective which reveals the presence of mind to imagine life feel the need to believe his health.

T. The world is the physical and life combines the first place and other manifestations of heat and starts applying the principle of my mind it is necessary (the principle of the attic) view (that the cause of each accident) and hence the principles of mental public the basis for all scientific facts, as well as the principle of proportionality.

The conflict philosophical determines the direction of the secret of life, while requiring not be confused with the scientific material and article philosophical, philosophy can not and does not think the split unity scientific instruments and means which owned rights, this case of the right of the flag alone.

And that the philosophical concept of the material consists of material and image, he has the scientific material can not be the first principle, because it involves the installation of a dual between the article and the picture. And where they can not each of the picture and article that exist independently of the other, that there should be an actor get ahead of the process of installation and according to that we are going to the relationship between art and some scientific systems according to this perception:

•

The physiology of man as one of the branches of the science of life refers to the amazing facts explaining actor reason that reflect the greatness of the Creator and accuracy of the details and secrets, then the digestive system (the greatest chemical plant in the world) what is creative by the styles of different foods analyzed chemically surprising and distribution of food Valid equitable millions of living cells.
art and science of life

The science of life when the last of the big secrets, a mysterious secret of life, which fills the human conscience, satisfied the divine sense of fear and faith and established it. Then attempts were made to switch to a purely material energy, any electric charge. Any disarmament physical character of the final element in the light of the theory of relativity of Einstein. As it decided that the body mass relative and not fixed it increase the speed up and says Einstein equation energy = mass of the square × speed of light and mass = energy ÷ speed of light squared. As a result, corn, including of protons and electrons is not really, but equal energy.

What follows from that put the original material to the world the reality of life and one common appear in various forms, and the physical properties of compounds cross. The properties of simple elements themselves, are not self-rule but are the qualities of a cross between common article all simple elements.

The material such as mentioned has become the light of past facts recipe occasional Also, they are not transgressed be from energy colors and philosophically, the assumption in the world of art life as the reason for the denials are higher, such as lack effectiveness as well as him.

Genetics

Mendel discovered the key principles of heredity and passed him by subsequent scientists. As it concluded after Tsoaj successive generations of split peas that back plant inherits the characteristics of advances in accordance with a mathematical formula that can be secretly for life and named after then Mendel's laws and then genetics was born at the beginning of the twentieth century after it was re-expose the principles of Mendel renamed hereditary Mendelian. Followed by many variables changed from the traditional qualities of the science of life to taxonomic qualities and stabilized Mendelian genetics.

The decline then Darwinian concept that depends on the basis of the theory of evolution that the changes and attributes that can be obtained as a result of practice or with ocean can be transferred by inheritance to his descendants.

The trend towards the emergence of species hypothesis by mutations some aspects of the sudden change in the number of cases that called on the assumption that the animal diversity of mutations grew up like that, and some of these changes may be inherited.

After that he moved the science of life of the traditional formula, formula description and tradition, classification and manifestations of organic evolution and modalities of the cell and in its entirety to the attention of the microscopic life that focused on exploration in the nucleus of molecules and chemical structures. • Material development

There are different views on the development of the mismatch that the organisms in all its forms and types of fixed and does not change, but some of them is not satisfied with the validity of this opinion, and expressed the possibility of change in living organisms, or that living beings are not static but constantly changing, depending on natural conditions prevailing. In follows a number of views on evolution with theories that have been presented each gained continuity and the other stopped until:

## H. Scientific Aspects and education

This deals with the foundations

• continuous updating of the curriculum.

• Adoption of flexibility in curricula and textbooks.

• employ educational techniques for various Saitha to promote and strengthen the school curriculum.

• improve the teaching of science subjects.

• Conduct amendments to the curricula and textbooks in the light of the reference data of the concepts of environmental education and environmental problems such as pollution and the effects of weapons of mass destruction and dangerous epidemic diseases.

• confirm the practical side in writing research reports and in textbooks.

• Enhance Alknulogi dimension and improve the performance of teachers using coaching skills after all.

Educational policy in Iraq, according to the proposed educational philosophy

• formulation of educational policy principles in Iraq

• educational goals

- Kindergarten

- Primary education

- Intermediate level

- Secondary education

- The preparation of teachers

Educational policy in Iraq, according to the proposed educational philosophy

Educational policy is the set of principles, norms and standards that define the process of education and the key trends that define the interface movement in society towards the main goals during a specific period of time and include educational policy several axes, including:

• premises of educational philosophy, including:

- Intellectual and cultural premises.

- National and humanitarian premises.

- Social and economic perspectives.

• educational goals and preferential areas of educational policy, which included:

- The organization of public education.

- Infrastructure for education.

- Curriculum.
- Educational Measurement and Evaluation.
- Human skills.
- School buildings and equipment.
- Central administration.
- Planning and Research.
- Educational Development.
- Missions, grants and training.
- International cooperation.
- Funding.

There are enormous changes in the various fields in most communities as a result of the high level of the individual and the community education and as a result of the evolution of scientific discoveries in the field of ICT and e satellites led to the transition of society from an industrial economy to the age of the digital economy and the information society and knowledge, which has become a power affecting the development and growth of any society It was accompanied by changes and transformations of political, economic and social enormous at the international level and of the collapse of the socialist camp system and the rule of the capitalist system and reverse the role of the state and the expansion of the role of the private sector and the emergence of what is known as the phenomenon of globalization.

The rapid transformations that we have mentioned is its leader and its tool of Education to change and the development of society and the nation in the preparation of generations and qualify to lead the future and the development of life. It was therefore necessary good selection of the various education programs, so as to achieve the educational goals based on the doctrine of the country and its culture and needs of the political idea and trends in order to achieve this several methods, including follow:

- adoption of new technology in the educational grades.
- development and modernization of laboratories.
- use of educational media and technology.
- The development of the technological potential teachers.
- Coordination with the scientific research centers.
- Provide advanced educational techniques.

And it can use the computer in the evaluation of educational education policies and put them in a scientific technical grounds because contemporary culture is influencing the education system both in terms of the contents of education and its link with the changing needs of the labor market quickly and respond to the changing technological quotient in the contents of the competencies needed for careers and methods of their performance or in terms of tools Education itself known dimensions of Technology Khola quickly. It can also assess some of the projects, including the educational technology project (Open Economics of Education) and the diagnosis of structural weaknesses in the education system and thus develop advanced educational policy has taken into consideration all the requirements of the present and the future as well as the most important foundations of the new Iraqi educational philosophy.

Education study and phases in which regulates the students that had previously been achieving its goals and set plans and curricula and trends legislature approved including laws and regulations and instructions that operate under the general rules for which are going in the light of the educational process and represent educational policy also the general framework that directs the technical and administrative work for the educational system and finally they represent the general educational goals emanating from the educational philosophy that had previously put founded and sources, which represent:

- Islamic faith.
- Iraqi national.

• multiple intellectual values of Arabism and Arab nationalism.

• contemporary trends in educational thought in the world.

• Iraqi society Properties.

A new formulation of educational policy

In the light of practical educational practices of the Iraqi regime and the sources mentioned, it was felt a new formulation of the principles of educational policy and the basic features are in:

• Education credible national aspirations response, which you'll see the contents of educational and measured accurately and development of the national sense.

• Establish educational system on the basis of solid originated from the Iraqi national targets and follow-up the contents of those goals in the reality of field work (the educational system architecture, integration stages and organization and management).

• The importance of the renewal of learning and teaching process through the provision of programs raise the mental capacity of students to expand the horizons of knowledge and diversification of education and to consider the management of cultural, social and economic transformations and a means to instill human values.

• supervision of the state on most aspects of the educational process to ensure national unity and social development allowing for some decentralization in educational policies and implementation of plans and decisions with a delicate balance between the two mode.

• integration between devices of different education levels and specialties and qualities and Masat between production and services on the other.

• mandatory education and expand learning opportunities for all components of the people starting from primary education and access to its application to include intermediate and secondary education.

• Develop education sector a top priority when reconstruction as the right road to accelerate the stability of security.

• Achieve Education and the consolidation of democratic principles of participation, according to the principle of equal opportunities for citizens.

• Adoption of the educational, strategic and integrated planning and development of the necessary so as to ensure the development of a comprehensive educational plan for the different stages of study and the follow-up, implementation and evaluation methods setups.

• Adoption of formulations educational strategy requires further development and is characterized by clearly landmarks, objectives and translate them into the possibility of accommodating different variables so as to ensure the development of the educational system targets and structure, structure and Ssaliba educational plans.

• renewal of the educational system and the development of a radical in terms of its objectives, structure, content and methods and means and methods of the calendar where departments of education and supervision and to highlight aspects of renewal through Mayati:

- Contribute to the provision of practical possibilities and means and mechanisms provided educational programs to prepare teachers and support competencies and specialized leaders.

- Increase the external efficiency of the educational system to document its links to society by emphasizing productivity and documentation between educational plans and policy work and its implications and development plans and Mttalibadtha.

- The employment of modern technology in the educational process and the use of educational technology and the use of modern technology such as computer and the World Wide Web and take advantage of distance education and media service.

- A brief review of the literature of educational policies and comparing models and improving the quality and effectiveness of education programs to shed light on the basis of the principles and objectives of the administrative organization in the education system and which ones to highlight priorities and unify policies and improve program efficiency.

- Develop and discuss and analyze some of these models and policies used to improve the quality and effectiveness of education programs in the development of administrative system for education and organization to keep pace with the times and the requirements of development and the needs of the community.

- Experimenting with new models of education and the establishment of these experiments on sound science and evaluate the results and dissemination of what works for them in the context of the adoption of technical innovations.

- Raise the efficiency of workers in the field of education, especially some members of the teaching staff, supervisors and specialists educators and officials from the educational administration in terms of their numbers and in-service training to develop their efficiency and increase their expertise and keep them up to developments in the various fields and the educational process.

• Seek in the educational system reform with the diagnosis and evaluation of educational reality and discover the strengths and weaknesses of it through:

- Develop the ability to analysis and creativity and positive dialogue.

- Promote the values derived from the Arab, Islamic and human civilization.

- Consolidation of the scientific method in the educational system.

- Expand the patterns of education commensurate with the potential.

- Pride in the status of scientific and social teacher.

- Hold courses and seminars and conferences in the field of education and to encourage research and studies.

- Need to focus on Arabic language and attach importance to teach living languages (English, French and the language of human conversation and book).

- Preparing educational materials print and audio-visual and translated into different languages.

- Permanent development of curricula.

- Prepare a data base for the production of individuals and institutions for the various stages of education.

- Centralized planning and follow-up and decentralization of management to achieve.

- The development of secondary and vocational education and the introduction of the principle of the smooth flow of graduates from middle schools according to the needs of the development plan.

Availability: the elimination of leaks and rumor-educated lifelong learning.

- Quality: the best response to the needs of the labor market and the requirements of sustainable development and access to the developed countries.

- Good management: performance evaluation, decentralization and anti-corruption.

- Confirm human values that religions and let them make important tributary to the virtues of morality and the rejection of intolerance and discrimination.

• consolidation of national unity through:

- Develop the spirit of good citizenship and patriotism and loyalty to him and the development of the spirit of cooperation between citizens.

- Deepen understanding of Iraqi society and its characteristics and its environment.

- Emphasis on national unity and social cohesion.

- Emphasis on joint Arab action, especially education and culture issues.

- Relying on the principle of the rule of law and equality of all in front of him and equal opportunities.

• scientific target considered a pillar of the educational process and the new educational system through:

- Personal Development balanced and integrated control and obedience and order and preserve the secrets of the country in accordance with the format of awareness need Iraq to develop its power to fulfill its need to meet the challenges.

- Understanding of the Arab-Islamic civilization and take care of the Arabic language and bring awareness of issues facing Iraq and the development of the forces of creativity and innovation.

- Opening And national policies for science and technology Iraqi (7)

• look at the science and technology policy in Iraq

• indicators and trends in science and technology policies in Iraq

- Scientific and technological reality

- Scientific and technological institutions with the Arab countries, foreign countries and international organizations

- The transfer of technology Activity

- R & D activity

• Assessment of science and technology policies in the world and some Arab countries and foreign countries and international organizations

- What is the science and technology policy

- Science and technology policies in some Arab countries and foreign countries

• science and technology policy in Iraq

- The formulation of science and technology policies

- The relevant authorities in the formulation of science and technology policies

- Proposed creation of a Higher Council for Science and Technology

- The Supreme Council for Science and Technology

- Financing and activities of science and technology plans

• policy for scientific research in educational institutions

• biotechnology policy in Iraq

- Development of Biotechnology in Iraq

- Future of Biotechnology

- A proposal for the development of biotechnology in Iraq

• towards the formulation of science and technology policy for the twenty-century atheist

look at the science and technology policies in Iraq

- Indicators and trends in science and technology policies in Iraq

- Scientific and technological reality in Iraq

- Scientific and technological institutions in Iraq

- Scientific and technological cooperation with the Arab countries, foreign countries and international organizations

- The transfer of technology Activity

- R & D activity

Look at the science and technology policies in Iraq

That science and technology play a crucial role in the formation of the challenges facing individuals, organizations and nations constantly, (discoveries of genetic engineering, industrial human Parties, mobile, biotechnology) and the challenge facing the human race at the beginning of atheist-first century is how all countries can benefit from the power of science and technology.

But Iraq has not clearly could improve the use of science and technology available to them, share the Arab countries despite the availability of thousands of consulting firms and construction companies and millions of university graduates and about a million Arabs engineer and hundreds of industrial companies.

Located challenges facing Iraq in the two groups first major caused by developmental problems (food security, health, housing, human rights, education and transportation) and difficulty caused by the absence of political culture required the second is in the nature of cultural require finding systems and national science and technology take on the situation science policy and technology to Qatar.

Indicators and trends in science and technology policies in Iraq

Iraq has made over the past decades, significant progress in several areas, it has allocated to education, social services and infrastructure resources have increased, which had a positive impact on the average per capita income and quality of life, followed by things calls for the underdevelopment and closing (during the nineties) has decreased the per capita GNP product and led blockade to halt efforts to diversify the economy and the adoption of the main sources of national GDP on non-renewable mineral wealth.

There are scientific and technological policies in Iraq hypothesis limited in scope and effectiveness as the strategies it is effective in the best of circumstances. So it was felt to submit a proposal for the development of new structures after the submission of some experiences of some developed and developing countries.

Scientific and technological reality in Iraq

This can be illustrated according to the following:

• The availability of financial resources in Iraq, especially before the nineties of the twentieth century boosted the great efforts that have been made in the field of manufacturing, especially in the field of military industrialization and succeeded Iraq in building the industrial base independent military through technological mastery in the processes of manufacturing modern, but the strategies and policies of manufacturing ago of measures effective to ensure the development of local capacity and providing appropriate incentives for local people to be able to modern industrial technology.

• In spite of this there is no Iraq still it needs to follow innovative ways to meet the daunting for a large number of other sectors of production and services, as well as challenges to the development of local scientific capacity and technological renovation of traditional industries and addressing a variety of social problems and Alaguetsaddaah.

• The scientific competencies in Iraq Once upon innovative ability and has a very important role in facing the challenges in spite of standard conditions they face and leave many of them outside Iraq.

• The scientific limited successes that have occurred in Iraq dating back to some of the science and technology institutions (Academy, the House of Wisdom, higher education institutions) and other words that important achievements have been made in the areas mentioned institution-building as well as in human resources development, Guerin mentioned institutions and others are still far from to play a role in enabling the field of development, and that the linkages and interaction between scientific institutions and governmental, technological and business world is still weak in spite of some formats, including contracting.

• Spending on research and development in Iraq at best less than his counterpart in the Arab countries and the picture more frustrating with respect to the output of science and technology, so that the scientific publications and techs in specialized areas and the number of patents granted to institutions and individuals, much less than the average of the corresponding figures in developing countries Other.

• As a result, the case of science and technology in Iraq needs a great deal of attention, Fmahrat scientific inputs, outputs and technological point to deficiencies in information systems, computers, advanced devices, scientific research.

• It is due to this fact can be determined according to the following reasons:

- The bad conditions in which the research taking place.

- Lack of clarity in the criteria used functional upgrade.

- The case of destitution that have suffered and suffered universities and research and educational institutions.

- Bad physical condition to the researchers.

Scientific and technological institutions in Iraq

These include institutions (universities, research universities and centers, educational institutions, the Atomic Energy ... etc) The palaces in Iraq, one of the reasons that led to the absence of a scientific policy and technological and recognition of the palaces, which represents the ineffective management practices and the presence of the aspects of structural deficiency is a recognition of the symbolic the need to develop their capabilities and be out in the next decade to work in recent years different from the environment environment.

These institutions are currently on:

• the production and dissemination of scientific research.

• Knowledge rating includes a range of disciplines and areas of application.

• Training of some researchers.

And therefore it requires a complete reform of these institutions and revitalization and to identify and support high-level priorities by increasing competitiveness and compatibility and interoperability create an effective system of funding policies and linked to industry and social and economic activities.

Scientific and technological cooperation with the Arab countries, foreign countries and international organizations.

The current levels of scientific and technological cooperation between Iraq and the Arab countries is very low, as well as with foreign countries, which is almost to be non-existent as evidenced by the lack of joint scientific research and the sum of joint publications. This refers to the need for Iraq to the quest to strengthen the scientific and technological capabilities to strengthen scientific and technological interdependence.

To strengthen communication and cooperation between the tender in Iraq with other Arab countries is a prerequisite in determining the scientific problems better and get acceptable returns from the available knowledge.

Among the basics of scientific cooperation to avoid delivery free of any test of Technology, which are almost the priority that scientists participate from two or more countries from the Arab countries and increase the possibilities for researchers in the country to attend scientific meetings and in more efficient use of international aid from international organizations key contracts.

Activity technology transfer

The term refers to technology transfer that technology is gained through learning by doing, not just through the transfer or importation of goods or services, technological and therefore they are not the manufacture of pre-prepared removable and cause technological deficit.

The Iraq experience in the transfer of Altknololjia are diverse, including what has been accomplished with foreign companies, which provided a comprehensive and complex technological transactions in the framework of the international market strategy, as the country has suffered from indiscriminate transfers that took place in the absence of any sound domestic policy to create an independent local base in various fields techs.

Thus, Iraq is facing two problems related to the first search for modern technology and its transition and assimilation, development and improvement and the second related to technology development.

R & D activity

The research and development two activities vital for them strategic importance in maintaining the quality of scientific personnel and in ensuring access to advanced science and in promoting the transfer of technology, as well as providing an early warning system in preparation for technological change which reflected the results on the technological and industrial, agricultural and medical progress is, in fact, investment is guaranteed.

That the efforts made in the field of research and development in Iraq may vary on the progress similar efforts in other countries in terms of sum and it is still inadequate to meet the challenges posed by developments in science, technology and the globalization process in spite of the allocation of funds in this area and was the last twenty years the best proof tended to him .

And it is supposed to take over Iraq's great attention to research and development in all fields as a base progress in all fields and allocate good proportions of their national incomes for this purpose and that is to develop the contribution of scientists and creators Iraqis and technicians as well as research centers in the state institutions and advisory offices in universities, taking into consideration that the benefits of R & D activities can not be performed without complex networks work and relationships.

Evaluation of science policies and Altknlologia in Iraq and the world

• What is the science and technology policy

• science and technology policies in the Arab countries

• scientific and technological policy in some countries of the world

- United States of America

- European Union

Evaluation of science policies in Iraq and the world Altknlologia

After review of scientific policies and technology in Iraq and other countries is evident she found work by institutions and centers and independently as well as in Arab countries and the review on the assumption that many developing Albuldn, including Iraq not have an actual capabilities for the use of science and technology in the development process and this will help in assessing the effectiveness of scientific institutions and technology as well as create and activate the national capacity to coordinate the contributions of those policies in the development and rely according to the following criteria when evaluating scientific Alsasat.

• presence or absence of a scientific policy.

• The existence of a scientific policy is effective (the inability to use science in development).

• evaluating the effectiveness of scientific institutions and centers generally.

What is the science and technology policy

• touted science and technology policies being range of sectors dealing with each other directly scientific and technological activities and its relationship to the development of infrastructure construction in the national science and technology. The wide range of investment and financial policies and laws relating to intellectual property rights and trade policies, export and transfer of technology and research and development.

• involves policies adopted by government departments and institutions in general to contribute to the national innovation, both based on local technological capacity or to the imported technologies. However, these policies may be unrelated, but may be sometimes contradictory Therefore, the main task of the policy of science and technology to find Alsasat to coordinate and increase the collective impact framework.

• can determine science policy requirements as well as the analysis of existing systems and the required policy through research a number of indicators and dimensions of these indicators include the following:

- The link between science and the market

- Budget policies between local activities and international links in the field of science and technology

- The needs of knowledge (the book production and dissemination of knowledge and elaborate tangible way Kalpraat, machinery and so on) is a collective experience.

- Analysis of policy approaches in the country concerned

- Evaluation of the links between the various activities the institution and bodies

Science and technology policies in the Arab countries

After reviewing the scientific and technological fact in some Arab countries illustrated as follows:

• There is no scientific and technological policies and clear.

• Do not include the formulation of policies and the subject of wide discussion include science and technology producers and users.

• Lack of public and private sector institutions participate in the formulation of the elements of science and technology policies.

• Follow-up research and development is not serious.

• There is no discrimination of priorities such as food security and health services.

• The specific implementation modalities, especially from policy to strategy to operational planning

• make use of international and regional initiatives is limited.

• efforts in scheduling of science and technology sporadic and incoherent strategies.

Scientific and technological policies in some countries of the world

The United States adopted the modern scientific and technological policy in the United States:

• the face of new global and national challenges directly.

• Government to impose a major role in helping private companies to grow.

• constitutes the focus of economic growth policies:

- Ability to competitiveness in the industry.

- Create jobs.

- The creation of technological innovation environment.

- Coordinate the affairs of technology between government departments.

- A contract of employment partnerships between industry and the federal government and universities.

- Focus on new technology (information, communication, manufacturing technologies).

- Regarded as basic science is the rule upon which all technological advances.

• scientific policies toward the user guide.

Scientific and technological policies in the European Union:

The following are the main goals of scientific and technology policies in Europe, the general framework:

• do collectively play an active role in the field of science and technology:

- Promote the use of research facilities better.

- Confirm the international role of European research.

- Strengthening the scientific capacity and European technology.

- Knowledge Base development.

• enhance the competitiveness of European industry at the international level.

• narrowing the gap between the technical areas and underprivileged areas in Europe.

• meet the social and economic needs of the EU.

• innovation with the participation of small and medium-sized enterprises.

• the development of human potential through the training of researchers.

Science and technology policy in Iraq

• formulation of science and technology policy

• the relevant authorities in the drafting of the world and technology policy

- Education Institutions

- Research and development of administrative institutions

- Academy

- Scientific institutions, professional associations and other

- Institutions, industry and techs

• proposal for the introduction of a higher council for science and technology

- The Supreme Council for Science and Technology

- Science and Technology Council variety

• scientific and technology policy

- Proposal scientific and technological policy

- Drafting the supplementary policy decision

- Financing and activities of science and technology plans

Science and technology policy in Iraq

There are no scientific policies and technology is evident in Iraq may be a form of development strategy and include long-term lines for the development of science and technology indirectly may be absent, and dealing independently within the development process, where invests Iraq private institutions of science and technology (Higher Education, education, education, etc.) in the the establishment of a scientific infrastructure and technological.

The scientific and international cooperation in Iraq is limited and weak at the present time, and in light of scientific blockade reduced budgets drastically trimmed institutions expenditures strongly led to the non-continuation of the previous levels of innovation capacity and the inability of the infrastructure, science and technology to do basic functions.

**The formulation of science and technology policies**

Iraq remained practices so far, far away from any real interest in the development of science and technology policies, and is due to many reasons previously we listed above. It is worth mentioning here that any scientific policies and technology in particular, depends on the quality of Iraq's exports of Petroleum Exporting Countries and Iraq, one of them a natural resource is renewable and therefore Iraq faces an urgent need to find alternative sources of income and exports reflected thus on science policy formulations.

And accordingly it requires a scientific decision-making in determining private channels Aloabedh between the scientific community and society and political leadership to deliver proposals to the decision making process. A departure from the indiscriminate issuing decisions under the pressure of time and these channels are scientific institutions and research and development of scientific and industrial associations, institutions and structural proposal to link these channels and their participation in the drafting.

As well as Zllk requires determining when drafting the objectives of science and technology policies and suggests in this case:

• promoting the growth of specialized in manufacturing activities and services to local companies.

• the advancement of specialized institutions.

• Increase the movement of workers in the field of science and technology.

• scientific and technology policies overlap with a range of social and economic activities other.

• develop the institutions that will be based on science and Altknlologia and work in a different environment.

• strengthen the structure of science and technology department.

• raise the level of resources.

• increase the effectiveness of research and development institutes.

• Development of technology transfer.

• strengthen international cooperation.

• improve the researchers' wages and working conditions.

• it is based on the visions of long-term scientific and technological development.

• that take into account the needs of all relevant institutions.

• determine effective roles of government and the interests of the private sector and non-governmental organizations and professional associations.

• The priorities include local resources and strengths and weaknesses in the scientific capacity.

• delete duplication and overlap and inconsistency.

• The monitoring and evaluation and include effective at all stages of drafting.

The relevant authorities in the formulation of science and technology policies

• propose here different views and a variety of related institutions in the formulation of science and technology policies, including institutions of higher education and research and development institutions and the complex scientific and scientific institutions, professional associations and other institutions as well as for industrial, technological and educational.

• that these institutions should have to work in different from recent years, environments environment is not limited as currently based on the production, dissemination and application of knowledge include a range of customizations and application areas, but to evolve and that interfere with a range of other social and economic activities and to cooperate with the private sector. Moreover it requires these institutions to propose concepts and scientific input according to a specific policy and included discussion.

• and it should involve the largest possible number of the relevant authorities in the preparation of the scientific and technological policies, including the representative of the professional economists and government departments and chambers of agriculture, industry, trade and non-governmental organizations and scientific societies.

- Executive Circle fourth.

- Department of Education institutions

That the performance of education institutions in Iraq, the weakest performance of the countries have the same resources and the level of development and is due to reasons including its organizational structure and in the prospects doth science policy as well as in its independence and isolation as well as the conditions of the blockade.

That all the efforts that have been made in the past to link universities scientific strategies proposed were not successful for the same reasons therefore it requires the involvement of these institutions in Dmnaat channels that suggests scientific policies and the development of new councils where scientific policy and scientific issues are discussed.

research and development of administrative institutions

Academy or the Supreme Council consists of:

- Executive Chamber Information and Documentation and scientific and technical publishing.

up to the peoples and conScientific bodies

• develop scientific bodies by the Technology Council, including, for example:

- Scientific Authority for Pure Science

- Scientific Authority for Informatics

- Scientific Authority for Medical Sciences

- Scientific Authority of Agricultural Sciences and veterinary

- Scientific Commission for Energy and Water

- Scientific Authority of Engineering Science

- Scientific Commission for Human Sciences

- Scientific Board for Economic and Administrative Sciences

• each headed by a member of the scientific Science and Technology Council body.

• Scientific Authority is made up of a number of specialists selected by the Science and Technology Council.

• meets every scientific body once a month at least.

• General Manager of one of the executive departments have planned for scientific body and one in the least.

• assume all scientific body following duties:

- The preparation of science and technology policies in the area of competence and updated annually during the first quarter of each year for submission to the Council of Science and Technology.

- The trade-off between the proposals research projects to grant funding available in accordance with the priorities of the Science and Technology approved plans emanating from it and recommend adopting policies during the fourth quarter of each year.

- Recommendation to contract the preparation of the technical and economic feasibility studies for investment projects related to the development of human and material potentials in the fields of science and technology under the specialization.

Executive departments

• linked to the Secretariat a number of executive departments responsible for implementing and following up the implementation of the Science and Technology Council resolutions and decisions of the Secretary-General and scientific bodies, according to the powers and terms of reference in the approved work mechanisms.

• develop circles as follows:

- Executive department first.

- Executive Chamber II.

- Executive Chamber III.

- Executive Circle fourth.

- Department of Foreign Relations.

- Department of Information and Documentation and scientific and technical publishing.

• each headed by the Chamber of the Secretariat General of the rank of director appointed by decree employees.

• The Director General of all planned executive service for body and one or more of the Council's scientific bodies.

• assume all executive department prepare the agenda of relevant scientific body work schedule and implementation of its decisions.

• assume all executive service announcement science policy and technological details related to its work in the third quarter of each year and set a date and recognizes the research projects of technological and scientific institutions in the country proposals.

• holds the executive service contract on the implementation of projects and scientific and technological research and studies of technical and economic feasibility of the relevant follow-up work and so on according to schedule

• specialized committees

- Linked to the Authority the following specialized committees:

□ Commission for Basic and Applied Science.

□ Committee of Medical Sciences.

□ Commission for Agricultural and Veterinary Science.

□ the Committee on Energy and Water.

□ Commission new technologies

□ the Commission for Social and Human Sciences

- Each of the number of specialists form (not less than six nor more than ten) is the full-time director nominated by the Commission and selected by the Council are appointed from among them president for two years, renewable once.

- Be one members of each committee degree full-time counselor and be scheduled for associated administrative director of the Commission.

- Council by developing or may be canceled and Hcesp need specialized committees.

• circles

- Linked to the Council following Chambers

□ Affairs Department bodies.

□ Contracts and funding.

- Each department employee heads the rank of director general.

• directorates

- Linked to the Authority three directorates:

□ Directorate of Finance and Administration.

□ Information Directorate, documentation and publication.

□ Directorate for External Relations.

• the goals of Science and Technology Council

The Council aims at achieving the following:

- Identify trends and national policies for science and technology.

- Monitoring the movement of science and technology in the world and keep the country secure and touching her active and influential in them.

- The advancement of the movement of scientific and technological research in the country to serve the national development plans.

- Create the social environment stimulating scientific and technological work.

- Explore creativity and sponsorship revenues and employment.

- Scientists and researchers care in all sectors and provide support and backing them possible.

- Develop coordination, collaboration and integration between science and technology institutions in the country mechanisms.

- Strengthen the link state of Arab, foreign and international organizations in the field of science and technology.

• Science and Technology Council tasks:

- Approve the proposed science and technology policy body of science and technology and its committees specialized in the second quarter of each year.

- The introduction or cancel specialized committees as needed.

- Choose a president and members of the specialized committees.

- Action on the recommendations and proposals submitted by the Commission on Science and Technology.

- The adoption of cooperation and coordination between the Commission and institutions of science and technology in the country mechanisms.

- Agheraralnsp percentage of how to distribute the resources available on the details of science and technology-based plans.

• tasks of science and technology body:

The Agency shall achieve the objectives of the Council and the implementation of the following tasks:

- The preparation of science and technology policies and presented to the Council.

- The introduction of recommendation or cancel specialized committees as needed.

- The nomination of president and members of specialized committees.

- Preparation of cooperation and coordination between the Commission and institutions of science and technology in the country mechanisms.

- Capable body to supervise the affairs of the public and the work of specialized committees

- Participate in the Qatari elves to cooperate with Arab and foreign Countries in the field of science and technology.

- Development of action between the specialized committees and departments and directorates of the body mechanisms.

- The preparation of quarterly reports to the progress of work and displayed a semi-annual reports to the Board or as needed.

• tasks of specialized committees

Each committee holds a specialized implementation of the following:

- Prepare a paper science and technology policies in the area of competence and updated annually during the first quarter of each year.

- Showing a proposed policy paper in an extended seminar attended by specialists for the purpose of participating in the development of policies, plans and details, which is one of the basic principles for effective planning and then reviewed in the light of the discussions in preparation for submission.

- The trade-off between research projects and received from the implementing agencies to compete for the available funding and recommend their approval during the fourth quarter of each year proposals.

- Recommendation to contract the preparation of the technical and economic feasibility studies for investment projects related to the development of physical and human resources in the field of specialization.

• circles tasks

- Department of Public bodies assume the following:

□ implement the decisions of specialized committees after ratification.

□ the collection and tabulation of projects and research proposals for submission to the relevant specialized elves.

□ study and analysis of the progress of work reports and research projects contracted to implement them

- Holds the Contracts and funding Mayati:

□ declaration of scientific and technological policy details the end of the second quarter of each year and set a date recognizes projects and research proposals and compiled for submission to the specialized committees during the fourth quarter of each year.

□ enter into contracts for the implementation of projects and research adopted and following up their implementation.

□ collection and tabulation reports of the progress of work and research projects adopted and forwarded to the Department of Public bodies.

• financing plans and activities of the body and style of exchange:

- The Authority's budget resources consist of the following:

□ proportion (2%) of the profits of public companies in the state.

□ central grant from the state budget.

□ allocations for the implementation of projects in a deliberate investment budget.

□ proceeds of funds and the proceeds Ntegadtha and publications.

□ donations mixed and private sector companies.

- Exchange body holds on its activities under special instructions.

- Commission adopts the style of a grant to the hand contracted by the implementation of the specific activity and give to those within the gates of the power exchange is restricted to work but are subject to codification and audit Alaousola.

Scientific and technological policy

Proposal science policy and technology

Higher Council for Science and Technology proposes scientific and technology policies of Qatar and raises to the Cabinet for approval in accordance with the following tasks:

• Efforts to formulate scientific and technology policies in diameter coordination at the institutional level.

• clarify the methodologies used in the formulation of scientific and technology policies.

• modernize and integrate science policy with development policy.

• analysis of scientific and technological capabilities in Iraq with a focus on research and development.

• the need for informed policy makers enough on the methods used in the planning and programming of research activities.

• Follow up the implementation of these policies by the Secretariat and the mechanism to provide advice on issues.

• Other regulatory matters related to:

- These things include input and output scientific and technology policy with an emphasis on the role of different institutions, including industrial, educational, scientific and own the complex.

- As confirmed in this area on the regulatory issues that link between these institutions and how to send the receipt of proposals and guidelines for scientific policy.

- As was confirmed on the availability of specialists to conduct an ongoing assessment of scientific potential and technology.

- Should involve the largest possible number of parties involved in the preparation of scientific policies as well as the interaction with the sectors of production and services by formal mechanisms.

- Also require the participation of the private sector in the financing of scientific capacity building.

- It should be taken into account their response to the requirements of development policies.

How to formulate the resolution of scientific policy

After collecting the proposals for scientific and technology policies of the various channels according to the directions of higher authorities, sent to the Council in accordance with the hard-mentioned proposes in this case scientific policy objectives (competition, trade barriers, the environment, new technologies) and send the upper bodies for approval in accordance with the special leadership policy political and adopted.

Therefore, and according to this perception it is emphasized on the need to start work for the establishment of institutional arrangements and the completion of the proposal in question accompanied follows and according to:

• Rating and scientific and technological forecasting.

• assess market demand for science and technology.

• planning and management of science and Altknlologia.

When scientific policy decision that the industry is taken into consideration:

• be based on local capacity and the scientific progress elsewhere and to balance them.

• regional and international programs contribution to the development of science and technology.

Financing and activities of science and technology plans

• requires in this paragraph to highlight the role of financial institutions and the mobilization of resources with respect to the need to allocate sufficient resources to invest in technologies that have been obtained or that have been developed locally by financial mechanisms to implement policies, such as allocating fixed resources for scientific activities and technology and increase the budget for research and development in the public sector and private.

• can finance activities and science and Altknololjia different channels, including plans:

- The state budget (current and investment).

- Public sector companies.

- Mixed and private sector companies.

• You can do a feasibility study with a view to the creation of a national fund for science and technology devoted his 0.25 per cent to 0.5 per cent of the country's budget to fund technological projects implemented with priority implemented through contracts awarded on the basis of competitive bidding and to allocate greater than the gross national product part to set up a process and technology capabilities and upgraded.

• The financing system in the private sector, which relies on the so-called risk-based adoption by industrialized countries ranging from tax incentives, grants and loans to encourage private investment in innovative activities, sharing policy.

Public policies for scientific research in educational institutions (83-84

• based mechanisms

• The use of the Arabic language in the scientific and research activities

• mechanisms set up a database for research and development

• The role of science and technology in the preparation of human spammers

Public policies for scientific research in educational institutions

Based mechanisms

The linking scientific research plans and programs to development plans and needs of the community and closer cooperation with the private sector require multiple mechanisms including:

• periodic survey of the needs of scientific and technological capabilities and implemented by the Secretariat of the Higher Council for Science and Technology and to submit periodic reports as needed for periodic survey results.

• constitute a technical working groups to carry out its tasks for the development of scientific research on the plans which represent the private Aalqtaa relevant.

• issued a planning committee of scientific research in coordination with the General Secretariat of the Supreme Council for Science and Technology Bulletin include priority research in institutions of higher education programs axes.

• The planning committee of scientific research in its attention to the following:

- The priorities of national R & D approved by the Higher Council for Science and Technology.

- A database of projects already implemented and scientific centers and educational institutions.

- Projects that are approved within the National Plan for R & D proposals.

Also, in coordination efforts carried out by education institutions in the field of scientific research requires other mechanisms including:

• The higher education institution to submit an evaluation report on their experience in the areas of scientific cooperation with scientific research and regional and international donors to support the aim of the scientific mainstream interest centers.

• The education institution to provide an annual report on research activities.

• the establishment of a joint body for the management of laboratories and specialized services in higher education institutions and universities in Iraq are functions:

- Inventory of common use specialized devices.

- Develop instructions to determine the modalities for the use of qualified researchers and programming devices used.

- Determine the types of new equipment required for shared use.

• the establishment of a comprehensive electronic library shared by the Iraqi education institutions and universities and your managed jointly by the library include electronically scout patrol and searchlights, manuscripts and historical documents and databases all.

• formation of a joint coordinating body Masssat of Education and educational institutions for the purposes of regional and international conferences that each institution and the University of supplying this body proposed its plan for annual conferences, seminars and workshops.

• The education of all the scientific data for the production of Foundation Tdresen and researchers Masters and PhD theses, and all messages and be this database available on the electronic library network.

Either needed to support scientific research in the education institutions and financial resource development also requires a number of mechanisms "

• the possibility of allocating 0.5% of the profits of large companies, which increases its capital for an end to be agreed upon.

• endowments local, Arab and foreign grants.

• grants from Arab and foreign sources of foreign aid.

And human resources development, also requires the following mechanisms:

• assigned to the joint coordinating body the following tasks:

- Identification of priority and resource mobilization scientific conferences to ensure the active participation of universities and qualified researchers.

- Evaluation of cultural and scientific agreements between educational institutions and universities, as well as agreements between the Ministry of Higher Education and Scientific Research and the Ministry of Education, governments and other agreements in order to activate and maximize utilization.

• encourage researchers and Altdresen to do the research carried out by independent research teams or joint, by giving it priority in support and reconsider the foundations of upgrade in Iraqi universities so that the value of the largest research carried out by research teams.

• Encourage faculty members on applied research.

• the exchange of information and knowledge with scientific research in the Arab world and in the world institutions and the use of the latest scientific revolution produced what techs such as the International Internet network and e-mail.

The use of the Arabic language in the scientific and research activities, the mechanisms supported:

• promote the use of Arabic numerals in the appropriate fields.

• Increased interaction between the complex scientific and educational institutions and institutions of scientific terms used for the purpose of dissemination of the Arabized terms and unification in general use.

• encourage scientific research in the field of Arabization of Science.

• encourage scientific research in Arabic.

• Prepare bulletins means specialized scientific terminology in cooperation with Arab institutions and distributed to the scientific research institutions, locally and regionally.

The establishment of a database for research and development is done by monitoring the output of the research and development of the Iraqi universities and measure the extent of their interaction with the industry and services sectors

Mechanisms:

• Each education institution job database on the outcomes of scientific research in terms of numbers published research addresses in court and abstracts presented at scientific conferences, journals, books published author, and the number of researchers at the university level of doctoral and master's degree, and patents registered.

• Each education institution Bhoudr number of scientific research and patents, and the Ministry of Education, establishment of a network between educational institutions, data centers and building a central database in the ministry for scientific research in educational institutions and updated annually.

• Each education institution to report on the envoys PhD in the areas of knowledge.

• Each higher education institution to limit the available scientific potential of the library, scientific periodicals, information networks and scientific equipment.

• Education is offered Masat research outputs by what turned them into the development of industry and services operations.

The role of science and technology in the preparation of human spammers

It reflected the role of science and technology in achieving the goals of the Iraqi educational policy in the following areas:

• universal primary education and the expansion of democracy in education at all levels and stages and comprehensiveness using technologies such as the means of telecommunications and improve the school building.

• modified curriculum, developed and constantly updated and linked to local needs and what is happening in the world of change in professions structures and needs of science and technology.

• adopt the ways and means and through the educational quality of teachers and their working conditions and to provide assistance to teachers and teachers as he can from the use of technology in the classroom.

• adopt modern methods to provide assistance to the teacher and the teacher Ptkhalis of the burden of the examinations and the patch also give him the means to measure the knowledge and achievement in a variety of content and potential.

• provide modern methods in planning, administration, evaluation and follo-up methods and implementation at the school level.

Towards the formulation of science and technology of the twenty-century atheist policies

After the creation of the Council, it may be exercised put science and technology including things related to the Horn of atheist twenty in Iraq, where requires the creation of some of the things that have been reviewed in the preceding paragraphs, including a brief analysis of scientific capacity and techs with a focus on research and development as well as aspects of change that took place in institutions policies scientific in the country and the extent of the involvement of Arab scientists and technologists in the scientific and technological activities, International, leading to the achievement of the following A_khasais and balanced socio-economic development:

• The science and technology development based on local capacity and to aspects of scientific and technological progress made in other countries with a balance between these two aspects.

• Private investment in science and technology in the creation of infrastructure, scientific and technological and scientific institutions.

• international scientific cooperation.

• promote national technological capacity development in the private and public sectors alike.

• efficiencies gained by scientific Altknuoger national systems and coordinate and integrate them with a huge number of competencies in the areas of social and economic activity.

• integrating scientific and technological policies set of national policies and programs that meet this purpose.

• continue the production and dissemination of scientific knowledge, technological and maintain an effective level of innovative capacity.

• continuous documentation when formulating scientific and technology policies.

General trends

Can be characterized by scientific and technological policies General trends are to:

• to maintain the balance between any directed towards the development of local science and technology programs and the establishment of linkages with external sources of technological knowledge.

• take place in several stages:

- Continuous assessment of the current status of science and technology.

- Develop an ongoing prospects for the development of science and technology during the year.

- Develop a strategy based on specialized studies related to Bahidv development policy.

- The preparation of a detailed five-year plan accompanied by executive programs.

Biotechnology Policy in Iraq

Scientific and technological policies is not a new issue, but it presents itself today unprecedented sharply because it has become crucial in the life of peoples. Many governments have been taken when developing scientific and technological policies to adopt general concepts shared another of their own based on the degree of economic, social and political development, and that these policies require confrontations unconventional and radical solutions, and these starting points, the preparation of scientific policies and technological of Qatar is one of the necessities of the advancement of the national economy and optimal use of natural and human resources through building on the scientific and technological base and advanced solid. In accordance with this concept it proposes the formation of an institutional structure determines the national policy for science and technology trends and monitor their movement in the world and believes in keeping the country touching her and active and influential in order to achieve this has to be that this institutional structure supreme region of possesses the vision inclusiveness to accommodate the political and economic system and future outlook philosophy To achieve this, can these associated Enterprise (FDA) that operates through committees specialized characterized by its members efficient and estimated in scientific fields, technological and enjoy a broad spectrum of knowledge and with insight into the edges of science and the ability to develop and formulate scientific trends and appropriate Altknololchih for the benefit of the country and aspirations and these committees propose formation of a committee of biotechnology is preparing own policies in this area.

Development of Biotechnology in Iraq

That biotechnology applications can be of far-reaching consequences in Iraq, which suffers from:

- Increase in the number of population.
- chronic food shortages.
- malnutrition.
- ill health.
- environmental problems.

The biotechnology and genetic engineering of the areas in which the country can achieve rapid and meaningful progress, particularly in access to food security and enhance the pharmaceutical industry.

There are various activities carried out by many governments, academic institutions and non-governmental organizations in the fields of biotechnology and genetic engineering, particularly in agriculture. And that progress in the areas of biotechnology and genetic engineering highlights the importance of investing in basic science that form the backbone for continued progress in science and technology, especially because of the height of research and development in biotechnology in the country. The importance of the human genome sequence organizing an event can match a man land on the moon, and is described as a landmark in the history of science will work to expand research into human biology that focuses on widespread diseases such as cancer. The Universal Declaration on the Genome and Human Rights adopted by the General Conference of UNESCO in 1997, and major breakthroughs in molecular Alpajuloggio genetic engineering has raised many legal, ethical and social issues.

The internal reasons for the positive development of biotechnology in Iraq is the following:

• Lack of long-term biotechnology policy at the national level in Iraq.

• the limited number of students and graduates of distinguished graduate in subjects related to biotechnology in addition to the low level of teaching in most institutions.

• lack of infrastructure and adequate for biotechnology research in Iraq.

• lack of coordination between the various Alheiat involved in biotechnology research and its applications.

• lack of modern curricula in Altknololjia vital in all stages of scientific teaching in addition to the small number of faculty qualifications of the underlying motivation.

• divert more resources to conduct research in the highly sensitive areas in biotechnology.

• establish suitable for the protection of national genetic resources mechanism.

The general principles for the development of biotechnology in Iraq

• definition of the objectives of national development in the fields of science and technology, especially biotechnology.

• definition of biotechnology strategy in Iraq, combined with the proposed Science and Technology Policy.

• the introduction of a private biotechnology in various stages of scientific educational outreach programs.

• to promote the introduction of appropriate aspects of the pharmaceutical industry legislation.

• create links and partnerships with the Arab and Islamic countries in the areas of biotechnology and genetic engineering in order to facilitate cooperation between these countries on the part of governments, industry and academic cadres.

• continue to deal with the developments in basic science and not to marginalize the backbone for the development of science and technology.

The special principles of this development mainly includes:

• avoid any exploitation of technology that could lead to finding unexpected material that could be harmful to health.

• the need to avoid the exploitation of any technology that may have environmental impacts can not fix it, but to prove that products not to cause any harm to the environment is noteworthy.

• It is not appropriate to expose people and the environment even less risk to note that the current levels of genetically engineered products are not of little value.

• it is not right to justify the risks of exploitation of technology because of the principle of withholds on the basis of sound scientific grounds that this principle may result in useful products in the future.

• The application of biotechnology to agriculture must be based on a scientific basis away from the foreign trade trends (genetically modified organisms).

• postpone the permission to publish Alorathia modified organisms and the use of genetically engineered foods to be done to reach a level of knowledge that enables to judge the safety that can bring to human health and the environment in order to exploit this technology.

The methods proposed for the development of biotechnology can be mentioned in accordance with the following:

• encourage and support biotechnology in Iraq pharmaceutical and industrial projects that have bases in the Arab and Islamic countries.

• Development of private databases relating to human vendor biotechnology and genetic engineering in the country to facilitate the evaluation of the strengths and weaknesses of the national.

• Support research and development based on the combination of various branches of knowledge in the relevant fields of biotechnology and to ensure the development of required human resources.

• saving money and government support for medical diagnostic applications as well as therapeutic in biotechnology as well as gene therapy.

• encourage and support the deployment of good quality research materials.

• Establish a fund of biotechnology in order to transfer technical expertise from other countries.

**Future Biotechnology**

When examining the technical issues relating to food and health to meet the basic human needs in Iraq development, occupies the subject of food and safe in the advanced position the diameter of the concerns and increase the importance and severity of the seriousness of the subject, taking into consideration the blockade and the lack of renewable resources and the deterioration of productivity and technical backwardness as well as other conditions.

The need for biotechnology to meet the basic need related to food and health of important things can be applied and the review of some of these things, those related to nutrition, especially vital. For example, there biotechnology applied to cattle, production and those related to face food shortages develop where it has exacerbated recent eerily and tragically. And that the food crisis in the country threaten the security and the future of the people and can not be addressed only by working hard and put a sharp plan to improve the means of agricultural production.

Biotechnology plays a distinct role in this area and in the development plan and the proposed plan can be summarized in biotechnology, which consists of components:

• Propagating minute through tissue culture.

• genome and suggest molecular characterization of all living species.

• bioinformatics include the collection of data from genomic analysis and easy assembled forms and hence the future plans of biotechnology can be formulated several priorities:

- Food secure.

- Increase and improve agricultural production and raising food production Supreme species as well as get-resistant species of insects and diseases and the protection of plant genetic diversity.

- The production of active pharmaceutical plant material vital.

- The production of vaccines and antibody monoclonal.

- The use of repetition and agricultural product for the production of ethanol and acetone and butanol.

Food secure

Agricultural biotechnology applications are characterized by being in Iraq in the future is promising for the purpose of providing the required requirements in agricultural production, food security and then carrying such resistance and stress is the neighborhood (drought and salinity) and to provide options to best rotation of my life in order to preserve natural resources.

Iraq is not the location of the use of genetically modified products to the fact that many of the advanced technologies are not clear in the mentality of farmers' fields, but can be expected in the future to improve crops in accurate and fast methods. The use of a flag functional genes to address the complex traits help preserve the genes and to improve the nutritional quality and management of natural resources using drags Almagbh efficient and is supposed to Iraq to be an active participant in this region in order to be to obtain the necessary needs for food security through scientific research Tail is done in cooperation with the bodies International, regional and, in particular, the uses of biotechnology in food security.

**Agricultural production**

Despite the importance of the industry and health sectors, the proposed priorities must take into account the agricultural biotechnology and to two main reasons:

- The research being done on the plants to improve crop directly related to the specific environmental conditions prevailing while those related to health and industry are more difficult in the country.
- preliminary data indicate that most of the research activities related to agriculture in the country.

The following themes agricultural biotechnology could be proposed to be used in Iraq in the future:

• the transfer of genes that Tafr of transgenic plants resistant to many of the nurses and techniques insects (insects sick), insecticides as well as thermal resistance levels, drought and salinity.

• control of natural resources and environmental technologies.

The additional proposals for plants include:

Plant Biotechnology

• Develop vital institution for treatment of sewage and use of this water in agriculture and processing in determining agricultural sites.

• develop means the productivity of vaccines using efficient methods and the use of agricultural waste such as beet molasses and rice straw and maize.

• Conduct for the use of the production of bio-fertilizer research to increase rice production in Iraq due to the bio-fertilizers caused great hospitality when used as an alternative for certain types of fertilizer because of its great benefits of the most important reduce severe pollution of groundwater as a result of the use of chemical fertilizers.

• use methods of biotechnology and the improvement and development of pesticides to control insects on plant as the main alternative to chemical pesticides in order to avoid risk, but the recent work on the production of these pesticides devices require special materials.

• increase the protein content of rice, which is characterized as a choice crop for the Iraqi people and the most important crops in Iraq and is characterized by the unique susceptibility to stress conditions.

• introduced handouts resistance to disease and the application of tissue culture and to improve the output and compensate the significant shortage in the preparation of the palm.

• Conduct future research to overcome the difficulties related to early flourish and the lack of consistency of cloned plants.

• the production of secondary materials using tissue culture Kalgulwaat example and use in medical, industrial and other areas by testing plant cellular lines in order to bear against salt stress and drought, as well as virus-free potato production as well as the minute of hybrid plants.

**Animal Biotechnology**

The introduction of private biotechnology transfer of embryos in cattle, and fish, for example, the adoption of the transplanted gene to improve domestic animal breeds there is a need for Qatar to their importance in the propagation of the number of animals and the election of the quality of high specifications in production can also avoid diseases.

**Microbial Biotechnology**

Biotechnology refers to:

• microbial biotechnology for the production of ethanol from sugar by-products and the production of methanol from industrial and agricultural wastes.

• microbial genetics and use in the removal of cracking and conversion of pollutants to the stabilizers and the formation of nitrogen Alsllowesah yeast isolates from the "S. cervisiae" subject to the consumption of cellulose or lactose.

• the use of appropriate biotechnology to convert biomass into biofuel and gas vital and transform agricultural biomass into biofuel and fertilizer adoption of various wastes for the production of biomass and rice crust one of the potential applications.

• bio-conversion of cellulose waste to Maud fermented protein-rich, followed by the production of microbial biomass product Sulaillowesa which decompose hydraulically.

• use of microbial treatment to remove oil and metal chromium in the process Alnturnh or remove ammonia.

Vital health technology

It requires vital health technology in Iraq generally great efforts in order to shrink the gap between developed and diagonals that country to get the desired goal in health care.

• establish a bone marrow transplant center in a hospital Ohz Attab provide a number of specialists, as well as equipment and materials.

• start research projects in the field of processing genes in tumors and other genetic diseases.

• the expansion of the pharmaceutical industry in the country and its development in order to meet local requirements as well as the modalities for the use of biotechnology.

Environment

According to the Environment:

• use of natural organisms (yeast, fungi, and plants) to convert hazardous materials in the soil.

• use of pollutants in the drainage systems for sewage micro-organisms as well as cleansing industrial sites.

• the use of biotechnology to avoid the resulting from the use of bioreactors products hazardous materials pollution.

• Aladlaat (forensic medicine) techniques to forensic science applications.

Informatics Life

This new scientific direction includes life science, computers will be the core of biology in the atheist and the twentieth century as the way to measure and control the computer with the help of thousands of genes and thus stimulate future developments in the pharmaceutical industry.

Cooperation with international agencies

This includes:

• cooperation with the Islamic and international agencies are required in the present time.

• Requires trained scientists from the Islamic and Arab countries are directly involved in the workout and the transfer of various biotechnology.

• Develop short training courses for graduate students Taatbntha international organizations such as the UN agency educational and cultural (United Nations Education Science and Culture Organization and UNESCO).

• Training of health personnel, especially doctors in the field of bone marrow transplants and to assist in gene therapy, especially in cases of cancer of the blood and lymph gland tumor.

• Expansion of biotechnology base as a necessity for the characterization, collection and preservation of Germblazem available mainly in the gene bank globally and provide information for all genes.

• training programs for joint cooperation with various countries in the following in the field of bio-control development areas:

- House information exchange.

- To produce a comprehensive insect host.

- To put out the vital crops and insects that develop joint projects.

- The computerization of information and the development organization to link networks in different countries.

- Training in the various features of biological control.

Foundation Center for Biotechnology in the light of Islamic feet can be concluded that the biotechnology center in Muslim countries has become a necessity those that have deeper experience in biotechnology that can help a lot in the establishment of such a center and private.

Hot issues

Muslim world needs a serious views on various current issues related to bio-technology and genetic engineering such as genetics and human cloning and living modified organisms (GMOs).

A proposal for the development of biotechnology in Iraq

Applications in plants and agriculture

Interference by applying the basics of modern genetics genetic engineering and the development of improvement by moving operations to allow farmers to produce new varieties in different forms more quickly and at lower cost and in contrast to traditional methods that require time and effort Here are some genetic engineering applications in the field of plant improvement:

• improve the quality of food, especially protein and other materials stored.

• production plants have the ability to nitrogen fixation and private non-leguminous.

• the production of new crops resistant to:

- Pesticides.

- Drought and therefore can be grown in deserts.

- Various pests and plant diseases (bacteria and Alfrusat).

- Salinity and therefore can be grown with sea water.

- Frost in the country and in the cold winter nights.

• the transfer of genes between plants for the production of new types of private and gene transfer between distant species ratios for the development of new varieties.

• production plants with extraordinary ability in the photosynthesis process.

• plant tissue culture in hybrid types of long descent through Albrootobalast Union after the collapse of the cellular walls and even on the mating different ways.

• reach a vegetable products by micro-organisms, through the transfer of plant genes and entered into the micro-organisms such as bacteria.

• biocides production "Production of Biocides".

• the production of bio-fertilizers "Production of Biofertilizers".

• tissue culture "Tissue Culture Techniques".

- The production of potato tubers and seedlings free of viral diseases.

- Mass production of date palm.

- The production of fruit crop.

- The production of secondary materials.

Applications in the animal

Build on the progress that took place in conjunction with new technologies such as artificial insemination and the use of the female incubator and other techniques, especially genetic engineering to improve the descendants of beneficial animals by producing new strains characterized by the following:

• increase the number of offspring.

• to contain the largest of red meat and less fat Kimat quantities.

• ability to der great deal of dairy cattle with respect to types.

• the great weight and an increase in the number of eggs produced with respect to poultry.

• excellent wool specifications with regard to sheep.

Genetic engineering may contribute in the field of events changes in the animal in some of the following specifications:

• the possibility of transferring nuclei from one object to another.

• The use of cattle as a factory for the production of certain hormones and proteins that can be sorted in milk and through:

- Animals that carry genes which are known movable animals Aberganah production.

- The other method used to insert genes into an animal is to use a viral vector known as the virus can I get bounced back inside the desired gene transplant is after the introduction of this virus in animal cells to merge with the animal dyes and gene begins to express himself inside the animal.

- And can take advantage of this method is the introduction of medical value to humans essential genes.

- Hormone synthesis in individual cells in dairy cattle only.

- The production of some of the coagulation factors (factor 9) in charge of the bleeding disease.

- The production of many drugs, such as substance interleukin-2 (to treat certain types of cancer) bunny.

- The production of human growth hormone (Adal gene in the mouse body).

• Animal increased productivity through:

- The animal hormone injection.

- The introduction of genes to increase the amount of hormones.

- Increase the amount of wool.

- Reduce the amount of fat.

• Build genetic maps for cattle, poultry and fish.

• purification implant embryos in cattle.

Some applications in the industry

Of the most important applications in the industry to use objects allocated genetic engineering in order to perform the following tasks:

• the production of antibiotics (the amount of the largest, best quality, cost less).

• the production of enzymes or different drugs have medicinal value.

• combating oil pollution by cracking chemical compounds or decontamination.

• transform human waste into food with high protein content for use in animal nutrition.

• draw some trace elements (uranium).

• Effective fermentation.

• industry and the production of materials for energy such as ethanol, methanol and acetone.

In the field of oil-contaminated soil treatment can be a highly efficient technique to help the oil sector to improve its products and expand its markets and increase the development of prices and features of this technology as a clean environment and ensure the safety of their application contributes to the economic turnaround.

The project includes:

• intensify and isolate bacterial strains that have the ability to remove sulfur.

• Increase the ability of bacteria to remove sulfur and is relying at this stage on the genetic engineering techniques where the transfer of certain genes carrying the desired traits and grown in the fast-growing bacterial plasmids.

Applications in human

The study of genetic diseases, either as a result of the number of chromosomes change or change in composition or change at the molecular level of the gene, and is recognized as the presence of the gene that causes genetic disease through:

- A family record.

- The use of modern techniques.

• Medical Consultations

- Family history is a ground for diagnostic tests.

• genetic scanning and diagnosis of genetic diseases

- Method of ultrasound.

- The use of drainage Amenyusa to identify the presence of chromosomal changes or chemical.

• modern technologies

- Diagnosis and the presence of certain genes in individuals before childbearing (the individual may be a carrier of the disease as a result of the presence of recessive gene without the appearance of satisfactory marks on it).

- The presence of the gene in the fetus before birth.

• Development of a sophisticated system of diagnosis.

• The development of PCR techniques.

• Develop the use of transgenic animals in medical research.

Environment

• Use of the vital signs "Bioindicators" and viewings vital "Biomonitors" to detect levels of contamination.

• bio-cracker and re industrial and agricultural waste cycle.

• the development of resistant plants.

Other areas of research

• enzymes technology

- Enzymes for diagnosis.

- Clinical biochemistry.

- Quality control for food.

- Industrial Operations.

- Environmental control.

• industrial enzymes

- Design and analysis of bioreactors.

- Waste water treatment.

- Vaccine technology.

The transfer of biotechnology

The use of an information base for the diagnosis of "STR (Short Tandem Repeats).

• Design prefixes "Primers Design".

• PAL PCR amplification for the purpose of the STR.

• The STR analysis.

• PCR amplification of Bal "(SNPs) Single Nucleotide Polymorphism".

• diagnosis of SNPs enzymes unequivocal.

• diagnosis of SNPs, respectively.

• PAL PCR amplification of D.n.o. Almetukondra.

• succession D.n.o. The mitochondria.

• STR network and data analysis, respectively.

• DNA "Finger Printing".

The formation of a Higher Council for Science and Technology

A detailed proposal

Education strategy in Iraq

• Provide educational and calendar first of the situation in Iraq

- Education and Human Resources Development

- Look to the beat of Education

- Education funding

• implications of Education's strategy in Iraq

- Strategic Plan for Education in Iraq

• knowledge to be the same in return for direct services to the community.

• respond to the requirements of the career system pasture versus the reorganization of the reality of work.

• access to education to avoid the quality gap between academic institutions in the world.

• Linking Iraqi educational institutions with their counterparts for the sake of academic excellence.

**The Iraqi educational philosophy**

Accommodate the doctrine of Islam and his laws and

• Education in Iraq derives its view of man and the universe and life according to the faith of the Islamic faith main pillars.

• develop education targets in Iraq are Iraqi patriotism and brotherhood of Arab and Islamic compatibility of Islamic and national pride and national levels.

• Iraqi education right for all the sons of the Iraqi society, regardless of its origins, gender, religion and membership of a countryside or cities and nationality.

• Education in Iraq contribute smelting different groups and assist in the formation of a national figure one.

• The Iraqi education to create good citizenship without compromising the identity of the individual and uphold all the rights of citizenship.

• Iraqi education an effective contribution to the Iraqi human breeding contribute include his body and his mind and social growth and the absorption of facts and concepts, theories and harnessed to the service of man.

• Iraqi education are the achievement of the ideals in practice in all the individual or collective behavior (the individual and society).

• Iraqi education is to raise the level of general education outcomes.

• Education in Iraq based on the Arab and Islamic heritage of the purified and developed with the nature of the change through which the societies in the entire world and absorb the heritage elements.

• Education in Iraq keep pace with scientific and technological development and interact with the development of techniques to process and conscious assimilation of technology and skill.

• To achieve harmonization with the labor market contributes to the Iraqi education.

• absorb the facts and concepts of the environment and the investment environment and take advantage of the technology.

Iraqi educational goals and mechanisms

Iraqi Education aims to:

• creating and preparing Iraqi citizen to be valid, free and responsible and contribute to education in the care of physically and mentally, emotionally and socially.

• Build Iraqi man as a driving force for the development of his country and its assistance to build multiple values and a hierarchy of criteria it used to his behavior.

• understanding of the natural environment and social aspects.

• Iraqi human educational, cultural, health and sports knowledge development.

• Adoption of the scientific way of thinking and research of the Iraqi people and raising the economic level.

Iraqi educational curricula and philosophy

It is known that the curriculum and founded linked to educational goals emanating from a clear educational philosophy and complement each other with the fact that the sources and one with a difference in the nature and mechanism of each and thus always requires the unification of the aims of education and the foundations of the curriculum for the construction of the new curriculum and this requires many studies and research sources to derive goals as well as identifying the general principles for the construction of the curriculum.

Researchers have differed as to the lay the groundwork for building the curriculum and in general are:

- educational foundations
- scientific foundations
- social foundations
- psychological foundations

And Iraq in general involved in the foundations above when it is put curriculum is contained within one region is characterized by the importance of a signed and Mstrkath, including:

- Religion
- Language
- frame of mind

And when it approaches, especially where the humanitarian situation that adopts the principle of unification of similar elements and useful properties in the planning, and educate students that Iraq is the general framework, social and moral frameworks and controls must be dominant and common law always master of the situation. Moreover, it is not logical methodology cloning experiments and actual other, but must Alastvadhmma is thinking fits in adequate education. According to these perceptions bear the Ministry of Education, the burden of guiding the generations as an educational institution stripped for shapes partisan and intellectual. And bloom for everyone to become a school form of multiple communities trends and stripes and the kind of pluralism oppose the independence of the educational curriculum.

**Psychological foundations**

There are a number of ideas that are consistent with the psychological foundations and contribute to the construction of the curriculum and fit with the educational levels of the learners and teachers, including:

• affected by its environment and Iraqi learner generated when it has distinctive characteristics resulting from it.

• the impact of a number of distinguishing factors on the growth of the Iraqi learner and the learner, including the situation in Iraq and his family and his school and community.

• There are many special features of growth, which is a continuous and integrated process vary from one individual to another with overlapping and distinct phases (of the Nativity and early childhood, and subsequent Alrahqh childhood and maturity stage).

• The growth requirements and the needs of the individual reflected on building curriculum.

**Social foundations**

Iraqi society is characterized by the presence of a number of common elements, including:

• Iraqi society Islamic society that believes in the Islamic faith.

• Iraqi society part of the Arab country.

• Iraqi society is the history of Old New track achievements.

• Iraqi society suffers from illiteracy and problems related to it.

• Iraqi society is a component of economic and geographic distinct.

• affiliation of teachers in primary schools to the same social environment as much as possible and that this social Privacy reflected on the educational process, including the curriculum.

• Re for the study of history and religious and social education curricula and Iraqi geography teaching in proportion to the situation in Iraq and regional and international contemporary curricula.

**Educational foundations**

The educational foundations it may many dimensions, including:

- educational philosophy
- Education and theories
- Modern Trends for Education
- Knowledge

It can be addressed through educational foundations:

- Creating public awareness of the concepts of moral and educational and commensurate political and social situation.
- not subject to the educational curricula Atjhat any intellectual or political or narrow partisan.
- Baltnci attention since the early stages (kindergartens and nurseries).
- teacher and teacher social and intellectual stature and benefit.
- schools and institutes, each of which is considered sacred campus of an educational and not the seat of partisan conflicts.
- Focus on educational spammers sophisticated and traditional.
- Strengthening of national belonging concepts in curricula and textbooks during the review to prepare for printing, especially the National Library.

Scientific bases

This deals with the foundations

- continuous updating of the curriculum.
- Adoption of flexibility in curricula and textbooks.
- employ educational techniques for various Saitha to promote and strengthen the school curriculum.
- improve the teaching of science subjects.

• Conduct amendments to the curricula and textbooks in the light of the reference data of the concepts of environmental education and environmental problems such as pollution and the effects of weapons of mass destruction and dangerous epidemic diseases.

• confirm Alttabiei side in writing research reports and in textbooks.

• Enhance Alknulogi dimension and improve the performance of teachers using coaching skills after all.

Develop a national policy for science and technology

• educational policy in Iraq

• look at the science and technology policies

Educational policy in Iraq

There are enormous changes in the various fields in most communities as a result of the high level of the individual and the community education and as a result of the evolution of scientific discoveries in the field of ICT and e, satellites and led to the transition of society from an industrial economy to the age of the digital economy and the information society and knowledge, which has become a power affecting the development and growth of any society It was accompanied by changes and transformations of political, economic and social enormous at the international level and of the collapse of the socialist camp system and the rule of the capitalist system and reverse the role of the state and the expansion of the role of the private sector and the emergence of what is known as the phenomenon of globalization.

The rapid transformations that we have mentioned is its leader and its tool of Education to change and the development of society and the nation in the preparation of generations and qualify to lead the future and the development of life. It was therefore necessary good selection of the various education programs, so as to achieve the educational goals based on the doctrine of the country and its culture and needs of the political idea and trends in order to achieve this several methods, including follow:

• adoption of new technology in the educational grades.

• development and modernization of laboratories.

• use of educational media and technology.

• The development of the technological potential teachers.

• Coordination with the scientific research centers.

• Provide advanced educational techniques.

And it can use the computer in the evaluation of educational education policies and put them in a scientific technical grounds because contemporary culture is influencing the education system both in terms of the contents of education and its link with the changing needs of the labor market quickly and respond to the changing technological quotient in the contents of the competencies needed for careers and methods of their performance or in terms of tools Education itself known dimensions of Technology Khola quickly. It can also assess some of the projects, including the educational technology project (Open Economics of Education) and the diagnosis of structural weaknesses in the education system and thus develop advanced educational policy has taken into consideration all the requirements of the present and the future as well as the most important foundations of the new Iraqi educational philosophy.

Educational policy renewable according to the Iraqi educational philosophy updated

Representing the educational policy of any society and as we mentioned beforehand to determine the overall shape of the structure of the educational process intellectual cover that provides them with the necessary immunity and principles under which are going as well as the overall shape of the allocations and qualities of Education study and phases in which regulates the students that had previously been achieving its goals and set plans and curricula and trends legislature approved including laws and regulations and instructions that operate under the general rules for which are going in the light of the educational process and represent educational policy also the general framework that directs the technical and administrative work for the educational system and finally they represent the general educational goals emanating from the educational philosophy that had previously put founded and sources, which represent:

- Islamic faith.
- Iraqi national.
- multiple intellectual values of Arabism and Arab nationalism.
- contemporary trends in educational thought in the world.
- Iraqi society Properties.

In the light of practical educational practices of the Iraqi regime and the sources mentioned, it was felt a new formulation of the principles of educational policy and the basic features are in:

- Education credible national aspirations response, which you'll see the contents of educational and measured accurately and development of the national sense.

- Establish educational system on the basis of solid originated from the Iraqi national targets and follow-up the contents of those goals in the reality of field work (the educational system architecture, integration stages and organization and management).

- The importance of the renewal of learning and teaching process through the provision of programs raise the mental capacity of students to expand the horizons of knowledge and diversification of education and to consider the management of cultural, social and economic transformations and a means to instill human values.

- supervision of the state on most aspects of the educational process to ensure national unity and social development allowing for some decentralization in educational policies and implementation of plans and decisions mode.

- integration between devices of different education levels and specialties and qualities and Masat between production and services on the other.

• mandatory education and expand learning opportunities for all components of the people starting from primary education and access to its application to include intermediate and secondary education.

• participation of all parts of society to establish effective educational process of citizens and parents of organizations, students, teachers and administrators.

• Develop education sector a top priority when reconstruction as the right road to accelerate the stability of security.

• Achieve Education and the consolidation of democratic principles of participation, according to the principle of equal opportunities for citizens.

• Adoption of the educational, strategic and integrated planning and development of the necessary so as to ensure the development of a comprehensive educational plan for the different stages of study and the follow-up, implementation and evaluation methods setups.

• Adoption of formulations educational strategy requires further development and is characterized by Brodouh features, objectives and translate them into the possibility of accommodating different variables so as to ensure the development of the educational system targets and structure, structure and Ssaliba educational plans.

• renewal of the educational system and the development of a radical in terms of its objectives, structure, content and methods and means and methods of the calendar where departments of education and supervision and to highlight aspects of renewal through Mayati:

- Increase the internal efficiency of the educational system in its operations and all of its aspects.

- Contribute to the provision of practical possibilities and means and mechanisms provided educational programs to prepare teachers and support competencies and specialized leaders.

- Increase the external efficiency of the educational system to document its links to society by emphasizing productivity and documentation between educational plans and policy work and its implications and development plans and Mttalibadtha.

- The employment of modern technology in the educational process and the use of educational technology and the use of modern technology such as computer and the World Wide Web and take advantage of distance education and media service.

- A brief review of the literature of educational policies and comparing models and improving the quality and effectiveness of education programs to shed light on the basis of the principles and objectives of administrative organization in the education system and which ones to highlight priorities and unify policies and improve program efficiency.

- Develop and discuss and analyze some of these models and policies used to improve the effectiveness of the varietal and education programs in the development of administrative system for education and organization to keep pace with the times and the requirements of development and the needs of the community.

- Experimenting with new models of education and the establishment of these experiments on sound science and evaluate the results and dissemination of what works for them in the context of the adoption of technical innovations and the expansion of their use in order to ensure the development of the educational process efficiency.

- Raise the efficiency of workers in the field of education, especially some members of the educational and training bodies, supervisors and specialists educators and officials from the educational administration in terms of preparation, training in-service for the development of their efficiency and increase their expertise and keep them up to developments in the various educational and practical fields.

• Seek in the educational system reform with the diagnosis and evaluation of educational reality and discover the strengths and weaknesses of it through:

- To achieve the harmonization of continuing education and investment patterns of parallel education.

- Develop the ability to analysis and creativity and positive dialogue.

- Promote the values derived from the Arab, Islamic and human civilization.

- Consolidation of the scientific method in the educational system.

- Expand the patterns of education commensurate with the potential.

- Pride in the status of scientific and social teacher.

- Hold courses and seminars and conferences in the field of education and to encourage research and studies.

- Need to focus on Arabic language and attach importance to teach living languages (English, French and the language of human conversation and book).

- Preparing educational materials print and audio-visual and translated into different languages.

- Permanent development of curricula.

- Continuous assessment and evaluation of the educational process and the preparation of standards for evaluating the performance of educational institutions, supported and developed.

• pursue and continue educational planning and increase efficiency and continue the practice of regional educational planning at the provincial level devices through:

- Put the priorities of work in the sector and diagnosis test Asahhristih tools for intervention points.

- Cooperation in the educational and management of educational institutions in various stages.

- Prepare a data base for the production of individuals and institutions for the various stages of education.

- Centralized planning and follow-up and decentralization of management to achieve.

- The continuation of work on the preparation of new curricula based on the contents of the Iraqi educational philosophy.

- The development of educational legislation as a framework within which the administrative and technical bodies to determine the powers and responsibilities.

- The development of secondary and vocational education and the introduction of the principle of the smooth flow of graduates from middle schools according to the needs of the development plan.

• The development of the educational system and according to the demands of comprehensive development with the introduction of new formats of education through various means to address the challenges:

- Availability: the elimination of leaks and rumor-educated lifelong learning.

- Equality: equality between boys and girls in rural and urban areas and among various ethnic Aswal different economic conditions.

- Quality: the best response to the needs of the labor market and the requirements of sustainable development and access to the developed countries.

- Citizenship (independence of Education and disconnected from politics and dissemination of human rights and respect for freedom of thought).

- Participation: community participation in the skip and evaluate the educational system and coordination with higher education.

- Good management: performance evaluation, decentralization and anti-corruption.

• faith in the educational system and to consider the humanitarian aspect of the Iraqi man supreme value and work on the development of his personality to interact with society and reflected Bmayati:

- Valuation and assessment of the Iraqi human value and enable the development of the Iraqi personality all the features of the spirit and body and thought and ethics, taste and administration and integration and balance and their interaction with society.

- Enable the Iraqi citizen of knowledge and understanding of human rights and duties and commitment to the advancement of civilization out.

- Attention to the Iraqi family as the foundation for the right upbringing and close cooperation among its members with child care and active role for women.

• Attention to the Islamic faith and other religions and find important channels to get to it through:

- Consolidation of faith in God and heavenly Brsalath.

- Confirm human values that religions and let them make important tributary to the virtues of morality and the rejection of intolerance and discrimination.

• consolidation of national unity through:

- Develop the spirit of good citizenship and patriotism and loyalty to him and the development of the spirit of cooperation between citizens.

- The study of the history of Iraq and depth Bmaziah civilization and pride Bmoroth.

- Deepen understanding of Iraqi society and its characteristics and its environment.

- Emphasis on national unity and social cohesion.

• educational system's dependence on Arab concepts and Arabism, nationalism and considered sources of educational policy through:

- Deepen the interlinkages that combine sons of the Arab nation as a language, history and other unit interests and the administration and determination.

- Pride in the characteristics of Arab culture and business of the Arab heritage revival and renewal.

- Depth specifications of Arab societies and their environments and the promise of wealth and human values thoroughbreds intellectual, moral and join forces.

- Emphasis on joint Arab action, especially education and culture issues.

- Prepare the intellectual and equal to the challenges of Alalomh and other Arab countries and try to find an atmosphere of common human labor.

• To achieve democracy, freedom and equality of the new educational system through:

- Emphasis on the principles of democracy in the relevant curricula as well as rationality, flexibility and a spirit of tolerance.

- Relying on the principle of the rule of law and equality of all in front of him and equal opportunities.

- The participation of everyone, especially the owners of opinion in various community activities.

- Adopt the methods of democracy in educational institutions (administration, teachers and learners ...).

• scientific target considered a pillar of the educational process and the new educational system through:

- Relying on scientific concepts and content, intellectually and practically, and address the issues and problems of modern scientific ways.

- Draw scientific heritage in the Arab and Iraqi and Islamic civilizations and adoption in the curriculum of scientific thinking.

- Follow up its development of contemporary scientific revolution and the absorption of its achievements in enriching and supporting scientific research.

• Develop updated concepts of the educational process and the new educational system such as work, power, construction, originality and innovation and continuous education of all humanity through paragraphs:

- The adoption of employment, vocational training and closer cooperation and coordination between educational institutions as elements in basic education.

- Personal Development balanced and integrated control and obedience and order and preserve the secrets of the country in accordance with the format of awareness need Iraq to develop its power to fulfill its need to meet the challenges.

- Understanding of the Arab-Islamic civilization and take care of the Arabic language and bring awareness of issues facing Iraq and the development of the forces of creativity and innovation.

- Opening up to the peoples and contemporary human thought and civilization.

- The human capacity for education and harmonization of education and the capacities of human life and civilization on the face of change and development for Amadrsa system for adult education development.

Educational goals for kindergarten

• Kindergarten is complementary to the first educational institution parental care in the family and the child's personality development aspects physical and mental, linguistic, emotional, social, spiritual, national and upbringing optimally.

• kindergarten programs be based on the philosophy of the state and that these programs include intellectual awareness for the children of Riyadh and is the follow-up Mtdmanatha in organizing educational and social activity.

• taken into account in the development of programs Riyadh must be at the basis of modern scientific citing childhood characteristics and needs, and characteristics of the environment and the nature of society and develop educational scientific studies for Childhood and trained teachers and founded on the principles and applications is the relationship between Riyadh and house documentation.

• are health care for children in Riyadh, both aspects of preventive, curative and are to cooperate with those responsible for securing such care in Riyadh itself and work hard to provide awareness Riyadh.

• Expansion in kindergarten be a response to social and economic need which enable women to work and contribute to the production process.

• be a kindergarten field of action is limited to women after preparation and rehabilitation educationally and professionally appropriate.

• Coverage of Riyadh educational play aimed at children that leads to the growth of a grave and growing dynamically balanced, giving them a strong and bold patterns of behavior and provide them with simple methods of mind through which they see the practical and technical progress in the world.

**Educational goals for kindergarten**

Child aged between fourth and sixth development of his personality in all its aspects physical and the spiritual, mental, psychological and social national which can translate format targets could be clarified according to the following needs:

• language development

They can be summarized by:

- The development of a child's ability to concentrate and listen and meta.

- Allow and encourage the child to express and safely correct vocabulary.

- Encourage the child to pay attention to the book.

• physical and mental growth

This is done through:

- The development of a child's ability to kinetic expression.

- Basic information about the body and the senses and functions of the child and the basic details of the food for the purpose of growth and play sports and other movements by the characterization of the child.

- Acquistion child development skill control his muscles for the purpose of balancing the consistency of his body movements and muscle include large and fine muscles.

- Definition of the child behavioral patterns that enable it to maintain the integrity of his body.

- Create a child to learn to read and write Bacassaph skills.

- Create a child to exercise its role in maintaining a clean environment (home, kindergarten, street).

- Mental capacity at the Child Development (observation and attention and excellence) and impart some simplified scientific concepts (space and time and counting and weights and lengths, etc.).

- Encourage the child to acquire scientific leanings, interests and express an opinion and take the appropriate decision and encourage him to express an opinion.

• social, spiritual, national and national growth

Help and train the child to:

- Sound development of social relations with the ocean.

- Growth and self-defined duties and rights.

- Instill a sense of belonging when the baby to his family and a group of his peers.

- The development of a child's ability to social conditioning.

- Know the child to employers trades and professions (Haddad, carpenter, doctor ... etc).

- Respect for public property and the application of the system.

- Savings and consumption, good governance and the development of moral courage.

- Faith in God and the Oneness of the child and the acquisition of behavioral concepts of Islamic values and the development of the child's respect for religions.

- The aesthetic and artistic taste and sense of music when the child's artistic expression and the free development (graphic representation).

- Instill patriotism and sense of belonging and loyalty to him.

**Primary education**

• continuity in curriculum development and preparation of textbooks in accordance with the educational goals inspired by the philosophy of the Arab Socialist Baath Party and national wealth.

• develop an awareness of the importance of folk crafts and industries through the primary targets and embed it their curricula.

• coordination of the activities of the school within the same school in terms of its activities and annual programs, whether with respect to the educational aspects of the social aspect and Sports in collaboration with the Libyan and professional organizations and social institutions in such a way as to enhance the principle of objectivity competition between one school and the rest of all schools.

• continue to develop teaching aids in quality and quantity and give it a major role in the educational process to contribute to raising educational efficiency and improve the quality of Almtalemen.

• Develop annual programs for school activities at the country level in the provinces include the participation of students in the collaborative work of collective, popular, and the exercise of their activities and various hobbies and raise them to respect the manual and craft work.

• evaluating its different levels of school activity, per school level, the level of joint activities, the county level and the country level.

• Include school activity programs of activities leading to instill the spirit of soldiering and characteristics in the hearts of the disciples Kalillt and obedience and order and preserve the secrets of the homeland and the patience and grandfather and self-confidence, cooperation, courage and sacrifice for the sake of the homeland.

• Provide adequate preparation of teachers to cope with the expansion of primary education in quantity and quality.

• Work on the development of teacher training institutions achieve the goal of preparing the teacher as part of a university response to global trends in this area.

• continue to provide opportunities for teaching and Ma'lemn parameters in parts of the country during the recent service by methods to improve the quality of education by raising the level of professionally and scientifically teacher and private teachers of the first rows and rows ended.

• educate teachers of public and private educational goals stages of study and academic materials and accessories implemented training programs to ensure the translation of these goals into behavioral patterns and included.

• granting teachers material and moral incentives are appropriate enhances the teaching profession and its place in society and provide a decent life for members of the educational community.

**The objectives of primary education**

There are general objectives under which the educational system is moving in primary education and seeks to achieve include the following areas:

• physical growth

Definition of the nature of the physical changes of pupils and the requisite physical growth of health conditions,

Enable students to realize related to sleep, food, health rules and enable them to apply the means of prevention of diseases and methods of sterilization, as well as enable them to assess the role of the physician and health institutions in the treatment of diseases, love cleanliness, strength and activity instill mind and basic necessities in the life of the individual and society. Enable them to exercise training and sports activities and events available to them and allow them to invest their leisure time exercise activities and sports events.

• mental development

Enable students to improve to express their ideas to the safety and clarity of writing and speech and possess the skill of listening and enable them to master the basic skills of the calculations as well as reliance on facts and correct information to reach conclusions and make decisions, and enable them to express their self-growth as develop their abilities and the development of their potential and initiative, creativity, innovation and trust science and the adoption of his methods in the development of aspects of their personalities and enable them to use some modern scientific techniques appropriate to their level of Adawat and game consoles and Ihsan investment and benefit from.

• emotional growth

Enable students from the adoption of dialogue and discussion to reach a common understanding and allow them to taste the aesthetic and technical aspects of their private lives and public and raise them on frankness and boldness and to exercise self-criticism in their words and actions, enabling them to respond to the attitudes that characterize justice and equality of opportunity and equality of all before the law and enable them to rush to do good and cooperation with their peers, their families and their community.

• social growth

Enabling them to realize the fundamental principles and objectives of the Iraqi modern society and enable them to realize their money from the rights and their duties towards themselves, their families and their community and their homeland. Raise them to respect the law and to respond to him to build a society, develop and enable them to exercise different activities and hobbies to disclose their abilities and skills in creativity and innovation, to raise them to respect the work of being a duty and an honor for the citizen and the practice of manual and craft work and pride in its institutions and their employees, enabling them to participate in various activities in their environment or exercise discipline, enabling them to constantly work for the growth of their personalities growth balanced and integrated as essential for effective participation in community activities.

• Spiritual Growth

Strengthen their faith in God and His Messages heavenly and strengthen their sense of the need for religious belief and that faith is required and a source of strength and reassurance in human life, deepen respect for religious beliefs associated with the celestial Balrsalat that everyone is responsible for his or her liberty sworn own.

• National

Instill pride in them in Arabic, Islamic and Islamic civilization and to deepen their faith in national ties that pull them to their fellow citizens and their nation and make them latch their hopes and their fate common and enable them to realize the importance of human and natural resources in small and large home and the need to invest and rationalization to deepen the all-round development.

**Intermediate level**

Objectives of the medium stage

It stems from the overall objective targets public education system which is moving in the middle stage and seeks to achieve include the following areas:

• physical growth

Students discover the physical preparations that appear at this stage to exercise and development of sports activities in accordance with the scientific bases, training students to motor skills and consistency with the overall growth.

• mental development

Training students to practice the mental processes of remember and imagine, understand and Adarak to develop their skills and their ability to use scientific thinking to face the problems daily, revealed the special preparations when students and develop diversified activities and school events and link them to aspects of life for future operation, curiosity development and the desire to fact-finding when students to pursue scientific developments automatically in their present and future, reveal the creative capacity of the students, develop and direct them to the appropriate fields, helping students to learn on their own mental energies and work with appropriate in their present and future.

• emotional growth

Training students to catch them for themselves to achieve emotional maturity, technical development tendencies and aesthetic sense of the students and their practice in their own lives and the public, instill self-confidence of the students and develop their abilities to decision-making and achieve independence and self-assertion within the community.

• social growth

- Family

Acquaint students with the importance of the family in building the foundation Alatqat social ties established for the growth of the individual and to serve the nation and its progress, and the formation of concepts and trends and social values sound when students to enable them to take responsibility for participation in family and community affairs.

- School

The development of the positive trend of the students about the school and by the promotion of democratization, closer cooperation between home and school and the development of parent-teachers and other activities, the definition of students exchanged between the school and institutions of social community relationship, help students use activities and programs of educational counseling and career guidance to resolve their problems, students learn the language foreign or more along with the Arabic language tool of scientific and social growth, professional development students' skills and ability to work efficiently and contribute to meeting the needs of the community in the present and its future development.

• Spiritual Growth

Deepening of faith in God and the spiritual and moral values that are enshrined in the heavenly messages other, and help students apply the ethical, social and scientific values of religion, enable students to understand and absorb the values and religious teachings.

• National

Definition of the nature of the students' Iraq geographic, economic, and the definition of the characteristics of students and the needs of their environment and enable them to take advantage of them, integrity and development in order to achieve growth and individual happiness and prosperity of society and the consolidation of national unity and faith of students.

**Secondary education**

**Targets middle school**

It stems from the overall objective targets public education system which is going in the middle stage and seeks to achieve include the following areas:

• physical growth

- Enable students to adapt to accept the physical changes of adolescence through them aware of the facts and scientific principles.

- The development of preparations and physical capacities of the students.

- The development of physical fitness among students compatibility motor plague through sports training and youth activities and aspects of school activity.

- Strengthening the body of the students organize and invest some leisure.

- Acquistion students good health habits Ptoeithm preventive health culture and styles.

• mental development

- Enable students to acquire information and concepts, attitudes and skills Almnumeih public and private mental abilities.

- Development of scientific thinking is based on observation and the organization and the formation of concepts and relations doth students.

- Practice methods of scientific thinking in solving school problems and problems facing students.

- Identify the individual differences among students and talented diagnosis and care.

- The development of mental processes and thinking up proper incentives and stimuli and the use of Arabic eloquence orally and in writing.

• Spiritual Growth

- Consolidation of faith in God and confirm the principle of uniformity calling for human liberation and stuff.

- Understanding of Islam in terms of being a major cultural revolution Arab humanitarian mission came to bear interaction between peoples and published.

- Strengthen the faith of students spiritual values and moral high heavenly practices such as calling early to truth, justice, equality, tolerance and brotherhood, sacrifice and chastity and the rejection of intolerance and discrimination.

- The development of conscience will configure calling to do good and deterrence of

• emotional growth

- Enable students to achieve their independence and confirm themselves to strengthening confidence and responsible commitment by stepping up school and social activities.

- Enable students to achieve the emotional maturity and self-control by providing extension services.

- Enable students to disclose their abilities and their talents and energies to launch the initiative and creativity to provide the democratic climate in the work, organization and efforts.

- Tmikn students to understand and accept themselves and their emotions and increase their ability to practice literary, artistic and social activities.

- Tnmacain students to satisfy the need for success and complacency by diversifying educational opportunities for academic, professional and technical activities, taking into account individual differences.

- Increase the chances of satisfying the need for students to role models and model to identify the sources of the tournament and excellence in the Arab and humanitarian heritage.

- Enable students to develop the ability to artistic taste and enjoy the beauty through the intensification of artistic, literary and social activities.

• Alnamwalajtmai

- Prepare students to contribute to the bear family responsibilities as a nucleus of society and carry out their duties burthens and the like that understand what their role in the exercise of their duties and rights of their money.

- Increased interaction between the school and the community through social, scientific and artistic activities.

- Acquistion of knowledge on their community of students (the Iraqi and Arab) and functions and its institutions and its problems and its relationship to other communities and its place in it.

- Upbringing of students on moral values and social trends etiquette upon which the Arab society such as tolerance, cooperation and sacrifice.

- Enable students to exercise their democratic style in teamwork and decision-making through youth organizations and social activities.

• scientific growth

- To prepare students for admission to higher education stage according to Astaadadthm and orientation within the framework of the development plan and the demands of society.

- Enable students to get a general culture studies paved the specialized scientific, literary and Astaadathm and according to their preferences.

• professional growth

- Acquaint students with fields of work in the community.

- Students and guiding them towards the creation of productive Astaadadthm and work according to their preferences and giving them practical experience in various crafts and arts inside the school or outside.

- Acquistion students the ability to act in their economic affairs and familiarity with the kinds of raw materials and goods in their environment and in other environments and how to get it and use.

- Instill trend towards working in the hearts of students as a right and a duty and an honor, and behavioral habits related to different professions.

• National

- The consolidation of national unity and confirmation of the proper relationship between the bi-national council. Arab and other nationalities.

- Insight into the philosophy of the new students community and the origins of this philosophy and personal relevance to the Arab nation and the Arab elements.

- Confirm the concepts of freedom, democracy, freedom and democracy which takes the process by the community in the school.

- Acquaint students with the resources and their home and means of development and rationalization of investment operations.

- Highlighting the role of the country in the face of Iraq's crucial challenges facing the nation and of imperialism, Zionism and reactionary populism.

- Prepare students to defend their homeland and their nation to the development of the spirit of the soldier and order (control and self-confidence and endure hardship, courage and redemption and readiness for martyrdom).

• National

- Install the concept of Arab unity in the minds of students and their souls as a necessity nationalist Iraq is an integral part of the Arab nation.

- Enlighten students on the Arab nation and understanding of the Arab-Muslim heritage and its role in human civilization in order to ensure their absorption and his pride in it.

- Familiarity students fateful conflict between the Arab nation and its enemies and between imperialist and their awareness of the seriousness of the conflict and schemes enemies in Palestine.

• humanitarian

- Educating students characteristics humanity of Arab nationalism and racial represented between individuals and peoples, nations, and the emphasis on character

- Knowledge of students thought the humanitarian and progressive interaction between him and against the Islamic thought, racism, capitalism and reactionary ideas.

- Educating students to the idea of international synergy and highlight the role of the country of Iraq and the Arab nation and humanitarian relations and international integrated and balanced in order to ensure the independence of

- Students realize the importance and necessity of cooperation between the peoples of the order

The overall objective to prepare teachers

The overall goal is to prepare teachers as follows:

Create teacher educators lovers of their homeland believers in their nation and the unity of the spiritual and Qamta High Matzan heritage humanitarian, to uphold the principles and socialist behavior and democratic and moral high, shareholders in the progress of society and the events of social changes and Alhagafih and economic which aspires to achieve them, taking manner of scientific thinking, armed with the necessary degree of knowledge and experience and basic skills in materials they will teach the students well versed understanding of the characteristics and growth Astaadaadtha unaware of modern educational trends to help them perform their mission Matzan their profession and morality are able to fulfill their obligations and thus achieving educational goals.

**Goals prepare teachers**

It stems from the overall objective targets public education system which is moving in the preparation of teachers and seeks to achieve include the following areas:

• physical

- Enable students to continue their preparations gross and sponsorship development is enabling them to employ them in elementary school.

- To deepen students' awareness of the importance of sporting activities and the continuity of practice to ensure the safety of physical and mental and emotional.

- The development of the different physical skills that students need to be teaching profession.

- Enable students to accept Aljmsah primary and secondary and sexual changes that are accompanied by an appropriate culture.

- Training students to practice first aid and civil defense and methods of prevention of accidents and introduce them to common diseases in the environment and protecting the environment from the dangers of pollution.

- Increasing awareness of the importance of food in the student body building and the formation of healthy eating habits

• mental

- Continue to develop the students' mental preparations and employ them in positive behavior.

- Familiarize students with free reconciliation and self-education which helps to sound literacy in the framework of continuing education.

- The development of the talents of the students and their special abilities to the fullest extent possible in pleasing their community.

- The ability to develop scientific thinking among students and use that problems they face.

- Create a variety of opportunities to meet the students at Alfdrah differences in side

• spiritual

- Deepen the faith of students in God and the spiritual and moral values that

- Students values of Islam as a great humanitarian revolution carried the banner to the peoples of the world.

- Instilling moral, social and scientific values in the hearts of students and uphold the truth and tolerance and freedom from intolerance and arrogance, community service and his defense and immunize students from deviation and intellectual invasion of the students absorb the noble values and religious teachings.

• emotional

- Opportunities for students to achieve the emotional and psychological growth and increased

- Positive emotions in the hearts of students and the development of values

- Help students to avoid situations of different psychological crises.

- Instill self-confidence in students and develop their abilities to make a decision.

- The development of collective values and studied in the hearts of students, helping them to adopt her expressive activities.

• Social

- Students realize the importance of the close relationship between education and overall development plans.

- The bad conditions in which the research taking place.

- Lack of clarity in the criteria used functional upgrade.

- The case of destitution experienced by universities and research institutions.

Activity technology transfer

The term refers to technology transfer that technology is gained through learning by doing, not just through the transfer or importation of goods or services, technological and therefore they are not the manufacture of pre-prepared removable and cause technological deficit.

The Iraq experience in the transfer of Altknololjia are diverse, including the large margin completed with foreign companies, which provided a comprehensive and complex technological transactions in the framework of the international market strategy, as the country has suffered from indiscriminate transfers that took place in the absence of any sound domestic policy to create an independent local base in various fields techs.

Thus, Iraq is facing two problems related to the first search for modern technology and its transition and assimilation, development and improvement and the second related to technology development.

**R & D activity**

The research and development two activities vital for them strategic importance in maintaining the quality of scientific personnel and in ensuring access to advanced science and in promoting the transfer of technology, as well as providing an early warning system in preparation for technological change which reflected the results on the technological and industrial, agricultural and medical progress is, in fact, investment is guaranteed.

That the efforts made in the field of research and development in Iraq may vary on the progress similar efforts in other countries in terms of sum and it is still inadequate to meet the challenges posed by developments in science, technology and the globalization process in spite of the allocation of funds in this area and was the last twenty years the best proof tended to him .

And it is supposed to take over Iraq's great attention to research and development in all fields as a base progress in all fields and allocate good proportions of their national incomes for this purpose and that is to develop the contribution of scientists and creators Iraqis and technicians as well as research centers in the state institutions and advisory offices in universities, taking into consideration that the benefits of R & D activities can not be performed without complex networks work and relationships.

Evaluation of science policies and Altmnlologia in Iraq and some Arab countries and foreign countries and international organizations

After review of scientific policies and Altknololjia in Iraq and other countries, it is clear that I found it works by institutions and centers and independently as well as in Arab countries and the review on the assumption that many developing Albuldn, including Iraq not have an actual capabilities for the use of science and technology in the development process and this will help in assessing the effectiveness of scientific institutions and technology as well as create and activate the national capacity to coordinate the contributions of those policies in the development and rely according to the following criteria when evaluating scientific Alsasat.

• presence or absence of a scientific policy.

• The existence of a scientific policy is effective (the inability to use science in development).

• evaluating the effectiveness of scientific institutions and centers generally.

What is the science and technology policy

Described science and technology policies being range of sectors directly address some of the scientific and technological activities and their relationship in the development of infrastructure construction national science and technology. The wide range of investment and financial policies and laws relating to intellectual property rights and trade policies, export and transfer of technology and research and development.

And involve policies adopted by government departments in general and institutions to contribute to the national innovation, both based on local technological capacity or to the imported technologies. However, these policies may be unrelated, but may be sometimes contradictory Therefore, the main task of the policy of science and technology to find Alsasat to coordinate and increase the collective impact framework.

Can determine science policy requirements as well as the analysis of existing policy systems and the required search through a number of indicators and dimensions of these indicators include the following:

- The link between science and the market

- Budget policies between local activities and international links in the field of science and technology

- The needs of knowledge (the book production and dissemination of knowledge and elaborate tangible way Kalpraat, equipment and machinery and so on) is a collective experience.

- Analysis of policy approaches in the country concerned

- Evaluation of the links between the various activities the institution and bodies

Science and technology policies in some Arab countries and foreign countries

After reviewing the scientific and technological fact in some Arab countries illustrated as follows:

No scientific and technological policies and clear.

Do not include the formulation of policies and the subject of wide discussion include science and technology producers and users.

Lack of public and private sector institutions participate in the formulation of the elements of science and technology policies.

Follow-up research and development is not serious.

There is no discrimination of priorities such as food security and health services.

The specific implementation modalities, especially from policy to strategy to operational planning

Benefit from international and regional initiatives is limited.

Efforts in scheduling of science and technology sporadic and incoherent strategies.

Scientific and technological policies in the United States of America:

Modern science and technology policy in the United States relied on:

The face of new global and national challenges directly.

The government imposed a major role in helping private companies to grow.

Economic growth policies axis:

Ability to competitiveness in the industry.

Job creation.

Creating technological innovation environment.

Technology Affairs Coordination between government departments.

Convene working partnerships between industry and the federal government and universities.

Focus on new IT (information, communication, manufacturing technologies).

Considered basic science is the rule upon which every technological advance.

Channel scientific policies toward the user.

Scientific and technological policies in the European Union:

The following are the main goals of scientific and technology policies in Europe, the general framework:

Do collectively play an active role in the field of science and technology:

Promote the use of research facilities better.

Confirm the international role of European research.

Strengthening the scientific capacity and technology European.

Develop the knowledge base.

Enhance the competitiveness of European industry at the international level.

Narrowing the gap between the technical areas and underprivileged areas in Europe.

Meet the social needs and Alaqsadah to the EU.

The establishment of an information society.

Innovation with the participation of small and medium-sized enterprises.

The development of human potential through the training of researchers.

**Science and technology policies in Iraq**

There are no scientific policies and technology is evident in Iraq may be a form of development strategy and include long-term lines for the development of science and technology indirectly may be absent, and dealing independently within the development process, where invests Iraq private institutions of science and technology (Higher Education, education, education, etc.) in the the establishment of a scientific infrastructure and technological.

The scientific and international cooperation in Iraq is limited and weak at the present time, while accused of doing competencies gained by scientific institutions in Iraq in coordination with the existing competencies in the areas of social and economic activity is located in the community.

Under siege scientific budgets reduced substantially reduced expenses institutions and led to sharply reduced the non-continuation of the previous levels of innovation capacity and the inability of the infrastructure, science and technology to do basic functions.

**Uncle formulation and technology policies**

Iraq remained practices so far, far away from any real interest in the development of science and technology policies, and is due to many reasons previously we listed above. It is worth mentioning here that any scientific policies and technology in particular, depends on the quality of Iraq's exports of Petroleum Exporting Countries and Iraq, one of them a natural resource is renewable and therefore Iraq faces an urgent need to find alternative sources of income and exports reflected thus on science policy formulations.

And accordingly it requires a scientific decision-making in determining private channels Aloabedh between the scientific community and society and political leadership to deliver proposals to the decision making process. A departure from the indiscriminate issuing decisions under the pressure of time and these channels are scientific institutions and research and development of scientific and industrial associations, institutions and structural proposal to link these channels and their participation in the drafting.

As well as Zllk requires determining when drafting the objectives of science and technology policies and suggests in this case:

- promoting the growth of specialized in manufacturing activities and services to local companies.
- the advancement of specialized institutions.
- Increase the movement of workers in the field of science and technology.
- scientific and technology policies overlap with a range of social and economic activities other.
- develop the institutions that will be based on science and Altknlologia and work in a different environment.
- strengthen the structure of science and technology department.
- raise the level of resources.
- increase the effectiveness of research and development institutes.
- Development of technology transfer.
- strengthen international cooperation.
- improve the researchers' wages and working conditions.
- it is based on the visions of long-term scientific and technological development.
- that take into account the needs of all relevant institutions.
- determine effective roles of government and the interests of the private sector and non-governmental organizations and professional associations.

• The priorities include local resources and strengths and weaknesses in the scientific capacity.

• rely gradient methods in scientific capacity building.

• delete duplication and overlap and inconsistency.

• The monitoring and evaluation and include effective at all stages of drafting.

**Public policy for Scientific Research in Higher Education**

The Higher Education and Scientific Research formulate the general policies aimed at the development of scientific research in their institutions and based this policy on linking scientific research plans and technological development plot development and the needs of society and the private sector and the development of the necessary human resources, financial resources and also encourage Arabic language support in scientific research.

**The relevant authorities in the formulation of science and technology policies**

• propose here different views and a variety of related institutions in the formulation of science and technology policies, including institutions of higher education and research and development institutions and the complex scientific and scientific and professional institutions and other associations as well as for industrial and technological institutions.

• that these institutions should have to work in different from recent years, environments environment is not limited as currently based on the production, dissemination and application of knowledge include a range of customizations and application areas, but to evolve and that interfere with a range of other social and economic activities and to cooperate with the private sector. Moreover it requires these institutions to propose concepts and scientific input according to a specific policy and included discussion.

• and it should involve the largest possible number of the relevant authorities in the preparation of the scientific and technological policies, including the representative of the professional economists and government departments and chambers of agriculture, industry, trade and non-governmental organizations and scientific societies.

**Institutions of higher education**

• that the performance of higher education institutions in Iraq as the weakest performance of the countries have the same resources and the level of development and is due to reasons including its organizational structure and in the prospects doth science policy as well as in its independence and isolation as well as the conditions of the blockade.

• It is worth all the efforts that have been made in the past to link universities and scientific strategies proposed were not successful for the same reasons therefore requires the involvement of these institutions in Dmnaat channels that suggests science policy in accordance with the following:

- The development of new councils where scientific policy and scientific issues are discussed.

**Research and development of administrative institutions**

The efforts in the field of research and development in Iraq may vary on the progress similar efforts in other countries in terms of sum and has become inadequate to meet the challenges posed by scientific and technological developments and globalization. Also, the links between research and development institutions Latvian need, where they are usually not based on cooperation, and sometimes conflicting because of inflexibility and bureaucratic practices.

The things mentioned about research and development, requires institutions to evolve in accordance with the ideas that have been developed for it and of preserving the special quality of the scientific staff and access to renewable science in the world and Tazizaltnumeih industrial and technology transfer and contributing to the social and economic planning and accordingly proposes that the R & D institutions as well as special functions:

- submit proposals by developed channels when scientific policy-making.
- Contribute to the provision of early warning systems in preparation for technological change.

**Academy**

Proposes in this area have a complex role in scientific policy-making, and is considered one reaches the main proposals for science policy or it may be an alternative to the Higher Council for Science and Technology and with the participation of specialized committees, for example, can Academy offers projects related to environmental pollution, energy, natural resources and projects will give a priority in the near future in the areas of agricultural development, public health and natural resources.

**Scientific institutions, professional associations and other**

It can be played specialized scientific societies to propose several projects and comprehensive survey of national scientific, human and material resources in the different terms of reference also promotes working groups, which could be developed and networks of cooperation between scientists with specialties in question and the creation of databases for selected sectors networks based on the priorities set by the vernacular policies and could conduct evaluation of research projects and public awareness of the importance of science and Alocknulogia increased attention to scientific research.

**Industrial and technology institutions**

• requires that these institutions have available and there is no specific set of interrelated goals with scientific and technology policies, which are consistent with the vision for the future are interrelated.

• Responsibilities include peculiar to the views of specialized consultancy tasks and to strengthen links and flows of knowledge and the provision of technical services.

• It is also there functions are concentrated on services related to the deployment of any technology transfer of knowledge and technology and their application is not a direct creation of new knowledge.

Proposed development of a Higher Council for Science and Technology

Due to the fact that the country is facing various difficulties in its institutions to strengthen the scientific and technological capacity and future planning for scientific question and thus gain the ability to innovate and Aeetmkn achieved only within the framework of sound science policy correlated.

He supposed to do new ways of these scientific and technological policies and the development of integrated growth strategies take into account local conditions and external scientific and proposes to implement the introduction of scientific academies or higher council for science and technology to do so.

This is characterized by the Board (or academic) being the institution with legal personality and financial and administrative independence linked to the Diwan of the presidency and shall achieve the following:

• Develop scientific and technology policies in accordance with the demands of current and future science.

• actively contribute to the movement of internal and external scientific development.

• To promote studies and scientific research in Iraq to keep pace with scientific progress in the world.

• establish scientific links and close cooperation with Arab and international destinations.

Academy or the Supreme Council consists of:

• members numbering at least 35 and no more than 40 including the leader of the Council or academic.

• secretary-general.

• requires the user to be a scientist and a researcher in one of the branches of knowledge (agricultural, industrial, Pure, medical, engineering) and be well-informed on a branch or more of the branches of knowledge and has a product Thoroughbred in which the member shall be appointed by a presidential decree and enjoy a private in peace administrative degree.

• appointed as head of the Academy of between Aedaha by a presidential decree and has the rank of minister and his deputy or more.

• Academy as a full-time secretary-general appointed by a presidential decree from among its members.

• Academy of multiple committees specialist working in coordination with other scientific institutions in the country (Committee on Technology, Water Committee, the Committee on Energy, the Committee on Information).

• The Academy also specialized departments of scientific knowledge within the general framework of (agricultural, medical, engineering, Pure).

• Academy is working on a proposal of science and technology policies and then filed for the Presidency of the Republic for approval (after revision through committees and departments).

• is also working on a recommendation to grant material assistance to centers and individuals and approve the establishment of various scientific centers.

• Develop work between scientific institutions and the beneficiaries and the mechanism of cooperation between relevant institutions and scientific institutions and international contexts.

General of the Supreme Council and Technology

The Supreme Council for Science and Technology consists of:

• Science and Technology Council

• Secretariat General

• scientific bodies

• executive departments

The Supreme Council for Science and Technology shall achieve the goals as follows:

• the adoption of science and technology policies to serve the requirements of the development plans.

• effective contribution in support of scientific research and technology development movement.

• create an environment of scientific and technological discovered creativity and motivating him and conscious of the role of science and technology.

• care researchers and technologists in the public and private sectors and to provide forms of support and backing them, according to Alamancnah.

• adoption of modern systems for the exchange of information between science and technology institutions in the country and work on the development of coordination and cooperation mechanisms between institutions.

• strengthen the link Arab and foreign countries and international organizations and to benefit from the products of their research in the areas of science and technology to enhance the potential of these areas relevant institutions in the country.

• Develop activity specialist scientific and technological information to be among the tasks of scientific and technological documentation and publishing.

Presidency of the Council of Science and Technology

Heads the Science and Technology Council, Mr. President of the Republic and enjoys a head of science and technology powers determined by the President of the Republic by Council.

Council of science and technology and duties lineup

Science and Technology Council is composed of a number of ministers and secretary-general of the Council of Science and Technology and the heads of scientific bodies in the Council and others selected by the President of the Republic.

The Scientific Council of the duties as follows:

- Develop general trends and goals of science and technology development plans.
- Consider the proposed science and technology policies of scientific bodies and approval during the second quarter of each year.
- Develop scientific bodies in accordance with the priorities and requirements of the development of science, technology and national development plans.
- choose a president and members of the scientific bodies.
- decide on the proposals and recommendations of the scientific bodies and departments related to science and technology.
- approve the work mechanisms within the formations of the Supreme Council for Science and Technology and issue instructions related to the implementation objectives of the Council.
- approve the development of research staffs and techs inside and outside the country plans.
- approve the percentages of how to distribute available resources policies and plans proposed by scientific bodies which details

General Secretariat

• Secretariat General consists of a number of scientific bodies and executive departments.

• head is the Secretary secretary-general appointed by the President of the Republic takes the following duties

- Is scheduled for the Council of Science and Technology

- Responsible for the preparation of the work of science and technology and the implementation of the decisions of the Council of tables.

- Supervisor of the proceedings of the scientific bodies and executive departments.

- Coordination with the concerned parties to implement the objectives of the Council.

Scientific bodies

• develop scientific bodies by the Technology Council, including, for example:

- Scientific Authority for Pure Science

- Scientific Authority for Informatics

- Scientific Authority for Medical Sciences

- Scientific Authority of Agricultural Sciences and veterinary

- Scientific Commission for Energy and Water

- Scientific Authority of Engineering Science

- The intrFinancing and activities of science and technology plans

• requires in this paragraph to highlight the role of financial institutions and the mobilization of resources with respect to the need to allocate sufficient resources to invest in technologies that have been obtained or that have been developed locally by financial mechanisms to implement policies, such as allocating fixed resources for scientific activities and technology and increase the budget for research and development in the public sector and private.

• can finance activities and science and Altknololjia different channels, including plans:

- The state budget (current and investment).

- Socialist sector companies.

- Mixed and private sector companies.

• For the first paragraph can do a feasibility study with a view to the creation of a national fund for science and technology devoted his 0.25 per cent to 0.5 per cent of the budget of the country to finance projects implemented technological a priority implemented through contracts awarded on the basis of competitive bidding and to allocate the largest part of the GNP to set up a process and technology capabilities and upgraded.

• The financing system in the private sector, which relies on the so-called risk-based adoption by industrialized countries ranging from tax incentives, grants and loans to encourage private investment in innovative activities, sharing policy.

Public policies for scientific research in higher education institutions

(1): linking scientific research plans and programs to development plans and needs of the community and closer cooperation with the private sector

Mechanisms:

The periodic survey of the needs of scientific and technological capabilities and implemented by the Secretariat of the Higher Council for Science and Technology and to submit periodic reports as needed for periodic survey results.

Constitute a technical working groups to carry out its tasks for the development of scientific research on the plans which represent the private Aalqtaa relevant.

Issued a planning committee of scientific research in coordination with the General Secretariat of the Supreme Council for Science and Technology Bulletin include priority research in institutions of higher education programs axes.

The planning committee of scientific research in its attention to the following:

- The priorities of national R & D approved by the Higher Council for Science and Technology.

- A database of projects already implemented in universities and scientific centers.

- Projects that are approved within the National Plan for R & D proposals.

(2): coordination in the efforts undertaken by institutions of higher education in the field of scientific research

Mechanisms:

• The higher education institution to submit an evaluation report on their experience in the areas of scientific cooperation with scientific research and regional and international donors to support the aim of the scientific mainstream interest centers.

• that each institution of higher education and the University of Iraq to submit an annual report on the research activities.

• the establishment of a joint body for the management of laboratories and specialized services in higher education institutions and universities in Iraq are functions:

- Inventory of common use specialized devices.

- Develop instructions to determine the modalities for the use of qualified researchers and programming devices used.

- Determine the types of new equipment required for shared use.

• the establishment of a comprehensive electronic library share Vihaa Iraqi higher education institutions and universities and your managed jointly by the library include electronically scout patrol and searchlights, manuscripts and historical documents and databases all.

• formation of a joint coordinating body of Masssat higher education and universities for the purposes of regional and international conferences that each institution and the University of supplying this body proposed its plan for annual conferences, seminars and workshops.

• that each institution of higher education and the University of evidence base for the production of scientific Tdresen and researchers Masters and PhD theses, and all messages and be this database available on the electronic library network.

(3) the financial resources necessary to support the development of scientific research in higher education institutions

Mechanisms:

The possibility of allocating 0.5% of the profits of large companies, which increases its capital for an end to be agreed upon.

Endowments local, Arab and foreign grants.

Grants from Arab and foreign sources of foreign aid.

(4): human resources development

Mechanisms:

• assigned to the joint coordinating body the following tasks:

- Identification of priority and resource mobilization scientific conferences to ensure the active participation of universities and qualified researchers.

- Evaluation of cultural and scientific agreements concluded between Iraqi universities and universities as well as agreements between the Ministry of Higher Education and Scientific Research, governments and other agreements in order to activate and maximize utilization.

• encourage researchers and Altdresen to do the research carried out by independent research teams or joint, by giving it priority in support and reconsider the foundations of upgrade in Iraqi universities so that the value of the largest research carried out by research teams.

• Encourage faculty members on applied research.

• the exchange of information and knowledge with scientific research in the Arab world and in the world institutions and the use of the latest scientific revolution produced what techs such as the International Internet network and e-mail.

(5) the use of the Arabic language in the scientific and research activities

Mechanisms:

• promote the use of Arabic numerals in the appropriate fields.

• Increased interaction between the Academy and institutions of higher education and scientific institutions of the terms used for the purpose of dissemination of the Arabized terms and unification in general use.

• encourage scientific research in the field of Arabization of Science.

• encourage scientific research in Arabic.

• Prepare bulletins means specialized scientific terminology in cooperation with Arab institutions and distributed to the scientific research institutions, locally and regionally.

(6): the establishment of a database for research and development by monitoring the output of the research and development of the Iraqi universities and measure the extent of their interaction with the industry and services sectors

Mechanisms:

• Each higher education institution job database on the outcomes of scientific research in terms of numbers published research addresses in court and summaries of the research presented patrols in scientific conferences, books author published, and the number of researchers at the university level of doctoral and master's degree, and patents registered.

• Each institution of higher education and the University of Iraqi Bhoudr number of scientific research and patents, and the Ministry of Higher Education and Scientific Research, establishment of a network between Iraqi universities, data centers and building a central database in the ministry for scientific research in Iraqi universities and updated annually.

• Each institution of higher education and the University of Iraqi envoys to report on the degree of doctorate in the areas of knowledge.

• Each higher education institution to limit the available scientific potential of the library, scientific periodicals, information networks and scientific equipment.

• are served Masat Higher Education Research outputs as much as turning them into the development of industry and services operations.

The role of science and technology in the preparation of human spammers

It reflected the role of science and technology in achieving the goals of the Iraqi educational policy in the following areas:

• the absence of students

Universal primary education and the expansion of democracy in education at all levels and stages and comprehensiveness using technologies such as the means of telecommunications and improve the school building.

• Education contents

Modifying the curriculum, developed and constantly updated and linked to local needs and what is happening in the world of change in professions structures and needs of science and technology.

• Raising the quality of education

Based methods and teaching aids and through the quality of teachers and their working conditions and to provide assistance to teachers and teachers also can use the means of technology in the classroom.

• Examinations and Assessment

Modern methods rely to provide assistance to the teacher and the teacher Ptkhalis of the burden of the examinations and the patch also give him the means to measure the knowledge and achievement in a variety of content and potential.

• educational administration

Progress of modern methods in planning, administration, evaluation and follow-up methods and implementation at the school level.

Biotechnology policy in Iraq

Science policy and Altknololchih not a new issue, but it presents itself today unprecedented sharply because it has become crucial in the life of peoples. Many governments have been taken when developing scientific and technological policies to adopt general concepts shared another of their own based on the degree of economic, social, political and development status of these policies require confrontations unconventional and radical solutions, and these starting points, the preparation of scientific policies and Tknololchih of Qatar is one of the necessities of the advancement of the national economy and employment utilization of natural and human resources through build on the scientific and technological base and advanced solid. In accordance with this concept proposes the formation of an institutional structure determines the national policy for science and Altknololjia trends and monitor their movement in the world and believes in keeping the country touching her and active and influential in order to achieve this has to be that this institutional structure supreme region of possesses the vision inclusiveness to accommodate the political and economic system and future outlook philosophy To achieve this, can these associated Enterprise (FDA) that operates through committees specialized characterized by its members efficient and estimated in scientific fields, technological and enjoy a broad spectrum of knowledge and with insight into the edges of science and the ability to develop and formulate scientific trends and appropriate Altknololchih for the benefit of the country and aspirations and these committees propose formation of a committee of biotechnology is preparing own policies in this area.

Development of Biotechnology in Iraq

• Reasons

- Causes global positive

That biotechnology applications can be far-reaching in the country, which suffers from:

□ increase in population.

□ chronic food shortages.

□ malnutrition.

□ poor health.

□ tricky environmental problems.

The biotechnology and genetic engineering of the areas in which the country can achieve rapid and meaningful progress, particularly in access to food security and enhance the pharmaceutical industry.

There are various activities carried out by many governments, academic institutions and non-governmental organizations in the fields of biotechnology and genetic engineering, particularly in agriculture.

The progress in the areas of biotechnology and genetic engineering highlights the importance of investing in basic science that form the backbone for continued progress in science and technology, especially because of the height of research and development in biotechnology in the country.

The importance of the human genome sequence organizing an event can match a man land on the moon, and is described as a landmark in the history of science will work to expand research into human biology that focuses on widespread diseases such as cancer.

World Declaration on Ajinom and human rights adopted by UNESCO's General Conference in 1997, which is a universal document in the field of biology and medicine and genetics.

Major breakthroughs in molecular Alpajuloggio genetic engineering has raised many legal, ethical and social issues.

Involve abnormal to close a series of genetic codes to the regular sequence of these constitutions, which was its evolution over millions of years, which represents a profound interference results unpredictable.

- Causes internal positive

□ lack of long-term biotechnology policy at the national level in the country.

□ the limited number of students and graduates of distinguished graduate in subjects related to biotechnology in addition to the low level of teaching in most institutions.

□ lack of infrastructure and adequate for biotechnology research in the country.

□ lack of coordination between the various Alheiat involved in biotechnology research and its applications.

□ lack of modern curricula in Altknololjia vital in all stages of scientific teaching in addition to the small number of faculty qualifications of the underlying motivation.

□ divert more resources to conduct research in the highly sensitive areas in biotechnology.

□ establish suitable for the protection of national genetic resources mechanism.

• General principles

- The definition of the objectives of national development in science and technology fields, particularly biotechnology.

- The definition of biotechnology strategy in the country and its integration with the proposed policies for science and technology.

- The introduction of a private biotechnology in various stages of scientific educational outreach programs.

- To promote the introduction of appropriate aspects of the pharmaceutical industry legislation.

- Create links and partnerships with the Arab and Islamic countries in the areas of biotechnology and genetic engineering in order to facilitate cooperation between these countries on the part of governments, industry and academic cadres.

- Easing fears that the strict invention patents and intellectual property rights severe and harsh legislation.

- Continue to deal with the developments in basic science and not to marginalize the backbone for the development of science and technology.

• Special principles

- Avoid the exploitation of any technology that could lead to unexpected finding materials that can be harmful to health.

- Should avoid any exploitation of technology may have implications for environmental can not fix it, but to prove that the products mentioned to cause any harm to the environment.

- It is not appropriate to expose people and the environment even less risk to note that the current levels of genetically engineered products are not of little value.

- It is not right to justify the risks of exploitation of technology because of the principle of withholds on the basis of sound scientific grounds that this principle may result in useful products in the future.

- The application of biotechnology to agriculture must be based on a scientific basis away from the foreign trade trends (genetically modified organisms).

- Postpone the permission to publish Alorathia modified organisms and the use of genetically engineered foods to be done to reach a level of knowledge that enables to judge the safety that can bring to human health and the environment in order to exploit this technology.

• Proposed methods

- Encourage and support biotechnology in diameter pharmaceutical and industrial projects that have bases in the Arab and Islamic countries.

- Private databases relating to human vendor biotechnology and genetic engineering development in the country to facilitate the evaluation of the strengths and weaknesses of the national.

- Support research and development based on the combination of various branches of knowledge in the relevant fields of biotechnology and to ensure the development of required human resources.

- Saving money and government support for medical diagnostic applications as well as therapeutic in biotechnology as well as gene therapy.

- Encourage and support the deployment of good quality research materials.

- Set up a fund of biotechnology in order to transfer technical expertise from other countries.

**Future Biotechnology**

Technical issues relating to food and health to meet the basic human needs in the country development, occupies the subject of food and safe in the advanced position the diameter of the interests and increase the importance and severity of the seriousness of the subject, taking into consideration the blockade and the lack of renewable resources and the deterioration of productivity and technical backwardness as well as other conditions.

The need for biotechnology to meet the basic need related to food and health of important things can be applied and the review of some of these things, those related to nutrition, especially vital. For example, there biotechnology applied to cattle, production and those related to face food shortages develop where it has exacerbated recent eerily and tragically. And that the food crisis in the country threaten the security and the future of the people and can not be addressed only by working hard and put a sharp plan to improve the means of agricultural production.

Biotechnology plays a distinct role in this area and in the development plan and the proposed plan can be summarized in biotechnology, which consists of components:

• Propagating minute through tissue culture.

• genome and suggest molecular characterization of all living species.

• bioinformatics include the collection of data from genomic analysis and easy assembled forms and hence the future plans of biotechnology can be formulated several priorities:

- Food secure.

- Increase and improve agricultural production and raising food production Supreme species as well as get-resistant species of insects and diseases and the protection of plant genetic diversity.

- The production of active pharmaceutical plant material vital.

- The production of vaccines and antibody monoclonal.

- The use of repetition and agricultural product for the production of ethanol and acetone and butanol.

**Food secure**

Agricultural biotechnology applications are characterized by being in Iraq in the future is promising for the purpose of providing the required requirements in agricultural production, food security and then carrying such resistance and stress is the neighborhood (drought and salinity) and to provide options to best rotation of my life in order to preserve natural resources. Iraq is not the location of the use of genetically modified products to the fact that many of the advanced technologies are not clear in the mentality of farmers' fields, but can be expected in the future to improve crops in accurate and fast methods. The use of a flag functional genes to address the complex traits help preserve the genes and to improve the nutritional quality and management of natural resources using drags Almagbh efficient and is supposed to Iraq to be an active participant in this region in order to be to obtain the necessary needs for food security through scientific research Tail is done in cooperation with the bodies International, regional and, in particular, the uses of biotechnology in food security.

## Agricultural production

Despite the importance of the industry and health sectors, the proposed priorities must take into account the agricultural biotechnology and to two main reasons:

• The research being done on the plants to improve crop directly related to the specific environmental conditions prevailing while those related to health and industry are more difficult in the country.

• preliminary data indicate that most of the research activities related to agriculture in the country.

The following themes agricultural biotechnology could be proposed to be used in Iraq in the future:

• the transfer of genes that Tafr of transgenic plants resistant to many of the nurses and techniques insects (insects sick), insecticides as well as thermal resistance levels, drought and salinity.

• the transfer of genes to improve the efficiency of conventional plant breeding is biotechnology methods.

• control of natural resources and environmental technologies.

The additional proposals for plants include:

## Plant Biotechnology

• Develop vital institution for treatment of sewage and use of this water in agriculture and processing in determining agricultural sites.

• develop means the productivity of vaccines using efficient methods and the use of agricultural waste such as beet molasses and rice straw and maize.

• Conduct for the use of the production of bio-fertilizer research to increase rice production in Iraq due to the bio-fertilizers caused great hospitality when used as an alternative for certain types of fertilizer because of its great benefits of the most important reduce severe pollution of groundwater as a result of the use of chemical fertilizers.

• use methods of biotechnology and the improvement and development of pesticides to control insects on plant as the main alternative to chemical pesticides in order to avoid risk, but the recent work on the production of these pesticides devices require special materials.

• increase the protein content of rice, which is characterized as a choice crop for the Iraqi people and the most important crops in Iraq and is characterized by the unique susceptibility to stress conditions.

• introduced handouts resistance to disease and the application of tissue culture and to improve the output and compensate the significant shortage in the preparation of the palm.

• Conduct future research to overcome the difficulties related to early flourish and the lack of consistency of cloned plants.

• the production of secondary materials using tissue culture Kalgulwaat example and use in medical, industrial and other areas by testing plant cellular lines in order to bear against salt stress and drought, as well as virus-free potato production as well as the minute of hybrid plants.

**Animal Biotechnology**

The introduction of private biotechnology transfer of embryos in cattle, and fish, for example, the adoption of gene implanted to improve domestic animal breeds there is a need for Qatar to their importance in the propagation of the number of animals the quality and the election of high standards in production and can avoid imported from the origins of the outbreak of disease, as in the mad cow disease and rinderpest and foot and mouth disease .

**Microbial Biotechnology**

• microbial biotechnology for the production of ethanol from sugar by-products and the production of methanol from industrial and agricultural wastes.

• microbial genetics and use in the removal of cracking and conversion of pollutants to the stabilizers and the formation of nitrogen Alsllowesah yeast isolates from the "S. cervisiae" subject to the consumption of cellulose or lactose.

• use of microbial treatment to remove oil and metal chromium in the process Alnturnh or remove ammonia.

• the expansion of the pharmaceutical industry in the country and its development in order to meet local requirements as well as the modalities for the use of biotechnology.

**Environment**

• use of natural organisms (yeast, fungi, and plants) to convert hazardous materials in the soil.

• use of pollutants in the drainage systems for sewage micro-organisms as well as cleansing industrial sites.

• the use of biotechnology to avoid the resulting from the use of bioreactors products hazardous materials pollution.

• Aladlaat (forensic medicine) techniques to forensic science applications.

**Informatics Life**

• training programs for joint cooperation with various countries in the following in the field of bio-control development areas:

- House information exchange.

- To produce a comprehensive insect host.

- To put out the vital crops and insects that develop joint projects.

- The computerization of information and the development organization to link networks in different countries.

- Training in the various features of biological control.

- The production of secondary materials.

**Applications in the animal**

• increase the number of offspring.

• ability to der great deal of dairy cattle with respect to types.

• the great weight and an increase in the number of eggs produced with respect to poultry.

• excellent wool specifications with regard to sheep.

Having reviewed the developments in the country and the Arab countries and the countries of the world for various scientific policy and scientific fact illustrated as follows:

• The absence of a central scientific and technological policy hinders a lot to continue the development of science and technology.

• The Iraqi society was not fully familiar with issues related to access to science and technology.

• society in the country did not realize the value of science and technology in ensuring the future.

• that there is importance to the acquisition of specific technologies (Information for example).

• and that there is significance to make science and technology part of social values to the Iraqis.

• The legislative mechanisms focused on laws to develop a framework for science and technology available.

• The institutional mechanisms such as the establishment of the Council or the proposed Academy are available.

• mechanisms that are available for the development of manpower to promote global education and support graduate programs and the expansion of training programs abroad.

**Recommendations**

• secure the information flow in an orderly fashion.

• examine the role of the private sector in supporting the activities of research and development.

• Study inputs allocated to small and medium-sized enterprises local technological.

• Identify some of the priorities in (food security, health services, improve the competitiveness).

• the employment of national initiatives to matters concerned (research and development, technology transfer).

• take advantage of international initiatives to promote local scientific capacity and technology.

• formation of specialized bodies within the Higher Council for Science and Technology is working with:

- Scientific domestic policies and technology.

- Research and development strategies.

- Higher Education strategies.

- Policies and strategies of science and technology in various application sectors.

- Policies and strategies of science and technology in selected technologies (information, technology Alahiaah).

• broad discussion when formulating scientific policies include curved science and technology and their beneficiaries.

• The need to adopt different ways of strategic action.

• introducing the concept of the strategic combination of styles and strategies related.

• adopt different methods to continue.

• facilitate the formulation of science policy through the legislative level such as restoring the organization of scientific institutions.

• scientific capacity development and the use of research results are dilated to improve the social and economic Alaossaa.

• set up a special examination and development centers and mechanisms to ensure industry support for applied research capacity and competitive productive capacity and provide incentives for the private sector to encourage participation in the activities of research and development.

• find ways researchers and others can workers in the field of science and technology from which to see the developments in the field of science and technology policies, strategies and program level.

• Provide intensive training and specialization in the field of scientific formulation of policies and strategies.

**Education strategy in Iraq**

## Submitting

Education strategy in Iraq represents a comprehensive vision of the reality and prospects of educational work and its objectives, tools and trends prevailing among those in charge of educational work at both the planning and the central guidance level or in professional practice and practical application in schools has been articulated goals of the development process in this strategy as part of an integrated philosophical vision Tstcherq global and regional variables current and future challenges.

These strategies emphasize the religious and national Iraqi identity and the defense of cultural heritage and cultural excellence in the face of globalization with a realistic vision to follow up the developments of the times Moreover, this strategy is aimed at building the individual is able to take advantage of one of the scientific and technical achievements and the development of the compound, which houses the individual learner. Finally, this strategy is the fruit of a long history of hard work and persistent effort to get to the educational experience mixed with the scientific and cultural accomplished contemporary world religious and cultural values and philosophical assets of Iraq and the Arab and Islamic civilization.

## Calendar first educational status

Requires first when Calendar educational structure survey in Iraq and the potential financial and regulatory constraints, programs and experiences in the field of distance education, and to identify the necessary ingredients for the success of distance education, including the capabilities and potential beneficiaries of the programs and plans, programs, and expertise available locally and regionally, Problems of communication and the Internet and the availability of means of transport appropriate, prioritization of joint programs and projects in the field of distance education and how to start the implementation of distance education and the options and alternatives possible practical programs as well as developing the necessary standards to achieve the level of quality in distance education programs, and systems Calendar, testing and quality control and to identify alternatives and options for expansion in distance education, and submit it without being necessarily rely entirely on the Internet and investing in distance education and encourage the private sector and take advantage of regional experiences in the field of distance education.

**Ntnumeih proposed Education and Human Resources Strategy**

Of the most important strategic themes of human development in Iraq and its relationship to education are the following:

• achieve a balance between population growth and economic growth through the achievement of the population growth rate.

• the provision of health care and reduce the rates of death and injury of various diseases through:

- Provide basic health care to everyone in Iraq through a system characterized by cost-effective, efficient and encourage disease prevention and public safety.

- The provision of preventive health services and emergency services.

- Reduce the rates of death and injury of various diseases including equivalent levels in different countries.

• disseminate and encourage and nurture the development of knowledge and education through the following:

- Create a climate works on dissemination and promotion of knowledge and care and literacy.

- Give priority to the development and dissemination of basic education accessible to all system through a cost-effective and efficient so as to achieve equal opportunities for citizens.

• Building an educational system based on key disciplines needed by the national economy and provide necessary to conduct applied research in all areas, including social and economic facilities.

• provide a system of primary and secondary technical education and vocational training for education is able to prepare workers can adapt to the labor market needs of the balance of sharing different disciplines and skills and to achieve the income level commensurate with their performance and productivity.

• creating employment opportunities in the public and private sectors to the Iraqis, training and rehabilitation, including suit the needs of the market through the following:

- Create jobs for Iraqis who want to work to avoid identified unemployment among themselves, and with the need for training and rehabilitation commensurate with the needs of the labor market.

- Raise the efficiency of the labor market and that by bringing the benefits between the public and private sectors.

**Education strategy in Iraq**

• entrance and principles of the strategy of Education in Iraq

• background to the strategy of Education in race

• Strategies

Entrance and principles of the education strategy in Iraq

• entrance

• principles of the Strategic Plan for Education

• the general framework of the strategy

• new legal and academic vision

**It proposes to reform the human resources development system in Iraq to do the following:**

• lifelong learning.

• Respond to economic growth.

• access to information technology.

• quality education.

It is proposed to launch an ambitious and comprehensive reform program entitled (Education Reform for Knowledge Economy) and includes a long-term goal of making Iraq a center for information technology and build a society with a Tnaksah ability in the knowledge economy on a global scale information.

The software will change the education system at all levels from early childhood to primary and secondary stages to meet the needs of effective participation in the global knowledge economy.

**Entrance and principles of the education strategy in Iraq**

**Entrance**

The strategy document aimed at improving education through long-term plan for future (10-25 years) distinct scientific vision of an ambitious and clear and poignant message of values and standards to evaluate a variety of education, patterns and methods of financing system. This strategy is an inclusive which included several axes including acceptance and alignment with finance, infrastructure, management, scientific research and educational work market.

And accordingly what is required to create a scientific base able to absorb modern science and technology and work to develop and add to consistently and this effort is not done, but researchers and scientists distinguished it must be oriented towards the creation of a scientific structure in Iraq of researchers and research centers and laboratories able to create a climate to enable Iraq contribute to the modern and disposal of scientific progress gradually from scientific liability and technology on the capacity and expertise of foreign when the introduction of modern technology products to local markets In this context, it can not be for Iraq to benefit from the capabilities of scientists immigrants are not able to create a capable internal scientific environment lured back to Iraq or at least the employment of their science and their knowledge and scientific researches.

The creation of a scientific Iraqi society believes in the importance of science and centrality requires long-term planning multiple reflection and guidance patterns for the development of his community base is able to recognize the centrality of science and scientific research, and that the reform of education systems in Iraq, the development of teaching methods and curricula and by encouraging scientific research and creating team spirit through the collective research and the need to provide incentives for Nabgan scientifically as well as starting in the field of scientific research adopt policies sings of the value of science and scientists in the community and provide a decent life for researchers and scientists.

Planning traditional strategic future of education in Iraq includes the specification of which control administration on the resources available within the educational institution and control of the external environment of the educational institution and balance factors between the level of output and the labor market and continuous dialogue between the departments at different levels, flexibility and knowledge of the external characteristics of the environment and take into account all the internal environment of the educational institution and interaction aspects between planning, implementation, and are re-identifying steps for strategic planning and features the general characteristics of the educational institution and analysis of internal and external environment of the educational institution, but there are a number of defects that affect on education strategy, including the lack of awareness of the future in humans that lead to the problem of the weakness of the strategy practice and this is due to the weakness of faith Strategic work in education institutions and the weakness of the government organizational support to work.

The strategy document for education based scientific document sets education policies for years to come and tracks education for years to come, and also depends on the futures and on the efficiency of implementation of the strategy and deal with potential challenges, including the

Which had an impact on the content of education in general, and about all that, the question that stands out and the role that education institutions Asenbga to be played in the short and medium term is it? It has to be considered to community needs of present and Almstaqbilh institutions and ensure access to the level of the budget between the contradictory forces in the visible and those interactions by:

• cultural modernity in exchange for preserving the local culture.

• trend towards global exchange for a commitment to the local educational institutions.

• Development of individual versus social equality.

• knowledge to be the same in return for direct services to the community.

• Provide general skills in exchange for specialized skills.

• respond to the requirements of the career system pasture versus the reorganization of the reality of work.

• Provide training in exchange for a comprehensive and specialized training.

And you see the document that despite the difficulty of finding quick solutions to the interactions of these conflicting forces, but the term (Global Education) will be used as an umbrella for those programs, events and activities that feature a global course such as exchange of students and teaching staff abroad, study foreign languages, international studies, Altdrasat common and comparison and other in referring to that imposed by globalization and the growing need in the global education, which calls for reforms in the education curriculum in the area of universal values such as democracy and human rights and the environment in the area of global knowledge in the areas of global skills, such as skills and language computing. The main elements that should be given the most attention are:

• the principles and mechanisms of international cooperation (of Education - and in - the outside)

• access to education to avoid the quality gap between academic institutions in the world.

• Linking Iraqi educational institutions with their counterparts for the sake of academic excellence.

**The principles of the Strategic Plan for Education**

**Ensure that the strategic plan for education in Iraq:**

• institutional structure of education

It will decide the infrastructure of education? And who will be called educated? What kind of institutions will join under the banner of education? What are the standards that will govern the building and convert various education institutions? Wi standards will govern the organization of educational institutions and the rights of the Iraqi specialization in education.

• administrative system

What is required between government authority and institutions of learning balance? What are the principles and the environment that will shape the administration at the national level? And how it will design the form of management and leadership at the level of the educational institution? And how will the administration and structure of the research centers?

• organization and attract scientific staffs

Who will be responsible for finding and convert and manage jobs and scientific centers and how they can be planned to attract academic angel? And how the process will enable the administration to improve the functional conditions of the staff? And how you will evaluate and upgrade academic and stimulate angel? And how it will develop qualifications and competencies Iraqi angel? And how Sttmn professors working skills, rationing, administrators, and will train them and subsequent generations?

• acceptance and care for students

What are the social and economic changes that will have impact on education? And how the system will be optimized for the selection of students and standards? The required services available to students? What events are required to enrich the academic performance of students?

• qualitative aspect in training

How will the relationship between teaching and research, and how will be the future of academic programs? And what would be the hours of study and evaluation? What are the qualifications required to obtain certificates and how they will develop the content of training programs in the Iraqi education? And how the system will be built to control the quality and the recognition of academic? What are the criteria to be adopted in the granting of licenses to foreign donors?

• physical infrastructure

What needs of the buildings, appliances, supplies sports? What programs Altsamamh required for buildings? And how it will establish an effective system for maintenance?

• management and financing of education

It will be financed from the Iraqi education? And you can repair the current situation and how it can improve the financing of educational institutions allocations?

Globalization and meet the challenges of integration with the challenges while preserving the cultural identity and lack of dependency and that this strategy is not designed to delve into the circumstances of the events, but aimed at showing the outline of a strategy to re-structure and reconstruction, and in the renaissance recent being the education system in Iraq to different directions and beset by various reform movements had to be to put himself clear strategy agreed upon to lead his footsteps towards achieving short-term goals or Baidth. This document is novel for the near future, including moving him present Iraqi society is going through and what variables of Fords, as it is recognized that long-term planning requires a stable foundation provides the vision minute fortified Idonh stability and which can not be considered to extend to the farthest possible.

## General framework of the strategy

After what Aanha Iraq over the decades of war and siege and the resulting loss of life and material damage, exceeded the individual to include the community at large but even the ground, but the damage largest in the loss of opportunities for progress and loss and the emergence of many the most important phenomena who missed learning opportunities and low levels of growth and loss construction opportunities.

The document that is experienced by the State of the collapse of almost universal in the educational system, which not seen previously, as a continuation to leave the academics of the country's salvation from death threats or otherwise heavy damage made in the infrastructure However, the current education system is going through a major phase in the reconstruction, and indicated to the ministry's efforts over the constant changes in subsidiaries and increase the effectiveness of its institutions as well as the development of curricula and assessment and quality of scientific research and technological education and academic freedom and the decentralization of education and the welfare of teachers and students and the sustainability of funding and spread globally. Stressing that the challenges facing Iraq can not be addressed in part but must coexist with influential forces in Iraqi society perspective the near future as a force of economic reform and population change and contact with the outside world, globalization and the increase accelerated knowledge, in the face of these challenges have to be re-regulating the activity of the basic education environment in future vision and strategic plan landmarks light (in the short-term perspective targets fact) which target this document and prioritization in order to be available investment sources indicating the mental reality and is an active and influential.

**New legal and academic vision**

That any legal vision must be based on a balanced legal basis to take into account the preservation of the original features that should be kept as part of the educational and cultural heritage, taking Althompsreeth changes required by the new harmonious society in its evolution with education and make it more responsive to the general problems facing Iraqi society economic, cultural and closer to the needs and problems detailed geographical areas and diverse population Based on this should be seen as Iraqi universities and scientific institutes endoscopic following:

- It provides scientific institutions in its bid to enrich the contribution of socio-economic change and promise sustainable human development.

- It's scientific institutions provide its bid to contribute to the organization of a developed society and be relevant document seeking to combat poverty and protect the environment and improve public health and nutrition and to strengthen the principles of civil society and is keen to develop other levels and patterns of Education.

- It's scientific institutions respond to changes in the labor market and civic culture changes, and therefore is seeking to develop civilian capabilities in the community and put up for special groups Ksaiha for the development of individual competencies.

The strategic vision of the new Bnigi have to look at universities and institutes of higher education pioneers development bodes trends actors about the labor market and lead to the emergence of new opportunities for future development resulting exert more attention to the changes the major developments in the market and the adoption of curricula and education systems in line with changing conditions based on the principles of Kmsaddaqah quality and management of education institutions involved as to fulfill its role of creative in the community, as well as to give the practice of self-administration of these educational institutions, legal formulas practical reality, and certainly on individual freedom and university autonomy which would download these institutions largest in academic work responsibility congenital and its contents as well as exchange issues and spending.

The document refers in its presentation about finding outlets for dealing with these challenges and which has become inevitable is no secret that address administrative weaknesses important issue in this regard and it is in the interest of education (public and private) that takes into account the calendar issues and awareness, including issues of recognition of educational institutions as part important for building an administrative system Rezin Among the elements that should be given special attention in this vision is to renew methods of education and its contents.

But in order to reach those goals, according to the vision of the future and for the renewal of Education, the new vision that draws them to the education system in Iraq image shown that:

• System High-quality training enables students to handle and influential actor in the framework of civil and professional activities, including developments in diverse specialties.

• system allows easily reach sources of science and to interact with its programs so as to ensure Equality and social welfare.

• system of education is based on the quality and knowledge alone trends and encourages knowledge and commitment to keep its focus in the minds of the graduates of the future as well as the sense of responsibility for the development of training in the service of social development.

Background to the strategy of Education in Iraq

• Look at the reality of education in Iraq

• Education Strategy is building in Iraq

Background of the Strategy (70-75)

Look at the reality of education in Iraq

To improve education in Iraq, has become a necessity for the time being, which represents a new society called knowledge society "Knowledge Society", a society requires brainpower superchargers, potential and special skills, and faces challenges that require qualitative changes in education systems.

But the biggest damage lies in the loss of opportunities for progress and loss.

That Iraq remain in place based on what you had saved from the expertise at a time the world was in steady progress, it is a particular loss.

Education funding in Iraq

• the growing demand for education.

• struggling economy growth.

• Iraqi educational institutions need to increase spending to keep up with modern technology.

• inability sources of government funding for the face of increasing education requirements.

• is the effective use of resources in the education institutions.

• High student in the cost of education.

• Iraqi society is going through an economic crisis have a profound impact on education systems.

• extra funding for education in Iraq is not a substitute for government funding.

• diversify the sources of funding for education.

• rationalization of spending on education.

Diversification of funding sources

• institutions, companies and banks

• public financing

- Grants, donations

- Endowments

Rationalization of spending on education

Indirect ways to fund education in Iraq

The default input schools indirectly to fund public education because it contains:

Strategic principles

Develop the personality of the learner

- The rights of the original learner

- The achievement of laws in the human personality

- Human values

- Humanitarian fraternity

- The national spirit

- To stickBuild a strategy for education in Iraq

It required when building a strategy for education in Iraq and clarity of strategic principles and ways to implement these strategies and perspectives and sources and identify local and external challenges her and upon execution require clarity stages with priority criteria.

**Strategic principles**

Of the major tasks for the development of education strategy to be available principles transpires including detailed targets and procedural derived from the philosophy of educational rely on a clear social philosophy is built on the heritage of the nation and the past quarter and the current reality and global challenges and accordingly, the success of the strategy depends on the availability of key principles that can be adopted as principles previews of the philosophy of social and educational, these principles can be summarized according to indicators as follows:

• humanitarian indicators

- The status of human

- Develop the personality of the learner

- The rights of the original learner

- Reliance on self-help

• Faith indicators

- Establish faith in God

- The achievement of laws in the human personality

- Human values

- Humanitarian fraternity

• National indicators

- The national spirit

- To stick with soil home

Means of implementation of the strategy

Many of the procedures required when implementing the strategy which is called the ways to implement the strategy, including:

• open the floor for discussion and dialogue in educational institutions and personnel in the fields of comprehensive development and dialogue include:

- Intellectual aspects of the strategy.

- Practical aspects of the strategy.

• the introduction or development of existing institutions.

• adoption of models to guide for the implementation of special priorities.

• differing priorities about what is in the other countries (financial problems, saving manpower).

• inventory of financial and human potential and Almadahz

• Linking plans and projects the educational goals and plans and projects of comprehensive development.

Strategic perspectives

Premises education strategy in Iraq depends on:

• a comprehensive education system.

• Education system has links to the economic and social regimes, political and cultural.

• Iraq's glorious heritage interactive and influential in human life.

Local and global challenges of education

Education in Iraq is facing domestic and external challenges involved together with the countries of the world and is characterized by special challenges, including:

• scientific and technological revolution.

• Knowledge of production.

• economic changes.

• political changes.

• cultural changes.

• social changes.

• globalization.

Stages of implementation of the strategy

There is a basic stages of implementation of the strategy vary from one country to another never this number of stages required procedures followed by the evaluation process and review can be summarized according to the following:

- create the requirements phase.
- stage determine the procedural goals.
- Develop operational plans stage.
- the implementation phase of procedures and processes.
- phase of the implementation of the detailed plan.
- describe and analyze the output.
- Calendar and follow-up phase.

The development of specific time periods for the implementation of the strategy is necessary per country has been Officials faces they were unable to fulfill the requirements of this point in time, as it may lead to delays in the implementation or planning or speed up the aspects require precision however must be set a time indicators help to touch the road at implementation of the strategy.

Priorities

Must first determine the priorities of priorities and which prefers to begin its implementation and mind beginnings or the basic starting points for the implementation of the strategy can be summarized priorities and determined according to the following:

- Priority difficulty and suffering of them.
- made a priority and presence seriously.
- Post priority in all sectors of society.
- Priority link to provide manpower.

According to these standards can be identified priorities for example, diversification and development of technical education to the importance of the money supply in the labor market appropriate frameworks.

Intellectual underpinnings of the strategic plan

That this historic phase of Omar Iraq is witnessing construction of a new society and, of course, what happened in April 2003 (and also suffered from the events) had plunged Iraq into a fateful Winding down in history, it summoned the building (or rebuilding) all joints of life.

In the midst of political, intellectual and cultural philosophies conflicting and between Traffic cultures that opened Iraq them (or opened it), the scheme finds himself in front of it extremely difficult to choose or to reach a solid philosophical landmarks gathering Ka these trends conflicting often (and sometimes contradictory) to recline in drawing vision drills for the future, should Noting the strategic decisions that logically justify its allocation to an agreed constants, but we did not find a single resource can rely on him and to his inclusion in most of the conflicting trends reliably, but a new Iraqi constitution.

Pillars of the constitutional strategy

Constitution is that document characterized by relatively stable progress and acceptance, and therefore, this document will be taken as a source based upon the coming years education strategy being characterized by stability mentioned, and as well as legislative texts clear the document from which to infer many of the indicators, as Seeley Etbianh.

• data preambular Constitution

Although the preambular Constitution are not legally part of it as governor, but they can provide a comprehensive social Tracker reading of the situation which resulted from which the Constitution, have been received in the telecom submitted inter signals can Nstl_khas including the following:

- Iraq, home to the apostles and prophets and the cradle of civilizations.

- Iraqis makers writing and Agriculture pioneers and makers numbering.

- Iraqis witnessed the pains of oppression.

- Iraq consists of several components.

- Iraq Aljdid- Iraq of the future - it is not sectarianism, racial or zonal or discrimination.

This scheme provides the Muslim ground interpreted - in the least - psychological / social and cultural situation of the culture of the current stage. When hopes Education site in this perspective, we have to take into account the following principles and concepts involving predominantly corresponding Education System:

- The principle of intellectual freedom: that the suffering of the Iraqi thinker of authoritarianism and individualism and the siege of the intellectual must produce a strong tendency towards the academy, which should be enjoyed by all educational institutions.

- The status of education: Education is supposed to be at the forefront of society in the future and read Orientalism requirements and preaching the Palate of scientific knowledge. It must occupy the high position in the interest of the country, and enjoys his joints - whole - needed to achieve this care.

- Responsibility for education: Education bears a humanitarian and national responsibility towards the country and towards humanity and contribute Brphi Iraqi society needs appropriate human powers, and at the same time contribute to the sustainable Brphi humanitarian empowered by its knowledge.

- Education relationship in the world: Iraq contributes to the intellectual and cognitive accumulation taking and giving, it appears - then - modernization and international concepts of human knowledge and the dissemination of culture.

• Executive premises:

Based on all of the above, Based on the dictum of indicators, what is required of the Iraqi education at this stage, and subsequent stages of education is said to be in Iraq:

- Education accessible to all, (for all shades of society and all of its components)

- Educated in charge, (in front of the nation and to society and humanity)

- Educated respondents, (to the needs of the country and the world - in general)

- Education products, (of knowledge and counseling and community service)

- Educated Grzia, (purposeful and planned future lens)

- Educated systematic, (planned and strategically)

• initial public constants:

There are a number of constants that relate to the role of education foundation can be drawn from the following paragraphs of the Constitution:

- Article 2 (First: A): No law may be enacted contrary to the laws of Islam.

- Article 2 (b spas): No law may be enacted contrary to the principles of democracy.

- Article 2 (c spas): No law may be enacted that contradicts the rights and freedoms.

- Article 3: Iraq is a country Mtaad nationalities, religions and sects.

- Article 32: cares for the disabled and those with special needs and guarantees their rehabilitation.

- Article 33 (II): The State shall guarantee the protection of environment and biological diversity.

- Article 35 (II): The State shall guarantee the protection of the individual from coercion and intellectual ...

These include the principles that can be deduced:

- Constants: Islam, democracy and freedoms.

- Pluralism: creeds and nationalities.

- State responsibility: to the environment and people with special needs.

- The responsibility of the state: in nappy on the intellectual, political and religious freedom.

• texts relating to education:

Word about the education system directly in the new Iraqi constitution, which can be regarded as an initial basis for a limited strategy, particularly in Article 34:

- Education is a main factor for the progress of society, a right guaranteed by the state, which is compulsory at the primary level, and the state guarantees fighting illiteracy.

- Free education is a right for all Iraqis in all its stages.

- Encourage the State scientific research for peaceful purposes in order to serve humanity and it promotes excellence, creativity and innovation and the various manifestations of excellence.

Therefore, this strategy was taken in this content - Jmie- focused on: education for all, free education, scientific research in the service of humanity, and care for innovation and excellence.

But there are flaws in the constitution kept him somewhat for the considered core document and a primary source of educational policy in Iraq, beyond Iraq from its Arab surroundings and refer to Article 3 of the first door / basic principles that Iraq is a multi-nationalities, religions and sects country is a founding and active member of the League of Arab States and binding charter which is part of the Muslim world did not say that Iraq is part of the Arab nation, which lose educational policy an important source of resources situation Moreover, the educational policy has lost a pillar least, and an important source of the Iraqi National represented when the Republic of Iraq said in article 1 / Title I / principles Home a single federal, independent at the time, which requires at Iraqi national adoption to be the Republic of Iraq is a state one third and the source of the Arabic language has been weakened by the Constitution and share in Article 4 of the first door / basic principles of the Kurdish language and have become official languages of Iraq and this is strange in the constitutions of the world.

And Article 34 / I pointed out that education is a key factor for the progress of society and the right guaranteed by the state, which is compulsory in primary school and the state guarantees fighting illiteracy and secondly pointed out that free education is a right for all Iraqis in all its stages and the private and public education is guaranteed regulated by law in IV / Article 34.

It is therefore difficult to emergence of educational philosophy of the Constitution of him overcome the lack of clarity and decreasing as well as the lack of clarity of the principle of equal opportunities in education, employment and equality between citizens in rights and duties and compulsory education is defined and restricted to primary education.

And it is assumed the development of an advanced Law of Education describes the sophisticated educational philosophy defines its parameters of the general principles of the Iraqi people, such as:

• Respect for the dignity of the individual, freedom and appreciation for all the public interest.

• social justice and providing equal opportunities for education for all the sons of Iraq.

• Help each student to normal growth physically, mentally and emotionally.

• The importance of education for the development of Iraqi society within the framework of the Arab countries.

• Attention to modern technology and take advantage of them.

About spiritual and many other Iraqi strategy

You can focus on the following objectives when developing Iraqi educational strategy, which is built of education by:

• Integrated growth of the individual.

• configure the behavior on the basis of national values.

• Iraqi dignity and prepare a unified Iraqi homeland.

• economic and social development of renewable concept and dimensions.

• promote the concept of the main principles of democracy, equality, justice, freedom and respect for the individual.

• scientific advances in theory and practice.

• Work to meet the needs of the Arab community of manpower.

• improve the standard of living and the renewal of genres.

• enable individuals to continue to fight for progress.

• international understanding and world peace.

• Develop linked to the national plan for the overall development of educational plans.

• renewal of private educational objectives for each stage.

• Publish schematic awareness "Planning Awareness".

• rehabilitation of workers in the schematic area.

• taking a centralized planning and decentralized execution principle.

• interest in educational administration.

• provide material necessary for the operations of educational planning and human resources.

• focus on the provision of education.

Future strategic requirements

Strategic principles when developing the educational system in Iraq requires:

• Develop educational and teaching staff training and raise the skill.

• the adoption of quality and quality control system.

• revitalization of the quality of education.

• promote education methods.

• Prepare projects for educational reform NES includes educational institutions and schools.

• Develop level educational institutions.

• adoption of educational planning according to international experiences.

• renewal of departments.

• set up systems to evaluate performance.

General targets for investment

The preparation of a good citizen insured, including the following:

• foundations of the philosophy of education in Iraq.

• adhering to all the rights of citizenship and work to take their responsibilities.

• Achieve moral ideals scientifically in all fields of individual or collective behavior.

• initiative and follow-up work and positive behavior in fruitful cooperation with others and follow the democratic style in relationships.

• following Alasash skills development:

- Easily transfer ideas to others through oral expression.

- The transfer of ideas Bsaoh to others through written expression.

- Follow Moazavih in cash.

- To help the individual to normal growth physically, mentally, socially and emotionally, taking into account individual differences.

- Raise individual and collective economic level and increase national income by providing a equal educational opportunities for all.

Being drafted overall objectives of the Education and aims of education in the various stages of education in the light of the principles and rules that take place identified in the philosophy of education in advanced law, and that the goals of the curriculum and teaching materials, which are the derivation of the overall objectives of education and its details are taken into consideration, as the system architecture and public policy are the means to achieve those objectives, taking into consideration that the philosophy of education proposed is a set of principles of intellectual and rules that define the educational system architecture also sets educational policy which is in a range of legislation and decisions taken by the education system to facilitate the education and guidance and organization to ensure the achievement of educational goals.

Strategies •

Strategic Plan axes

• Strategic Plan for Education in Iraq

Strategic Plan axes

Strategic plan for education for the next decade in the hubs include seven specific represents in fact it's the most important and the Ministry of Education's activities, which in turn represent the educational process elements of these themes revolve mainly around human resources, physical and technological resources, management systems, finance, and quality.

And the inclusion of these themes can also be included and will be followed towards them questions Stjsab by strategic goals, and events and activities planned to achieve them, and as follows:

• organizational structure and institutional construction of Higher Education

- Determine who decides Education infrastructure and administrative power to shape?

- Identification of Education.

- Institutes that join under Education.

- Standards that will govern the association and the financing of various educational institutes.

- Standards will govern the organization of education relationship Iraqi society.

- Required between government authority and institutions of learning balance.

• Student Affairs

- The access of females and disadvantaged in education.

- The optimum system for the selection of students and standards.

- Required to be provided for students services.

- Events required to enrich the academic performance of students.

• Human Resources

- Planning to attract angel academic.

- Manage the process of improving conditions for career staff.

- Evaluation and Upgrade angel academic and stimulated.

- Development of qualifications and competencies of the Iraqi cadres.

- The development of skills of workers (professors, technicians, and administrators).

• school system

- The relationship between teaching and education.

- The future of academic programs.

- System of hours of study and Calendar.

- Qualifications required to obtain certificates.

- Academic recognition.

• physical infrastructure and infrastructure

- Needs of the buildings and appliances, and sporting supplies.

- Design programs required for buildings.

- Building an effective system for maintenance.

• financing and financial management

- Tmioal Iraqi education.

- Reform of the current financial situation.

- To improve the financing of scientific institutes allocations.

- Dimensions investment and the productivity of educational institutions.

• Quality and quality control

- The process of evaluating students.

- Performance evaluation and measurement of quality standards.

**The current status of education in Iraq**

• idea of the draft strategy

• The importance of strategic planning for education

• Strategic Planning Steps

• actors involved in the development of the strategy

• Strategic axes

• Strategic Setup

• the proposed strategic framework document

• vision of the future of education

• Education correlation model in Iraq and the Iraqi reality

• Recommendations

The current status of education in Iraq

• Education Institutions

- Planning

- Supervision

- Administrative regulation

- Public Education

- Private education

with soil hoThe current status of education in Iraq

• Education Institutions

- Planning

- Supervision

- Administrative regulation

- Public Education

- Private education

• Some of the strengths and weaknesses of the Iraqi educational system and supporting the process of developing a strategy for development

- The administration's commitment to the need for development.

- The availability of cumulative experience over the past years was characterized by continuity update according to the nature of each stage of development.

- Iraqi plan in the development of public basic education.

- Iraqi economy strategy 2020.

- Adopt the idea of electronic government investment knowledge.

- The availability of infrastructure, legislation and regulations.

- Development Plans and Projects in Education streak.

- The absence of a shared vision for educational institutions.

- The need to review the legislation and regulations relating to education in Iraq.

- Multiple supervision of educational institutions in Iraq views.

- The scarcity of institutional research that address the internal problems of the Iraqi educational institutions.

- Poor coordination, communication and dispersion of efforts between the Iraqi authorities supervisory.

- Take advantage of the lack of technology in the educational fields and management of educational institutions.

- Shortcomings in the accuracy and flow of goods and information system between institutions.

Factors influencing the need to develop a strategy for the development of education in Iraq

• social changes and transformations on the Iraqi society.

• Global economic changes and the trend towards a knowledge society.

• technological changes and the communications revolution.

• cultural shifts and new concepts related to globalization.

• the recommendations of international conferences related to education and nuclei.

**Challenges facing education**

• the creation of the Iraqi citizen to face the challenges of globalization while preserving the identity and system of the Islamic, Arab and Iraqi national moral values.

• Poor use of technology in education and not to adopt many of the educational institutions modern educational concepts such as pivoting on the learner and self-learning and diversify the teaching methods and strategies of education, and others as well as to the lack of technology and technocracy in educational institutions.

• modalities of higher education institutions with regular basic education and education outcomes deal.

• traditional systems in teaching methods are still prevailing in the education institutions.

• continuing to rely heavily on the state as the main source for the financing of education and limited support him and other sources.

• limited contribution to scientific research in the economic and social development and the absence of a clear national policy to guide and care for scientific research.

• multiplicity of agencies overseeing the education institutions, which requires the development of a framework for coordination among themselves or be placed under the supervision of a single point.

• Lack of Iraqi cadres educational experience.

• decline in the relationship between the educational institutions and the external environment.

• slowness in applying the mechanisms and controls related to ensuring the quality of education.

• Compatibility between the output of educational institutions and business needs of market disciplines and the quality of the competencies required.

The idea of the educational strategy in Iraq project

• directed the Council of Ministers the need to develop a strategy for the development of education.

• Raising the paper work for the distinguished Council.

• formation of a working group to study the reality of education and the challenges it faces.

• develop a draft strategy for the development of education hubs.

• prepare a document axes consisting of several axes in

The idea of the draft strategy

• The project presented to the legislative bodies.

• The Council of Ministers Decision strategy that includes education project.

• form work teams and write documents related to axes and conduct field studies and follow up the implementation of the idea after it became a reality.

The importance of strategic planning for the Talim

• develop a general framework to determine the future of the educational system Altojat.

• encourage the authorities supervising an education to work together and to participate in the formulation of a common and unified vision for education.

• visibility of future goals and objectives for all beneficiaries and those in charge of this vital sector.

• area open to the participation of a wide range of diverse segments of society in the formulation of the strategy.

• raise awareness of the importance of change and raise administrative efficiency to bring about the desired change.

• calendar during the previous phase of the Comprehensive Environmental Survey and stand on the strengths and weaknesses in the educational system and the challenges it faces.

• fruitful direction for the efforts and resources invested and better.

• strengthen the role of the government and institutions involved in setting priorities in accordance with the methodology of scientific study.

• help devise new ways and mechanisms of action of improved performance level.

• identify areas of change and the challenges facing the educational system and develop appropriate solutions to remedy them.

Steps of Strategic Planning

• Planning

- Environmental Survey

- Analysis

- Empowerment

• Documentation

• the application and follow-up

Foundations and basic sources of project strategic development of education in Iraq

• principles and values acquired from the sources of legislation Kalaqidh Islamic educational constitutions and laws.

• approved constitutions of the Iraqi state.

• principles and moral values of the Iraqi compound.

• own development plans in Iraq.

• vision of the future of the Iraqi economy.

• the views of educators and academics state officials.

• the views of the people of science, culture and ideology, economics and business in the Iraqi compound.

• global trends in the educational field.

Home-dimensional strategy to build

• the population dimension

- The high proportion of young people.

- Increase the number of the population of school age.

- The high number of the population in the coming years.

• The political dimension

- Determining the vision and objectives of education in the framework of the cultural principles.

• The economic dimension

- Vision of the future of the Iraqi economy.

- Diversify sources of income.

- Expand the participation of the private sector.

- The development of the national workforce skills.

- Attention to education and training programs.

• the cultural dimension - the social

- The change in the pattern of social and cultural values determine the vision of individuals and affect their attitudes toward their vision of education and its objectives.

• the global dimension

- Iraq within the global system affected by variables that take place in this system.

- There is great interest in education and development in line with modern technological developments in the cognitive domain.

Strategic axes

• axis visions and goals

• axis of Education and the external environment

• axis alignment in Education

• Quality axis in Education

• financing education hub

• axis of cooperation and partnership in education

• axis of facilities and infrastructure

• the focus of the legal and regulatory framework for education

• axis Department of Education

• axis linking public education and higher education

Vision and goals axis

Opinion directed about the current education goals

• there is a gap between the goals of public education and higher education.

• Current targets not be considered a student of the labor market.

• concentration on goal theoretical aspects and practical aspects of neglect and footwork.

• interest in traditional patterns of education that focuses on conservation and indoctrination and neglect of intellectual and creative aspects.

• Do not encourage the independence of the student's continuous self-education.

• To Ataatuaq with new developments in the world of knowledge.

The importance of vision and goals

• The basis for the development of education in Iraq strategy.

• include the general orientation of the education system in Iraq, which is based on social and cultural values of society.

• determine the attributes and features of desirable educational system in the future.

• expresses the ambition to face future challenges.

• help in the development of the lines and policies and educational projects.

• closely linked to other axes of strategy.

Axis Education and the external environment

This axis deals with all the factors that have to do with education, directly or indirectly, a natural resources, human resources, economic, and infrastructure of the state, population trends, and social and cultural community in different directions at the level of countries in the region and the world.

Axis alignment in Education

This axis to focus on:

• dedication and activating the principle of harmonization between education and the requirements of the next phase system.

• Almumh means compatibility and connectivity with various partners in the education system, the labor market, the other stages of education, domestic and foreign technology, and compatibility with the concept of lifelong learning in every time and place.

• Human Development Report 2005 in Iraq refers to the focus of education on theoretical knowledge and differences for vocational and technical education and high unemployment rates among graduates of general education and the lack of compatibility between education and labor market outcomes.

Quality axis in Education

This deals with the current reality axis of the educational institutions

• In the field of quality in terms of the process of education and teaching.

• methods of evaluating students.

• Academic Programs and patterns of Education.

• The quality of the preparation of teachers, supervisors and administrators and the quality of the public education system output compared to the output of basic education programs.

• Education and the quality of facilities and infrastructure outputs It thus includes inputs, processes and outcomes (outputs).

The focus of education funding

Of the challenges facing education in Iraq continued to rely heavily on the government as the main source of funding for education and public form and higher, with an average growth of government spending on education during Afattarh about 8.4%, in addition to the decline in enrollment for the age group rate reaching in about 12 0.6%.

The focus of cooperation and partnership in education

This includes cooperation

• between the various education institutions (programs and activities and events Taamah).

• bin education institutions and the community, the private sector and all relevant parties.

Axis facilities and infrastructure

It deals with this theme:

• Knowledge of current educational institutions and the requirements of their suitability for the educational process.

• the need for additional buildings that are consistent with the specifications of international standards and in line with the local environment.

• contribution of the private sector in finding an acceptable structure for education by building schools.

**The focus of the legal framework**

**Axis addresses**

• study and analysis of the current regulations and legislation.

• re-drafting of legislation to comply with the developments that have taken place on education forms, and types.

• agreements and resolutions of international organizations, regulations and regulatory decisions.

• the possibility of the unification of regulations on scientific and academic activities.

**Axis Education Department**

**This includes Hub**

• exposure to the importance of management in achieving the educational goals set by using:

- The best ways to invest manpower and resources available.

- Modalities for the reform and development of the programs of study so enjoy the flexibility to meet the challenges.

• reduce the multiplicity of agencies supervising, duplication and lack of coordination and do:

- Training and Rehabilitation administrative.

- Specialized courses.

Strategy for the modernization of education and training

When developing a strategy for modernization of education and training requires a lot of reforms:

• A description of how to develop a course for this type of education.

• Clarify how to prototype and evaluation and improvement before adoption.

• diagnose the conditions of education and training in their relationship to work and the economy.

• comprehensive reform of curriculum and technology.

• reform based on the school to develop the educational institution and enable each school to improve her performance.

• human resources and professional development program to provide qualified personnel capable of achieving the ministry.

• institutional rehabilitation of decentralization for the treatment of big government and a review of the organizational structure for workers program.

• administrative system program and a review of the organizational structure for workers.

• technological development and information systems program Astkham increase of technology in education, such as distance education and self-education.

• Calendar and follow-up program and the aim is to establish a complete system for follow-up, calendar and calculator in all aspects of the educational process.

• school buildings program to provide the necessary preparations.

• Develop kindergarten program.

• reform of the educational component for Basic Education program.

Strategic relationship of government plans

Strategic Setup

• hold educational seminars and community to view the strategic axes and glitter with all the views of the community groups and classes on strategic axes.

• formulation of the strategy document and according to the axes and amendments thereof by experts and community groups.

• General review of the preliminary drafts of the strategy document by a seasoned team.

• Showing a national conference to view the content of the strategy document and adjusted according to the findings and recommendations of the conference.

• formulate and document the development of education in its final form strategy.

• Showing document Alastrtejah the Ministry of Education and Education to provide opinion and comments and then submitted to the Cabinet for approval and accreditation.

• analyze the current situation of the educational system.

- Make some field studies.

- Development of strategic axes.

- Writing in the strategic axes.

- Takimm documents relating to the axes.

The proposed strategic framework document

• Provide

• strengths and weaknesses in the educational system

• Challenges and Change Dauaa

• vision and goals

• the general principles of the strategy

• Key strategies for achieving the vision of the future

• work and cost plan

The general principles of the strategy

• Develop a new framework for the educational system in Iraq.

• building capacity and efficiency of the system learning.

• give the powers and independence of the educational institutions.

• Develop a mechanism for measuring educational output.

**Strategies Home**

• strategies for students

- Student progress.

- The link between the public and higher education.

- Education for All.

- Absorption

• management-related strategies

- Department of Education.

- Facilities and infrastructure.

- Legal and regulatory framework.

• strategies for achievement

- External structure.

- Cooperation and partnership.

- Standards and quality.

- The quality of output in the context of globalization.

- Scientific research.

• strategies for funding

- Financial efficiency.

- To contribute to the funding.

- Capital investment.

• Technical Education Sector Strategy

- This awareness of the role of education in economic development.

- Provide technical education and vocational training to meet the modern labor market of labor requirements.

• public education strategy

- Raise the level of basic education to become comparable with the standards in developed countries.

- Give high priority to increase non-avoidable in demand for public education services in the allocation of government resources.

- Rebuild the academic curriculum focusing on informed consent charts such as mathematics, computer science.

- The development of the teaching of the English language.

- Cancel the dual school system or more.

- Make the lives of textbooks for a specific period.

- Encourage the private sector to take over the role in the education sector.

**The priorities of education strategy in Iraq**

• improve the quality of manpower and increase the suitability according to the following mechanisms:

- Admission policy.

- Educational process.

- Calendar and upgrading.

- Infrastructure.

• Develop a close relationship between education and research institutions and the needs of the labor market.

• To achieve flexibility between manpower and business that you make.

• especially in some fields priorities.

**Raise the quality of manpower and increase the suitability**

It supposed to be the subject of education constitute the most prominent problems in Iraq through the need to influence in the following areas:

• Admission Policy

- Enrollment rates (demand).

- Alantvae standards (or demand response).

- Display distribution (geographical and social map) and equal opportunities.

• educational process

- Teaching staff (teachers, teachers).

- Curriculum content.

- Teaching methods.

- The language of instruction.

- The education of science and technology.

• Calendar

- Curriculum calendar.

- Evaluate teaching methods.

- Collection calendar.

• infrastructure

- Patterns of structures

- Horizontal structure (terms of reference)

- Vertical structure.

- Education in traditional institutions.

- Distance education.

- Continuous education.

- Recurrent education.

- Rehabilitation.

- Self-education.

- The need to avoid the lack of human spammers in information and communications and biotechnology.

**Education institutions and the needs of the labor market**

This is done through the support of the relations between the institutions and Education:

• other scientific and technological institutions in the inside and outside Iraq.

• production and community institutions and professional organizations (Altkabat and associations).

**Strategy alternatives**

• There are a number of strategic alternatives for the development of Education in Iraq

- Linear continuing strategy.

- Partial reform of the adequacy of the internal strategy.

- Partial reform of foreign adequacy strategy.

- Comprehensive renewal strategy.

• There is a strategy for the dissemination of scientific and technological knowledge and the preparation of human spammers alternatives

- Alternative rationalization.

- Alternative Altotini.

- Alternative technological root.

**Alternative technological**

A community without a school or in the school community, using a number of means of access to and use of information and communication media or mass media education outside of school, depending on the variant forms of many of the technologies, including:

• Maomat user-friendly systems.

• unique self-learning systems.

**Science and technology strategy, education**

• require formatting and identify Futures Education and planning the following:

- Education and its relationship to society (social demand).

- To provide the country hand-eligible high (the need for labor).

- The rate of return account (another indicator of the surplus or rarities).

- The role of education in meeting the needs of the labor market.

• Technology and Education

Dealing with this relationship:

- Major technological changes and its effects on education such as information and communication and improve communication (Ansan- family).

- The effects of these technological changes on education through changes that challenged on the economy (the contents of the professions and methods of practice professions).

- The impact of technological change on education through change education tools.

- The impact of technological change on education through curriculum change

- The impact of technological changes on education through the improvement of educational research.

- The impact of technological change on education through computers, like the computer in education and the Internet in education and learning as a computer.

- An integrated policy for the educational systems interlocks Ktdakhl science and technology, education (curriculum, research, and scientists Ttaiwiralbagesan) and the overlap of science and technology and the labor market.

Strategic alternatives for the development of science and technology in the education sector

There are a number of strategic alternatives that pose a wide choice in the use of science and technology which is focused on the potential offered by science activities and technology, including the alternative continuity rationalization and alternative reformist and most importantly, where are some of the current school aspects of renovation and planning education and management, which depends on:

• Increased use of modern educational methods and technologies in education.

• the introduction of new forms of technology vocational training.

•Strategies

• General Education

• Prepare manpower

• Technical Education

• General Education

**General Education**

The purpose of the strategy

• raise the level of basic education to become comparable with the standards in developed countries.

• give high priority to increase Tnadea is not possible in the demand for public education services in the allocation of government resources.

• rebuild the academic curriculum focusing on the scientific disciplines such as mathematics, science and computer studies and principles of economics.

• teaching English since the beginning of the first year of basic education.

• Cancel with diets school system or three meals.

• Encourage your system to take greater role in the education sector through appropriate measures.

**Prepare manpower strategy**

**Purpose:**

• Preparation of human resources from various disciplines necessary for productive activities and the activities of science and culture.

• human element highlights the central theme of competencies as problematic.

**Technical Education and strategy**

Of the forces of change affecting the technical education, growth in global markets and coupled with intense competition and the emergence of service-based industries to knowledge and the implications of information technology on society in general and to work on the degree of respect and changes in methods of organization of work within the institutions market institutions, for example rearranging administrative structures increase the devolution of powers and the emphasis on teamwork and the multiplicity of per capita skills and changes the population of Alnmwalscane and increasing numbers of young people aspiring class to better jobs and social changes resulting in changes in the levels and lifestyles as well as the growing demand for consumers on a wide range and variety of services and at high quality levels.

Lies the strategic importance in the outlook that takes into account the changes that rapidly occur in the balance of sharing different areas necessitate a fundamental shift in the methods used titling manpower and technical skills required, and the challenge is to find a technical education creator and creative at least at the level for technological education in the countries of the world, but and developed and added to it as well.

Therefore, when developing the strategy for technical education is supposed to be taken into consideration:

• This awareness of the role of education in achieving economic and social development.

• Raising the competitive and support the country in the public and private sectors.

• Provide technical education and vocational training modern and flexible to meet the requirements of the labor market of skilled labor and semi-skilled manpower.

**Higher Education strategy**

Take the strategy into account the needs of the current Iraqi society and future harmonized with the development goals and plans, with the aim of graduating students are eligible are able to meet these needs, and through policy and determine necessary to provide an appropriate environment for the development of procedures, and through these themes around which this strategy (founded admission, curriculum, accreditation and the foundations of quality control and quality, encourage creativity).

**Target**

The development of a high-quality system capable of graduating qualified human cadres and specialized in various fields of knowledge.

**Mechanism**

The development and modernization of education sector to become more able to meet the different needs of economic, social, political and cultural activities.

**Overall objectives of the strategy**

• Prepare a human Kudar qualified and specialized in various fields of knowledge.

• Provide academic and psychological and social support for creativity, excellence and innovation environment.

• find the founders of the close link between the public and private sectors on the one hand and educational institutions on the other.

• improve the quality and efficiency of education to harmonize the requirements of society through the development of standards and foundations for the adoption of quality control.

• keep abreast of developments in information and communication technology and employment in the administration.

• taking into account the economics of education to ensure secure funding and the development of appropriate mechanisms for the distribution of the financial resources available.

**Measures needed to provide an appropriate environment for the development of the education sector**

• give the role of the private sector to participate in the future of education and industry by increasing its representation in the councils of the ministry of planning.

• Study of financing for public schools to ensure Balmurad supplying them with the necessary funding and the creation of fund for needy students.

• the introduction of the concepts of quality and quality control in the various components of the education system.

• Provide mechanisms to embrace students who have the ability to excellence, creativity and care.

**Study programs**

• reconsider the plans and programs of study to update it continuously at a rate of once every four years.

• Work on the establishment of centers of excellence in specific disciplines and a review of the disciplines that does not hold sufficient ingredients of good quality.

• the establishment of teaching staff development centers.

**Information Technology**

• Reload public computer courses continuously in the light of the increasing knowledge of high school students with these materials.

• The use of ICT in all the programs in terms of content and teaching methods, and methods of evaluation.

• provision of equipment and infrastructure necessary to enable the teaching staff and students of the use of technology in teaching and learning.

• the use of information and communication technologies in distance learning programs.

**Funding**

• the establishment of needy students and the allocation of funds ratio of the annual government support for these funds is increasing annually until it reaches 100% in 10 years to become fully dedicated government support her.

• Encourage financial institutions and other community organizations to set up a fund a study of students boxes.

• allocate a portion of the extra government support to finance centers of excellence and disciplines.

**Accreditation and Quality Assurance**

• creation of an independent body to evaluate and quality control, quality and in line with international standards.

• establish an independent body for approval.

• set up offices for accreditation and quality control.

**Creativity**

• the establishment of a supreme body comprising distinguished representatives of public education and the private sector and the Supreme Council for Science and Technology and institutions dealing with scientific research institutions are working on:

- The unification of scientific efforts.

- Set up a special fund for financing.

- Guide researchers toward the most useful scientific research to meet the needs of the community.

- Support for researchers and give them incentives and discretionary and moral.

- Closer relations with public and private institutions concerned with scientific research to undertake research to their advantage.

- Promoting the dissemination of scientific production and unify efforts to issue a specialized scientific journals.

• build a complete data base for scientific research.

• provide funding for scientific research, creating cadres chapman efficient and provide appropriate opportunities for them to acquire the necessary expertise.

• the establishment of centers of excellence in line with the strong disciplines.

• the establishment of an effective, regulatory and technical partnership between educational institutions and sectors of development and production and various services.

• Provide necessary to embrace and care for students who have the ability to excellence and innovation mechanisms.

**Administration**

• enable the Council and the Ministry of Higher Education and the bodies of the adoption, and quality control and quality of scientific research of information gathering and analysis in order to serve the purpose of making appropriate decisions.

• the use of information management systems in decision-making so as to include the following systems:

- Student Information.

- Acceptance.

- Financial affairs.

- Administrative Affairs.

- Human resources and payroll administration.

- Assets and warehouse management.

• adoption of efficiency and competition in the selection of university leaders.

• the adoption of decentralization in the administration, implementation and expansion of the delegation of authority in universities.

• the adoption of the principle of transparency and accountability in the management of the sector at the national and university levels.

• Improve the creative sector management and scientific research.

• set up offices in universities to follow Sean graduates and their employment.

**Legislation**

• Amend the Education Act to implement the formation of the proposed policies bodies.

• reconsider the legislation in order to ensure:

- Amendments to the Companies Act allocation of 1% of annual profits for companies and diverted to fund scientific research.

- Modifying the regulations and instructions relating to the rules of the promotion cadre education.

- Modifying the regulations and instructions Balantaat.

- Amend laws and regulations to increase private sector participation in the sector councils.

**First: Acceptance**

Figure who was responsible activity indicators time / Results

1 review of the foundations of acceptance and approval of the Ministry of Higher Education Council of the beginning of each academic year - reflect the merit of students and fairness and equal opportunities

- Ensure the quality of graduates

- Reflect the implementation of international cooperation agreements

- Achieve the greatest possible harmonization between the wishes of students and disciplines available to them

2 Identify the preparation of all students admitted to the University and the Ministry of Higher Education Council of the beginning of each academic year

3 to accept foreign students and distributed to public universities Council and the Ministry of Higher Education and the beginning of each academic year

4 to accept students with certificates of secondary non-Iraqi universities beginning of each academic year

5 accepting students in parallel and international programs and supplementary and evening universities beginning of each school year

6 accept students in the disciplines of Physical Education and Arts universities beginning of each academic year

7 admission of students in private universities and private universities beginning of each school year

Second: study programs

Figure who was responsible activity indicators time / Results

1 revision of school curricula and university programs once every four years - keep up with programs of scientific and technological developments and adapted to the requirements of science market

- Distinguish certain universities in specific disciplines

- Develop the capacity of faculty members and the development of teaching methods and evaluation of students

- Providing the universities with qualified staff

2 An assessment of the programs in each university to choose the best ones and strengthen them and supplying them with cadres and equipment to become centers of excellence universities constant

3 expansion of graduate programs continuously universities

4 set up to develop the performance of faculty members university centers 2003-2004

5 to send envoys to get a doctorate degree from prestigious universities in the disciplines required continuous universities

Update courses for computer compatible with the general secondary outcomes in this area and the Ministry of Higher Education and Scientific Research and universities annually - increase the likelihood of graduates get jobs

- Improve the efficiency of teachers and graduates in the employment of information and communication technology

- Increased communication and cooperation between faculty members at universities opportunities

2 the use of information and communication technology in all programs including distance learning programs continuously universities

3 faculty members develop the capacity of universities constant

4 provide basic infrastructure for the use of information technology and communications universities constant

5 develop operational plans for the use of information and communication technology in higher education and the Ministry of Higher Education and Scientific Research and universities

Fourth: Finance

Figure who was responsible activity indicators time / Results

1 reconsider university fees commensurate with the income of citizens and the cost of the study universities annually - enable academically able students from attending universities

- Improve the financial situation of the universities to enable them to maintain the quality

- Providing the universities with qualified staff

- Spreading an atmosphere of competition between universities to become centers of excellence and maintain.

2 Create boxes needy university students 2003-2004

3 boxes needy students finance the Ministry of Finance and financial institutions and other community institutions annually

4 distribution of subsidies between universities based on the Council for Higher Education students prepare annually

5 to allocate a part of Mozanat universities for the purposes of scholarship for a PhD degree in the disciplines required continuous universities

6 extra continued government support for the completion of the new infrastructure for universities and the Ministry of Higher Education and Finance annually

7 allocate a portion of the extra government support (it competes universities) to finance centers of excellence and disciplines. The Ministry of Higher Education Council annually

8 marketing of higher education to attract foreign students and facilitate enrollment in universities and residence of the Ministry of Education and Ministry of Interior continuous

Fifth: Accreditation and Quality Assurance

Figure who was responsible activity indicators time / Results

1 set up an independent body to quality and quality control of the Ministry of Education in 2004 - university's commitment to quality control and quality standards

- University's commitment to international accreditation standards

- Raise the level of universities and encourage them to compete in the graduation of qualified students

2 set up an independent body to adopt the Ministry of Education in 2004

3 set up offices for accreditation and quality control in the universities and the Ministry of Education in 2004

4 an efficient unified exams for graduates of the Ministry of Education in 2004

Sixth: creativity and scientific research

Figure who was responsible activity indicators time / Results

1 Create a supreme body for scientific research and the Ministry of Higher Education and Scientific Research in 2004 - raising the level of scientific research and standardization efforts

- Providing funding

- Directing scientific research to the requirements of development

- Cooperation and encourage the active participation between universities and the productive sectors

- Take advantage of the creative capabilities of students

2 provide funding for scientific research and the Ministry of Higher Education and Scientific Research and the private sector in 2004

3 establishment of centers of excellence in universities and the Ministry of Higher Education and Scientific Research

4 make way either to participate in the various committees and events for Higher Education and Scientific Research and the Ministry of Higher Education and Scientific Research of the private sector

5 care creators and talented students and encourage them and motivate them and the Ministry of Higher Education and Scientific Research is continuing

Seventh: sector management

Figure who was responsible activity indicators time / Results

1 to provide the necessary cadres of the Council of the Ministry of Higher Education and the accreditation body for the collection of information, analysis and Ministry of Higher Education and Scientific Research - and bodies to enable the Council to take appropriate decisions

- Increase the efficiency of the performance of the higher education sector at the national and university levels

- Ensure transparency and accountability to improve the efficiency of university administration

- To find a link between universities and graduates

- Study the reality of the labor market

- Help graduates get jobs

2 to provide training and equipment to these cadres of the Ministry of Higher Education and Scientific Research, and the private sector in 2004

3 building information management systems in decision-making in the management of the university and business use, and the Ministry of Higher Education and Scientific Research continuously

4 Select university leaders in a competitive manner through specialized committees and the Board of the Ministry of Education continuously

5 Use of decentralization and devolution of powers in the university administration and the Ministry of Higher Education and Scientific Research continuously

6 to provide the necessary cadres for creativity and Scientific Research of the Ministry of Higher Education and Scientific Research is continuing

7 set up offices in universities to follow graduates and employment affairs

Universities

Eighth: legislation

Figure who was responsible activity indicators time / Results

1 modification of higher education and universities for the implementation of laws for the formation of the proposed Ministry of Higher Education and Scientific Research bodies - the establishment of the supreme bodies of the adoption of quality control and quality scientific research policies

- Raise the level of faculty and academic titles unify ranks in universities

- Providing the universities with qualified staff

- A greater role for the private sector development and productivity in various sectors and the future of higher education industry

2 Companies Act amendment to allocate 1% of their annual profits for companies and converted to Fund Research Council of Representatives and the Council of Ministers

3 modifying regulations and instructions relating to the rules upgrade and a full-time scientific Tdrejan in universities and the Ministry of Higher Education and Scientific Research

Amendment 4 regulations missions and the Ministry of Higher Education and Scientific Research

5 amend laws and regulations to increase the participation of the private sector and the Ministry of Higher Education and Scientific Research

Terms of the success of e-learning

• adequate technological infrastructure

• underlying financial resources as this type of high cost of education.

• The existence of qualified human resources quantity and quality appropriate.

• partnership relations with foreign governments and institutes in unforeseen countries.

• cooperation with countries of cultures and in similar level of development.

• Starting small and limited group projects targeting to acquire the assets needed for subsequent expansion.

Equality: the elimination of disparity in the enrollment of learners between boys and girls in rural and urban areas and among various ethnic backgrounds and different economic conditions.

Quality: Improving the quality of education for a better to the needs of the labor market and the requirements of sustainable development and catch up with the level of high-performance in the field of education and upgrading the capabilities and efficiency and attitudes of educational and faculty countries in response.

Citizenship: Adoption of the independence of Education and disconnected from politics and dissemination of human rights and respect for freedom of thought and expression and to promote tolerance and social cohesion.

Post: promote community participation in the planning and evaluation of the educational system and strengthening coordination with higher education and other sectors and contribute to the development and promotion of the private sector.

Good management: orientation document on reliable evidence and data and evaluate the performance and decentralization and the fight against financial and administrative corruption planning.

Reconstruction of school buildings

The infrastructure of the educational system had been neglected two decades the situation has been exacerbated as a result of the large amount suffered by the education sector in the wake of the destruction and looting process that occurred in March and April 2003 and subsequent months. Digital and data indicate that about one-sixth of the number of schools in Iraq in 2751 schools have been stolen or burned or damaged.

More than 2,400 schools have been stolen and wounding 146 building during the military operations and the burning of 197 school also said 138 schools had been used stores of ammunition and 101 school stores of weapons, and based on surveys conducted in 2003, 80% of school buildings currently require repairs and rehabilitation of different degree These reforms between the school and the other and that the proportion of secondary school buildings, which are still acceptable condition less than 30% of the total schools while the vast majority of schools suffer from severe damage or moderate damage and there proportion of very poor buildings so you need to complete reconstruction or not be available in the vast Great schools of minimum health requirements.

At the beginning of the 2003-2004 school year on October Alaoln been Aammaroaslah more than 1,500 schools at the end of the calendar year arrived

**Patterns of Continuing Education**

**Education program**

Able student that knows himself and his effort and self-offering or follow the steps Nkos directly traced back to 1926 when he invented the Percy family includes a set of questions and correct her answer, knowing that education is missing in Iraqi schools.

**Correspondence Education**

And that the elements of good education to follow Vooqat vacuum prompted many to look for useful work they invest their time, they are not available in Iraq, this type of study.

**Open universities**

It receives students from different educational levels to provide them with types of studies required for the development of the individual and society, and suffer from the weakness of Iraqi schools to accept this kind of education.

**Communications and satellites devices**

Managed communications satellites that give prospects wan to enrich the educational process at all levels of education, especially in higher education was not this potential theories or hopes for longer and it was possible to lecture from school to home or transferred from another school.

It began the University of Hawaii Islands experiences on the use of satellite ATS-I transfer voice messages and print between its facilities in various islands in 1971. The university has set up floor plans for the transmission and reception of educational television and in the exchange between libraries and medical conferences and student discussions, teacher training and joint research. In 1971, use satellite ATS-I to provide educational programs for schools, rural medical treatment and to guide some medical lectures to some body of teaching staff in the Faculty of Medicine at the University of Washington as the other experiments at Stanford University in conjunction with Brazil using satellite ATS-6, as well as the experience of the Rocky Saunton province using the moon ATS 6. experience Ablastaian province using satellite

It has been in Ottawa, Canada, and Stanford University in California in the United States from the exchange of experiences between teachers via satellite and the exchange of distances where the school has been directing the school distances from Carleton University in Canada to the US Stanford University and vice versa.

The Arab satellite ARABSAT experience says Dr. Ali Mashat Arab Satellite Communications Organization earlier that the Director of the Arab satellite system will provide successful solutions to educational problems in rural areas will also help to offset and counter the serious shortage of teachers, and there is this kind of education in Iraq despite the his foot.

**Data banks**

Information banks are considered an outlet conventional storage problem and can retrieve information stored in these centers using one of the elements Description Albelograve such as document number, author, title search and objective information banks to link the centers and scientific institutions and universities by the light in front doors of knowledge and information they gather their international experiences, either in Iraq There are no techniques and even the terms of this education.

**Technological option**

No way in front of Iraqi schools only technological option to develop and change, specifically technology knowledge which is considered the most important factors helping to produce knowledge, which offers technological logic and then access to the production of more advanced technology.

Reinforced developments in technology and communications to take the open university education pattern, due to its reliance on knowledge and information technologies, and given the rapid development in the communities and the transition to a knowledge society, and will see the knowledge society, expanding greatly in taking this type of education, with the justification that we have mentioned previously, including:

- integrated use of other modes of technology.
- beating many of the obstacles to regular education.
- submission services to individuals of all ages.
- beating barrier place.
- done without meeting the teacher learner.
- Allow the teacher to work and learn.
- entrench a culture of continuous education.
- Develop opportunities to develop the performance of employees in state institutions.
- increasing social demand for education.

**Knowledge technology / concept and its relationship with the Iraqi university**

There is a clear relationship between knowledge and understanding of technology knowledge society are two sides of the same coin role of the knowledge society is evident and its dimensions are formed as a result of stunning advances in technology applications, where communications have developed a stunning view of the Ttoraltknlogi ,. Technology turned to a revolution represents one of the important creations done by man in the late twentieth century and early twenty-first century. Then this session widened in the areas necessary for humans, such as:

- chemistry and medicine

• Life Engineering

• Other Sciences

And it contributed to a substantial change in the pattern of human life and became Alnasr mind and thought is the basis of profit and investment, and therefore there is sufficient justification which emphasizes the importance of and the need to take Iraqi universities this type of education

• rapid technological evolution.

• significant increase in the preparation of educated and interested in education.

• democratic education and the right of citizens to education.

To meet the challenges of learning society, attention Petknlogia information and building rules Marafih and telecommunications networks modern and integrate technology in teaching and learning processes and research and that the success of these universities and turn them into learning societies depends on how much attention in Altdrassen professional development and the adoption of the principle of participation and planning and activation and the use of modern technology and transform the classroom to active and effective environments for learning.

As for the challenges of the knowledge society depend on the nature of the information and knowledge that are published daily in different parts of the world super-fast, find out about different types, including the globalization of knowledge and virtual knowledge and technology knowledge, globalization of knowledge basis of atheist-first century society part, called the third millennium.

**Scientific research**

Alahtma began scientific research after the change the concept of education, so that teaching is no longer the only function, but the look in-depth scientific research in general, it is clear that scientific research suffers from several problems, including lack of funding, as the teaching staff salaries what Iqri than 90% of the budgets of educational institutions and the remaining is distributed to all other aspects of spending, and thus reflect negatively on the requirements of scientific research equipment and other supplies, as well as it suffers scientific research than as just the performance of functional, doing educational Alkdar, to separate their own targets for his need.

Information technology in Iraq

Not Tstaa educational institution-building networks are sophisticated information which can provide Internet services for workers and students alike, but the ratio of the number of students to the number of devices available is still very high and where there is no general framework for the educational plans in information technology, he notes the existence of chaos in the development of this plans that are most often repeated to some random In light of these must be put general framework which defines the requirements and terms of reference to be developed.

Educational institutions are suffering from a severe shortage in the number of specialists from the campaign master's and doctoral degree related to science and engineering as well as computer software that facing many difficulties and challenges that limit their economic feasibility and benefits of future industry.

The computer entry in the field of education puts education planners in Iraq on the eve of social and economic changes as well as the educational changes that need to be absorbed and dealt with to create a modern educational systems can her keep up with innovation and development, and the number of computers in Iraqi universities up to (10000), where I lost most of them during and after the war and the lack of Internet networks are widely, and in some universities and colleges these advanced networks lost quickly and urgently, with a need to Anja (30000) Computer

**Continuous education**

Appeared in the last quarter of the concept of the century (continuing education) did not features identified and clarified its properties, but since 1960, when the Second International UNESCO education conference was held adult "Aduct Education" in Montreal, Canada City and it put the first touches as it was decided that it is no longer enough to spend an individual specific years in school education to be able to behave that way in life may extend for more than fifty years, as what we have learned at a young age may become obsolete in today's need to add a new to him.

In 1964 and approved by 119 countries in UNESCO's General Conference on the recommendation of the effect that different forms of education school abscess adult and learning should be considered an integral part in the education system, to have the opportunity for males and females to continue in education and lifelong learning, and still patterns of Continuing Education in Iraqi universities suffer from the continuous breach and weaknesses in performance in spite of the long period

**Scientific prospects"**

**Humanities**

Science is divided into two parts of humanity in general and Pure Sciences, Humanity and social belong to all what for humans and the resulting behaviors, attitudes and therefore the human sciences dealing with man and society, man and nature and man.

The schools that dealt with human understanding are numerous, including the Greek school grabbed the attention of the Greeks from the likes of Socrates, Aristotle and Plato man and was the jurisprudence of these philosophers refers to the method of analysis was based on mental Rai greater than Lalai practical and scientific, either human site in Romania civilization considered the case of a continuation of the Science Greece literature and thought.

Subsequent manifold mismatch theories of Marxism, schools and other well-known scholar in the social sciences do not have to think about the concepts and vocabulary and methods of analysis to study the analyzes, this was logical to think from within the scientific perceptions that shall be given to the phenomenon of scientific sense and scientific phenomena.

And as we have mentioned, the rights in accordance with all interpretations positively and negatively is the man in its interaction with the other must be seen this relationship look integrated comprehensive approach in its relations requires thought out in the essence of man and his mind and his thought according to balance and positive double Valtnaiah looking man in symmetry and diversity.

Arab achievements of the Islamic and Chinese output and account Indian Roman and Greek discourse and dialogues have contributed, as well as Persian stories in the renaissance of science as well as the Sumerian educational system a year ago (3500-BC) and discovered writing system.

Played Egyptian civilization important roles in the progress that has got to start from about (300 years BC) Some believe that the Egyptians the top are the educational radiation source and the Egyptian civilization has made the Greek civilization and that education was based mostly temples (Ain Shams and the Temple of Karnak) and was teaching It is the ebony and ivory and papyrus pages. The system of book successfully Line hieroglyph as others believed that the concept of education in the contemporary scientific context of which were by the ancient Egyptians and teaching aids were carried out by them ancient Egyptian drawings engraved on the front of the walls of temples as a means to connect the sensory information.

The Greek history is represented two cities important (Sparta and Athens), which agreed to make the way of Education to prepare a citizen who serves his country, in Sparta, the educational strategy focuses on the ongoing military exercises and scientists distinct favored date their achievements of Hippocrates the father of medicine, who got him and discussed the medical issues in theory and in practice And still the Department of Hippocrates Hippocratic Oath of the oldest rare historical documents that show the ethics of the practice of the medical profession. Valeonaon one of the first people who began to study memory and all the functions of the brain and heart.

The Romanian civilization has played a role in the history of human thought, where he was Education focuses on reading and writing skills and learn the young in schools, and she was known as Allodoux primary school teacher Palmadb called educational ideas and still being prominent educational guidance.

As for China and its civilization was the farthest civilizations era education and the most famous male in history, was philosophical doctrines are Chinese Confucianism since the sixth century BC. Chinese civilization is represented in many sciences, including medicine, chemistry and still Chinese needles and means of treatment.

China Eastern are rigorous breeding and tradition of the past and estimate and a commitment to a series of teachings and governance humanitarian tradition and appreciation of the past and focused on:

• Hold tests for students.

• Save information.

• Spirituality.

• compliance with laws and regulations.

The Confucius (479 BC -55 BC), one of the foundations of Confucian movement, which used several means of socialization, including:

• take advantage of the political positions.

• skillfully take care of reading books.

• the study of language and literature.

• obedience to the father and submit to him.

• obey the ruling.

Accordingly, the Chinese education characterized by:

• moral education.

• the clarity of vision.

• Faith continuing education.

• Focus on the heritage of the former.

The Indian civilization Fmtnoah, the mismatch Buddhism which is characterized by full-time for worship and self-liquidation and all the stuff hedged difficulties and desires of the causes of suffering. Buddha's teachings have made wide acclaim in the world, either educational features at Buddha Vtmthelt its means in the dissemination of education and teachings through lecture and conversation, either concentrated his commandments not to murder, adultery, theft and alcohol. And it addressed the Indian civilization philosophy, medicine, and the study of stars, science and mathematics. It is cultural anecdotes book (Panchatantra), which is interested in reforming the souls and put the Indian philosopher Alipidba. The educational features of the Indian is represented spiritual, moral and social education.ample, the ground). the treatment of diseases (AIDS, cancFuture perceptions

Characterized anthropology limited and relative lack of absoluteness of knowledge required to be a non-comprehensive and non-fixed and It's a change and evolution and modification, and what human reach of scientific knowledge and the laws in a particular time may not be right and acceptable at a later time. And that human science progresses with time, and that the human mind does not detect does not understand the realities of the universe at once, but at different stages of life and the pursuit of knowledge is an ongoing process. But despite the limited human mind and its inability to often understand and interpret the many scientific issues, it is a permanent quest to understand the puzzles did not prevent him from seeking future after he became the tools needed to study and Orientalism future and predictive available.

Perhaps what distinguishes man from the rest of the creatures being created future, since ancient times was a man thinks his future, and A_i_rach and predictable. The scientific advances in science such as chemistry, life sciences, mathematics, computer sciences and other sciences increased annually and became the basis for most of the other sciences. For example occurred tremendous developments in chemistry and life sciences has been extremely clear and flourished new branches in chemistry and has seen tremendous developments in the study of the atomic structure and molecular such as the use of lasers and X-rays, high-energy and low-lying rays Kama and the use of sound and light together in the study of atoms and molecules and ions in the gas cases and liquid and solid and got great development in analytical chemistry included diagnosis and extraction and separation processes as well as in biochemistry have emerged, including molecular biology and chemical Alahiaiat aware Chemical Bionic, which has quite significant in the present and the future, also got tremendous developments in the science of enzymes either in the life sciences is no longer note one but I took comprising tens of Science Sub-Specialized Kaalom anatomy and tissues, embryos and cell physiology and genetics and the environment, plant and animal and human diseases, behavior, and meeting and fossils geographical distribution of animal and plant classification plant and animal and the date of ancestors and natural history of the neighborhoods and Sciences bacteria, viruses and parasites, worms, insects, fungi, fish, birds and mammals, genetic engineering, technology, biotechnology has been used Genetic engineering applications in medical, agricultural, industrial and vaccine and production of hormones and the treatment of incurable diseases that have no drug areas, In 1997, it was exciting scientific event, which was announced in the February 27, 1997 which is the birth of Dolly the sheep in a way transfer of somatic cell specialized to unfertilized eggs to the Shah again after removing nucleus and implanted in the womb of a third sheep, and the most important event of this scientifically is a specialized and stable cell genetic return to the situation after the loss of this status. With other developments in other Pure science it is clear that this tremendous progress was hiding behind him backward in human cognition for other things such as the production of sufficient quantities of food to feed the human in the world as well as problems related to the scarcity of resources and energy and the increasing pollution and population explosion problem.

### Outlook for education

Will help the spread of education in all countries of the world and cultures to build a community, there is no escape from the cultural relationship between education and society, and that education is only one element of a variety of factors, including the social, economic or political, religious or military or geographical, and that the conflict between the world powers and interests Local will be educationally and politically challenging.

Will remain challenge is the importance of the face of religious, racial and ethnic, cultural and national conflicts and structured learning ability in schools and universities to reduce these conflicts are governed to a large extent the political administration of the leaders in the community about the participation of serious and positive education and educators in such cases, but it can not reduce This rage levels of educational participation without military strategies will not be enough in itself.

And must be seen teachers starting from pre-school onwards that they are a means of cultural and democratic understanding, that means sterile democracy and this in turn depends on the development of mutual understanding between cultures or multicultural understanding.

Related challenges ahead of us to get into realistic ways to bridge the huge gap between the rich and the poor in the world, this gap exists in the same cities, or between cities and rural areas or in their own countries or between one country and another in the various regions of the world and including a boundary between the North and the south.

That educational interventions are important pivotal when responding to this challenge, and should provide the education students Palmeart academic and technical necessary for many developing countries seeking to move away from subsistence farming and a single can teach firm in the concept of participation that helps guide the citizens to make informed decisions on spending Thrift, investing in education and health care.

In order to address these and other challenges need to be on the societies of the world will revive its institutions and governments, and it should enhance community life and education Perhaps one way to achieve that participation must either be the subject of a major educational all at different levels. But the first important step which must be taken by governments is to provide the money necessary for the public education system, which starts younger siblings at an early stage of their age and extends to adult facilities. The public education system is under the expansion and reform must be centered primary and secondary schools effective.

In order to meet the challenges it has to be for all levels of education in a significant impact on students, which should be effective education, and effective education, which can be a society of confronting the challenges ahead should be on three major issues. Societal goals of the mission to create an effective education system for all students that includes students from underprivileged economic sense and that helps all students to achieve higher levels of academic achievement.

Should educators in different levels to put a plan for curriculum and programs can take into account the best they could be taken into account of individual differences in their interests and talents and motivations and cultures on educators to understand that academic achievements and intellectual activities in education can not be placed in any way on the social, emotional and moral developments.

The community leaders to make education contributes to an effective contribution in the face of challenges, including the creation of learning communities because the school alone is not enough with the role of head of public schools and the development of learning communities provide education for the entire population and build a more cohesive communities with a network of civil associations with community schools.

And it requires education in the future:

• new integrated curricula with effective interactive multi-media.

• multimedia interactive being developed by leading international scientists.

• appropriate communications and computer technology standards to the level of each student to stimulate innovation and research.

• Change the textbooks and a wide range of academic software and hardware PC and discs programmed educational television.

• new roles for teachers and a new in-service training and outside to gather knowledge.

• strong partnership between home and school.

• a new way to assess and identify their abilities and preferences.

• diversity education away from traditional forms.

The schools of the future require a balance between the general objectives and these requirements:

• equip students for the future of the product.

• care of intellectual abilities of students.

• trained to understand their culture.

And these schools are in:

• to decide priorities and programs with respect to the curriculum within the framework of the year and a strong curriculum and standards.

• to decide its priorities with regard to spending.

• carry out their own students at all levels calendar and prepare annual reports on the results.

Schools of the future includes five elements internationally proven it's crucial for the successful schooling:

• active focus on individual schools.

• Flexibility in Alastmaah to the needs of students of the same school.

• the local community's commitment to the school education system.

• realistic standards.

• accountability to the local community.

Future studies in science

Most of the current and future studies, which ended the year 2000 and some little when even 2025 and is required to conduct such studies:

• means of special programs for the forecasting process.

• need to many experts, technicians and programmers are not one of competence Among them mathematics, chemistry, physics, life sciences, scientists and others have been fired them knowledge engineers 0 complex thought) or tank thought think tank.

In light of this, it is possible that future studies in science is divided into three types, and this arbitrary division Snstamlh for simplification only, since it was difficult to separate the three types of single studies on this and other species:

• Studies that rely on to predict what will happen in a particular scientific development like that would happen in the field of genetic engineering, computers, chemistry, mathematics and physics field.

• comprehensive future studies that rely on intuition, which looks at the impact of the current operation achievements on the future of humanity, whether on one level or more.

• comprehensive future studies that rely on statistical information detailed within the sports programs or computer models are used in specialized areas such as forecasting energy consumption.

Studies that rely on scientific prediction

These future studies focus on several areas of process Kalkemiae, physics, life sciences, mathematics, sparked by actors on the scientific and industrial institutions and other such studies and indicate to the rule-based industries science within the terms of reference of renewable energy and genetic engineering techniques and life industries and electronic industry computers, transportation, telecommunications, space and materials science.

Researchers from these studies are expected, for example, the use of satellites for the transfer of power by converting solar energy into microwave waves can be transmitted to ground stations and then converted back to energy can be tapped. Either in the field of genetic engineering and remember a lot of scientists and future perceptions following in this area:

• clone of genius or disease-free rather than relying on coincidence the advent of births may not be a genius.

• production plant roots produce potato tubers, while the plant itself produces Tmata.

• producing strains Kruvilih produce any human being does not depend on its food on animals but on the solar energy and carbon dioxide, water and some inorganic elements.

The electrons in the field and in the computer industry:

• The computers in the future to help the doctor make the necessary in his examinations such as blood analysis and other tests and then diagnose the disease and provide necessary medication to patients.

• The entire computer in a private houses inhabited by the management of those with disabilities Kadah home and turn on the TV and cooking and catering by robots.

In the transportation and Telecom rights in the future will be able to do with:

• lapses without going to the university, but in front of his computer.

• attendance to the scientific conference thousands of kilometers away and participate in the discussion without having a physical presence at the conference.

Either in the field of space scientists Vitenbo do a great achievements on the level of outer space, including:

• the establishment of satellite settlements on Mars and the moon and the atmosphere that surrounded the ground.

• set up factories satellite many electronic components and produces medicines (outer space is very useful for the pharmaceutical industry is pure).

• freezing of embryos and put them on spacecraft to form generation and directed embryos from freezing conditions to form the second generation and how the human continues to send his grandchildren to the ends of the universe.

Either in materials science will be able in the future of human industry materials with distinctive characteristics can be used in clothing and the automotive industry as well as the use of carbon instead of silicon in the computer industry life instead of silicon computer.

In a study on the future of the world, after four hundred years of perceptions it may be referred to the future, namely:

• severe pessimism.

• pessimistic warned.

• optimistic warned.

• enthusiastic about the growth and technology.

Extreme pessimism

This perception is supposed

• The technology would have little effect.

• that man plays a big drain natural resources.

• income disparity between the developing and poor countries will increase.

• a decrease in food.

Pessimist warned

• growing population will lead to diminishing returns.

• resource depletion.

• Increased pollution.

Optimistic warned

For this model

• growing resources as a result of technical progress.

• rising standard of living.

• sanity in the consumption of resources and the preservation of important resources.

• the end of the problem of absolute poverty.

Future Shock

In the future will create new generations get the so-called future shock according to the following scenarios:

• Future Shock acute illness experienced by increasing numbers of people (disease inability to adapt to rapid change).

• Responses future Daalh act depends on what you know in the ability to adapt.

• proliferation of psychological and neurological diseases with a lot of people.

• Technological advances may create facts on our awareness could not be absorbed.

• human future will be Lansana expatriates without roots and without confidence amid the dunes of quicksand.

You need models to study the future of many things, including:

• database includes comprehensive statistics.

• a set of preliminary studies that rely population or resources.

• adoption of specific scenarios and serve the planned targets.

• Computers with a great capacity for storage and handling information.

Most of the current and future studies completed in 2005, where required for further studies:

• technical means such as computer to store and recall the many information.

• Use of special programs for the forecasting process.

• Many experts, technicians and programmers are not one of competence.

• adoption of the current concepts of the industrial revolution (the second) that emerge from science, who founded the various discoveries.

- Chemistry contributed and developed mainly for new technology production processes.

- The foundation of mathematics and physics in nuclear fission and the invention of the electronic calculator.

- Life sciences effectively have an impact in the fields of agriculture and medicine.

- Electronics industry associated with the computer industry.

And we see the scientific progress in chemistry, life sciences, mathematics, computer sciences and other sciences is increasing every year and became the basis for most of the other sciences. It is possible that future studies in science is divided into three types:

- Studies that rely on prediction.

- Future studies that rely on intuition.

- Future studies that rely on statistical information detailed within the sports programs or computer models.

In light of this, there are four scenarios may be referred to the future, namely:

- Severe pessimism.

- Pessimistic cautious.

- Cautiously optimistic.

- Enthusiastic about the growth and technology.

Psychology

Is the study of human and animal and regulations gradient in intelligence, and called on the science of animal behavior Ethology named Pallaithologi a branch of zoology (many animals, unable to move and respond to environmental changes, feed on plants and animals), and divides Alaathologi to a branch of which continue animals Animal Communication and learning Animal neurological Alaathologia Neuroethology knew self represent study of behavior and the mind, thinking and personal to humans and previously was a science that examines the soul, mind and brain function and subdivisions and sensations and the device visual on location and the cerebellum and medulla oblongata responsible for the body and breathing balance. There are several schools of psychology, including the Alaepoukrah and structural and Padua and Freudian psychoanalysis and modern existential analysis and others.

Philosophy of Science

And it represents one of the branches of philosophy that deals with the philosophical and virtual foundations and implications of various sciences (physics, Akemiae, life sciences and humanities and social sciences) and contents that have been roads to it, including:

- scientific theories
- formulation of various scientific methods.
- scientific method.
- the credibility of the scientific arguments.
- modalities of the production of science.

Philosophy (

It believes that the term philosophy Philosophy was a Greek origin, and consists of two words (Philo-Velia) and its meaning and loving (Sophie-Sofia) their meaning wisdom, either in the Arab-Muslim heritage has found her several expressions, including science ethics. Philosophy of the most important specifications that fields wide and several kinds of sequential steps leading towards the formation of thought essential to simplify the perceptions and understanding of the humanitarian dilemmas. But they are all looking for the nature of things by using the mind. Has spread in Greece a lot of contradictory intellectual movements philosophy idealism and realism is better see the values fixed and virtues do not change, while the second sees Care senses more important than focusing on the imagination and are consistent with the ideal in the fact that the virtues fixed, educational philosophy required that the education part of the fully human existence Calvin, science and language. The concept of education is different from, for example, according to the philosophy that deals with the type, Vsagrat points out that education is the formulation of the same human being announced and goodness is believed Plato that education is consistency between the soul and body, while Al-Ghazali believes that the education her priority is the spiritual and humanitarian atmosphere. And lead exemplary education to the highest degree of maturity of the children, either natural philosophy refers to the mental preparation of the child. While education at Vtaatsour existential philosophy that man is free and subject to the inevitable and philosophy of pragmatism as vision Dewey suggest that continuous education organization of experience and adapt to the social reality. And John Dewey, one of the philosophers who pointed out the principles on which the modern concept of education and the education of them small community which comes to life and forever continuing education and curriculum must keep pace with life and the task of education is to prepare the individual for life.

Among the Arab philosophers who were interested in the problems of education such as Ibn Sina and al-Ghazali and Ibn Khaldun. Son Khaldun (1332-1406m) has been limited educational principles should be in the gradient and the transition from the known to the unknown and from easy to difficult. It particles to colleges. The European Education, which Pflasvetha including Jean-Jacques Rousseau (1712-1778m), who called for equality and return to normal life away from corruption and meet the needs of childhood. And Herbert Spencer (1820-1903m), which focused on the psychological concept of education and private psychology Tql.

The subject of philosophy is knowledge () and knowledge of natural facts (relative), facts standard (values and ethics) is the philosophy according to perceptions of Dewey, for example, the general theory of education as some mention of it humanitarian features of human beings turned into intellectual trends Others believe that celebrities philosophy are the leaders of educational thought Ksagrat, Confucius and Plato and Jean-Jacques Rousseau, Fichte, where reflect the philosophy about the conditions of education and working on Tviem and criticism of the educational process. But the fields of philosophy and topics of our time metaphysics (metaphysics) or divine science and knowledge, values and philosophy of education are to apply philosophical outlook in the field of human experience which we call education.

In other words, are the educational outlook emanating from the theories and ideas of philosophical as part of a cultural and educational philosophy way not only consider abstract ideas, but consider how to use the ideas in a better way, and lead educational philosophy to the development of channels and the principles and foundations required for our daily work in the educational teaching and practice of education and how We are learning and human growth. The show traces of Greek civilization that they worked on the allocation of Investigation extensiIdealist philosophy Idealism

And it linked to this philosophy pioneered by Plato, which is the perception of the existence of two worlds, the world of ideals (hard) and the real world (variable). The community consists of two layers, one of them thinking and other works In other words, the first class is linked to the educational framework for the purpose of access to knowledge and that requires the mind, and thus Knowledge is that link the mind be true and unchanging. This philosophy and believe in basic principles of strong belief in the existence of an independent in a perfect world the real absolute ideas.

This philosophical school follow a curriculum N constant evolution. The idealism and teaching methods are based on the basis of mental training staffs. According to Applied features of this philosophy in the educational field accumulation of knowledge approach is clear and unchanged and adoption of tools learning as ways constant teaching and exams without the use of traditional means and focus on the mind and the lack of school trips adoption and the adoption of corporal punishment and finally that learning is the focus of the education process, as well as it sees Plato and teacher Socrates (470 BC -399 BC) to fixed values and virtues do not change and that education must take care of reflection and imagination and that human nature is made up of two-spirit and body and must take account of these bilateral upbringing of the individual in society as the True. According to these perceptions education focuses on realism must prevail without neglecting the spirit of the body. And it provided the ideal philosophy for education and training of the Socratic idea generation which is based on the mind and stir to pay for self-search, addition, Vehtm Education at idealists experimental spiritual philosophical issues with abstract thinking and keeping information practiced by high school.

**Realism Philosophy**

According to this philosophy of Braidha Aristotle (384 BC -322 BC) that there is only one world is the world of reality is characterized by fixed principles and the most important of the senses Care Turkao the imagination and are consistent with the ideal in the fact that constant virtues.

Follow the realists narrators approach stability depends on continuing this philosophy to discover the universe and the world and work to understand the laws in force, which includes all of the facts within a stable and constant attempts world. Accordingly Valoajpat placed on realism philosophy for Educational clear according to the following:

• follow the approach which accumulate natural, social and cultural knowledge.

• acquire the knowledge, skills and habits to prepare students for life.

• Prepare specifications teacher distinct process.

• extracurricular activities necessary realism of the school.

• the learner is the focus of the educational process.

• extracurricular activities is an important part of the curriculum.

• curriculum of a set of facts discovered by scientists consists.

• The use of programmed instruction machines.

And the fact that the realist school Qdj filed down the senses and inflicted philosophy and ideals of meditation and fantasy to reality and the senses so that scientific knowledge and curricula astronomical and mathematical Kalaom Square widened. And John Locke (1632-1704m), one of the pioneers of realism philosophy and endorsed it by his conviction gave critical thought ample room and the experience and reality and senses no example and abstraction based on knowledge and science he sees that the child a blank page draw Storha of fact, as well as it has added Luc need to study natural phenomena, along with other math and science.

The Komenus (1592-1670m) realist sensory Making of the image of the most important methods of child education in schools as well as facts enthusiasm and believed to be the founder of the first special education child of freedom. And the development of sensory approach to teaching and took care of the physical and moral education to both.

Finally, the real reached a multiple convictions, including:

• established curriculum experimentation and exchange of scientific and methodological doubt (the road to see the existence of God, the real life).

• encouraged the learner to observe natural phenomena in an orderly manner.

• called for acts of mind in the analysis and the independence of the senses.

• This school did not distinguish between the world of ideals and spirit and the world seriously and reality.

• did not elaborate on the mental meditation.

• focused on professional education.

• invited to link educational curricula Balhajiat life.

**Pragmatic philosophy Pragmatism**

The roots of this philosophy go back to ancient times and the writings of Hrakulais (535-475 BC) and Undtlaan (30-95m) and Harzubayrs and William

James (1842-1910). While contemporary pragmatism linked to the New World (the second half of the nineteenth century and early twentieth century) called pragmatism generally Baladhatih and pragmatism, development and operation.

Of the principles of this philosophy that education is linked to the life and education to community service philosophy with Alentalm give a measure of freedom. And the human organism adapts to the environment and that the relative biological world is in constant flux and the truth is absolute and variable society and democracy to decision-making. John Dewey has established the features of this philosophy.

Pragmatic philosophy role in the development of teaching methods and by improving the traditional way and in a manner trial and error, or experimental way. John Dewey School-based Albergmteh philosophy has historical intellectual revolution against traditional schools, which focused on the information nor firmly believes in values, but the relative morality is renewed according to the convictions of the community.

The teacher in the philosophy of pragmatism he does not teach traditional materials in a systematic way and the teacher moves from idea to another in a sequential manner and deal with all the idea on the grounds that it in itself and suggests future problems for his students. And that the values vary depending on what they put time and society and the convictions of the individual.

Education at the pragmatic philosophy is not broadcast knowledge to the student in order to process knowledge, but to help him cope with the social needs of the environment and was able to stir up the student as required by the forces of social attitudes. And rely perceptions pragmatism philosophy on building curriculum, including confirmed by the student on what he wants from him, and using the student Alqrah, writing and arithmetic as a means rather than targets, and the curriculum is interested in the facts relating to the nature of the child, also emphasizes the pragmatic philosophy to develop vocational education, natural sciences, while human studies and languages have secondary importance.

This philosophy has paved the Educational Progression some progressive educational and considers her application and was the most prominent pioneers John Dewey at the beginning of the twentieth century (1920-1945m), which came a revolution on a massive and cons of the traditional system produced a key concepts including:

• Coordination with the environment as the best environment for education.

• reliance on textbook and curriculum need to the principle of the freedom and flexibility and experience.

• teacher prompt administrative.

• take into account individual differences for each learner.

• requires the student to ask about the surrounding environment.

• the school has to cut off from the outside environment and that depends on the style of problem solving.

• variable values and science.

**Islamic philosophy**

There are major Islamic philosophy perceptions reflected on education, and these perceptions that the Quranic verses explain the nature of the universe and man, knowledge and values, and everything in the creature universe to God and the changes that occur in the universe is governed by paths and rights in Islamic philosophy has marked special was the gift of the Creator stature over the place that was given for all the other creatures in the universe create for his service.

Education is happiness in this philosophy and that the child has the bright page where there is no hindered Education In practice Education is starting science Quran and Hadith beyond are going to learn science has refuted Ghazali arguments philosophers in his book, The Incoherence of the Philosophers while defeated Ibn Rushd philosophy wrote The Incoherence of the Incoherence Canadian and longer of philosophers who emerged in Islamic history.

## Natural philosophy Naturalism

This philosophy believes that man is inherently good and that is what human spoil the society and its institutions, education, her goal is to provide an opportunity for the growth of the natural child, as the jam Ptsourath negative and positive learner childlike nature. This philosophy also believes that the senses sources of education and outlets for the development of thought and not the role of cognitive balance.

Natural philosophy appeared to calls by reference to the educational activities that are aligned with the natural laws of indigenous cared Rousseau (1712-1778) these ideas and Crystal natural philosophy and Allamadrsah cemented the role of nature in the development of children in terms of:

- frees the child from school and classroom activities.
- meet the child's needs and the need to remove obstacles facing it in accordance with environmental requirements.

Rousseau was one of the first to call for self-learning and the nature of the most important educational principles that goes with it, and that women found the man to please, and this philosophy advocated the need to respect the individual and the protection of children from societal pressure differences.

## Existential philosophy Existentialism

Kdeckart existentialists and Sartre believed seriousness and absolute freedom and human a philosophical vision of human existence, responsibility and human nature and the world, knowledge and values existential emerged in Europe after the First World War (1914-1918) in Germany and then in France. It is believed the existential education as perceived by them to human existential indoctrination and does not accept the promotion of education for the existence of rights and Arts, music, philosophy and the arts of the general requirements, dialogue and debate and the basics of teaching methods and programs stable unacceptable.

### Allamadrsah philosophy Deschooling

Each of John Holt (1992) and appeared to Cash (1994), the pioneers of this philosophy, which depends entirely on educational institutions Some believe the philosophical foundation of this school is Gandhi's theory of education, especially in environmental education and the foundations of this school:

- learner be close to its environment and interact with it.
- learner learns from his peers.
- To achieve the same learner.

It is believed the others of the pioneers of this school that limiting education within the school not feasible and therefore extracurricular activities requires an important role in our schools and to be environmental education general philosophy in all the curricula, not scholarship material separate philosophy Allamadrsah is the return of simplicity, inclusiveness and return to the educational roots

### The philosophical concept of the mystery of life

There are three philosophical concepts of the world and developed as a result of human intellectual effort concept of spiritual and physical concept realistic and unrealistic concept of the divine. It can be evaluated and try to rush to one of them or the formulation of the concept of the center between them. The conflict between the divine and the physical manifestation of the conflict between idealism and realism and that the philosophical concept of the world one of two things ideal concept of the physical and the concept does not correspond to reality at all, realism is not according to the materialist conception as the ideal is not the only thing that is opposed to the materialist conception, but no concept last realism Divine is a realistic concept. The concept of the divine world does not mean cutting out natural causes, or to rebel against everything from facts and sound science but it is a concept that God is a deeper reason. Physical back door is in an ideal spiritual claim to the area and either spiritual concept in the divine way of looking at it is the reality.

As for the scientific field, there is no God and material Valfelsov whether divinely or materially believes in the positive side of the flag, there is no issue in the scientific philosopher, my God another material, but there are two Filsvetan and in conflict when it was a matter of existence beyond nature.

Valalhei thought to the fact that just about art, any outside experience and physical deny it is believed to be natural causes revealed by experience and extended her hand science is the primary reason for existence. Divine and believes that the human spirit and (I) with abstract art and perceptive and thought phenomena independent of nature and art.

The nature of evidence that can be presented by my God is mind, not unlike the experience of direct material traditionally regarded as evidence of the experiment on your concept, it believes that the concept of divine or metaphysical issues in general can not be proved experimentally. Materialism need a guide on the negative side, which distinguishes itself from the divine and it's the direction of philosophical Kalalheih, because science alone does not prove that the material of the concept for the world to be a material process but whatever undisclosed knowledge of facts and secrets in the world of nature leaves room for assuming the highest reason above article. If we look at a set of basic concepts about life and the way of thinking which can then be first addressed to the theory of knowledge and secondly for philosophical concept of life and when studying the theory of knowledge can focus on the reliance on mental way of thinking which represents the necessary knowledge over experience as well as the study of human knowledge value on the basis of logic mental not physical.

Epistemology represent the mystery of life and perception of ratification which reflect the cognitive The former is a presence, such as heat, light and sound, while the second (ratification), which represents the recognition that the heat energy source, for example, the sun and other concepts. A number of theories have been dealing with these perceptions, including:

• recall theory

The (recall of past information and that the human soul exists independently of the body). They isolation from the article, and can correct some of the mistakes of this theory, which represents that the soul is not something that exists in the abstract before the existence of the body, but as a result of the core material movement.

• mental theory.

Put forward this view by (Descartes) and (was) pointed to the existence of exporters perceptions firstly sense (heat, light, taste, etc.) and the second instinct (the human mind has the meanings and perceptions did not emanate from the common but are fixed at the heart of instinct). Vlachtlavat with recall theory is the fact that the source of the sense of understanding of the perceptions and not the only reason. And drawbacks of this theory is to return the entire cognitive sense.

• sensory theory

Sensory theory is based on experience and common sense in this theory is the infrastructure that is based on the base of the human perception.

Therefore, the theory of knowledge can be used to follow up the mystery of life and access to him, in the perception of objective value of life as expressed in the presence of thing Mdarkina and certification is an objective which reveals the presence of mind to imagine life feel the need to believe his health. The world is the physical and life combines the first place and other manifestations of heat and starts applying the principle of my mind it is necessary (the principle of the attic) view (that the cause of each accident) and hence the principles of mental public the basis for all scientific facts, as well as the principle of proportionality.

The conflict philosophical determines the direction of the secret of life, while requiring not be confused with the scientific material and article philosophical, philosophy can not and does not think the split unity scientific instruments and means which owned rights, this case of the right of the flag alone. And that the philosophical concept of the material consists of material and image, he has the scientific material can not be the first principle, because it involves the installation of a dual between the article and the picture. And where they can not each of the picture and article that exist independently of the other, that there should be an actor get ahead of the process of installation and according to that we are going to the relationship between art and some scientific systems according to this perception:

• material and cell

The second half of the nineteenth century between the serious findings life witnessed by the theory put (cell) at the hands of Matthias Hlibdn in the plant and then to de Theodore Schwann in the animal. As the plant and animal bodies made up of cells, the evolution (cell theory), development is a critical stage in the advancement of the science of life Atomic Kalnzerah in chemistry.

The physiology of man as one of the branches of the science of life refers to the amazing facts explaining actor reason that reflect the greatness of the Creator and accuracy of the details and secrets, then the digestive system (the greatest chemical plant in the world) what is creative by the styles of different foods analyzed chemically surprising and distribution of food Valid equitable millions of living cells. Due to those of living cells to justify the issue of causal efficacy mystery of life that fill self astonishment and admiration cell while adapting to the requirements of place and circumstances.

• art and science of life

The science of life when the last of the big secrets, a mysterious secret of life, which fills the human conscience, satisfied the divine sense of fear and faith and established it. Collapsed when the depth of self-nascent theory, but the unequivocal scientific experiments, demonstrated the invalidity of the theory of self-generated. Article foundation has been studied in the science of life and the idea has spread mainly elements and the idea of atoms former core of the universe be second nature to the material elements consist of a central core revolves around the core electrons (negative charges) and the nucleus containing protons and neutrons. Then attempts were made to switch to a purely material energy, any electric charge. Any disarmament physical character of the final element in the light of the theory of relativity of Einstein. As it decided that the body mass relative and not fixed it increase the speed up and says Einstein equation energy = mass of the square × speed of light and mass = energy ÷ speed of light squared. As a result, corn, including of protons and electrons is not really, but equal energy. It appeared in various forms and multiple images. Article has been converted into energy and energy into matter.

What follows from that put the original material to the world the reality of life and one common appear in various forms, and the physical properties of compounds cross. Fajasih liquidity occasional water and not self-evidence that it consists of two simple as possible and in the secretion of these two elements from the other. Virgaan to gaseous condition and status of water disappear completely.

The properties of simple elements themselves, are not self-rule but are the qualities of a cross between common article all simple elements. The material such as mentioned has become the light of past facts recipe occasional Also, they are not transgressed be from energy colors and philosophically, the assumption in the world of art life as the reason for the denials are higher, such as lack effectiveness as well as him.

• Genetics and Article

Mendel discovered the key principles of heredity and passed him by subsequent scientists. As it concluded after Tsoaj successive generations of split peas that back plant inherits the characteristics of advances in accordance with a mathematical formula that can be secretly for life and named after then Mendel's laws and then genetics was born at the beginning of the twentieth century after it was re-expose the principles of Mendel renamed hereditary Mendelian. Followed by many variables changed from the traditional qualities of the science of life to taxonomic qualities and stabilized Mendelian genetics and then rolled up on the basis of discoveries that have been transferred from the traditional life science formula to formula description and classification.

The decline then Darwinian concept that depends on the basis of the theory of evolution that the changes and attributes that can be obtained as a result of practice or Balvaal with ocean can be transferred by inheritance to his descendants. The trend towards the emergence of species hypothesis by mutations Khdot some aspects of the sudden change in the number of cases that called on the assumption that the animal diversity of mutations grew up like that, and some of these changes may be inherited.

After that he moved the science of life of the traditional formula, formula description and tradition, classification and manifestations of organic evolution and modalities of the cell and in its entirety to the attention of the microscopic life that focused on exploration in the nucleus of molecules and chemical structures. They are a mass of spherical or oval material that looks heavier and thus around. Then it emerged that in the nucleus clusters of tiny objects membership figure they called chromosome or chromosomes containing genetic factors mentioned by Mendel he released on each factor word Jane. Valkhalih contain genes and each type of neighborhoods particular number of chromosomes in every cell of the human body and forty-six (46) except chromosomes in the female reproductive cell (egg) and sperm in the male sperm containing both of them (23) chromosomes. But there is one chromosome in a group of male sperm chromosomes decide the sex of the fetus generated from a fertilized egg, it may be X, Y, while the egg Vthtoa between Krouomusumadtha on chromosome X. not only effect of these two Alkromusuman Altanaslin to determine the sex of the newborn. But their genes also determine the genetic characteristics of a male and a female, then it found that the nucleic acids present in the nucleus of cells issued instructions for growth and break apart and there are two, one of them the D.n.o. The second R.n.o able Creek and Watson (1962) develop a model of acceptable for this installation is composed of two bands of units Nclaiutaydah (with four bases adenine, thymine, Kuanyin Saitosin) arranged in a corresponding sequential and such a specific genetic model conveys information to the R.n A., which controls the formation of proteins.

• Material development

There are different views on the development of the mismatch that the organisms in all its forms and types of fixed and does not change, but some of them is not satisfied with the validity of this opinion, and expressed the possibility of change in living organisms, or that living beings are not static but constantly changing, depending on natural conditions prevailing. In follows a number of views on evolution with theories that have been presented each gained continuity and the other stopped until:

• Aristotle

Showing conceivable according to the following:

□ that both humans and animals with a single installation consists of natural degrees and non-living plant and animal organisms and low-lying marine animals and finally humans.

□ that the high-end living organisms which represented human can not arise from the low-lying objects, but found originally on this image.

• Haraclt

This perception of the Greek philosopher builder dialectics that everyone and everyone is being changed.

• Safa Brothers

He said Safa Brothers to the evolution and the doctrine of evolution that the worlds of animals, plants and inanimate objects one separated by the borders of another coup minutes.

• Lamarck

Lamarck that the organic and inorganic world is constantly changing and says that the impact of the elements and the mating and the use of the member or neglected. And high-end living organisms originated from low-lying objects over a long period and that the low-lying and slowly turn things emerged from self-solids, in the period of development. This development is based on the use and neglect and transmission of acquired characters and genetically including the emergence of new members in the body commensurate with the requirements. Contemporary animals and divided into six grades according to the impact of the Creator.

• Darwin

Darwin believed according to his theory, the factors that prompted the organisms to evolve, heterogeneity (variation) of natural selection (survival of the fittest) industrial franchise and conflict to stay in heterogeneity, differences in returns even in members of the same species of plant or animal to the environment and self-willingness to shift and use negligent or members consider Darwinism that human represent the final episode of a series of episodes beginning of the only cell organisms, which has been grappling with the forces of nature in order to stay In order to access forms of the most complicated complex, life is as convinced Darwin for the strong and smart and death for the weak and the survival of the fittest. Through Darwinism launched many of the foundations for the development of new lifeve philosophical basis for the study of human behavior.er). Concepts and the possibility of life science and intellect to decipher the mystery of life

There are life concepts on the table in different formats reconsider the way to clarify the mysteries of life, including the cell and cellular technologies and techniques of textile giant molecules. It can also ask the secret of the emergence of life in this area where some scientists believe that life appeared on Earth around 2000 million years ago that the Earth has been more than 2000 million years of life free. Some have said that life must have originated from the water and others argue that the air over the water flexibility and shift and others recall that life is generated from mud, while Hrkulait that continued believed Nara universe has been able to turn into air change to the water and to the last crusty then return dry to water Vhoa lighthouse. The Ambdoukls have come to the theory of the four elements to ensure that all objects are out of soil, water, fire and air. Accordingly, the intellectual side of these theories can be deduced from the scientific perceptions that we have mentioned it can keep them after the four elements to the hypothesis by breeding that have been fabricated intellectually by Razi First and Second Pasteur.

Human has made several masterpieces and intellectual effort and a pilot confirms his position of this universe has tried to link the theoretical idea of experimental self-perception and trends was confirmed that the life does not arise only from life as well as that of divine truth still play a significant role.

Body study in its early stages depend on the microscope and chemical analysis and other techniques developed later The former allows the researcher that tells minute details, either the arrangement of atoms can not know this technology, while the chemical analysis refers to compounds that make up the body and the elements that comprise these compounds.

Later it developed several techniques including electron microscope where the Alusa became obtain precise details of the body for a few minutes and to reflect the composition elements in the body and the other technique called X-ray, which is studying the arrangement of atoms in a lot of life vehicles.

The radioactive isotope techniques have continued as a tool useless in the search for the secret of life which is in addition to a means for the treatment of researchers have made hundreds of radioactive isotopes, generated from non-radioactive elements in nature including sodium, sulfur, calcium, chlorine, copper, cobalt, gold, iron, mercury and silver. One of the important uses that have been adopted photosynthesis C14 and follow-up process as well as in the development of new analytical technique called immune radiation and in which the concentrations of compounds found in very small amounts account, especially hormones.

Several techniques, including chromatography separation and deportation electricity used The former used to separate many of the life of vehicles and purification and first uses and which is still the mainstay for the study of the structural composition of the first proteins and amino acids thanks to the use of the positive ion equivalents.

**Organisms secret life**

When studying living organisms and adopted a fraction of the mysteries of life can be touched on a number of concepts Kalmthalah that reflect the objects that this life is a facts exist independently of feeling and perception, or is the color of our thinking or our perception as far as it comes to realism philosophy can ask another question The position of the organisms on the borders of this philosophy as a sensible Vtkon is the general cause of all the phenomena of existence and the universe or bypassed to another reason the deepest represented by a physical field and another spiritual and last beyond the organism as to the cause over the Spirit is the concept realistic Divine (Divine password) and this concept does not mean dispensing natural causes or rebel against something Facts sound science, a notion that God is a reason he is called to a deeper knowledge of the wider field to explore the nature and continuation of which living organisms.

Valfelsov whether physically or divinely believes in the positive side of science Kacetkchav unit life of organisms with the knowledge that is not a scientific issue philosopher in God and another material.

Agassiz was introduced the year 01 858 m) opinion is that every kind of living organisms create special an act of creative power. This view is consistent with the opinion of all Razi and Pasteur and they settled on that every living creature that has to be generated from a living organism like him. Herman also said Erhard Brichter has stated that every neighborhood eternal nor generated only from the cell.

**Cell of the mysteries of life**

If watched for example the gastrointestinal tract in humans, which represents one devices physiology of man, we find the greatness of the Creator and accuracy in the multiple details of the various secrets, for example, we can imagine being a laboratory chemically developed where being different methods of analysis for food, then distributed the food fair distribution to millions of cells which make up the human body and involving the secret of life and that fill self astonishment and admiration which adapts according to the place and circumstances where these cells form different geometric textile technologies mixed organic cell, it is in the digestive system nearly two hundred thousand interaction within 24 hours, some of which makes the heart muscle shrinks millions of times and have fun during the whole year tirelessly and how to get Ateltaqh necessary to think, movement and speech, including what the disposal of waste and toxins within the body looking at the cell which is the approach one of the secrets of life perceived to be adapted according to the requirements of the position and the circumstances and noted it was one serious Almtkchwet life, which saw establish cell theory at the hands of Sheldon and Schwann then considered cell theory serious stage in the progress of the science of life Atomic Kalnzerah in chemistry, or that the cell in the body of the living Kalafrad object in the communities or the living cell behave especially with specific work ways that can cause a particular job as a technology, it has become a cell as a plant or a chemical plant accurate. Neural Valkhalih example behave as a system Kahrokemiaia Alkimacaiaiah energy into electrical energy and electrical energy into mechanical energy or kinetic energy. As the industry some cells produce hormones and other life defense articles attacking their products every intruder and there are cells in the process of purification and filtration cells secrete and absorb movement and other commands and signals that each cell as a plan of action does not deviate from them.

And originate from one fertilized cell differentiated tissues and different members of different functions and is embodied in the end in the bones, muscles, cartilage and twigs and skins and vessels and the blood.

The private living cell technologies came, for example, nerve Valkhalih Kahrokemiaia system can transform chemical energy into electrical energy and last into mechanical or motor. Or cell as a laboratory or may become a chemical factory in which hundreds of complex chemical processes take place and there are cells specialized in manufacturing and other hormones to produce weapons Baaologih cells filtration and purification.

It is clear from tissue that arise from the fertilized cell one then divided into thousands of similar cells is embodied in the end in the bones, muscles, cartilage and twigs and skins and vessels and the blood, where these tissues are formed in the early embryos and mutate to organs and systems stand-alone, but they are integrated in the performance of Rsalatha.

It is clear from the information about the cell that there is a strange force managed in living cells, the world (Walker), professor of Velsjh Plant says (that the cell components group in a strange way emerged through life. And still, the researchers are unable to cell making blood of know the exact components and this so-called the mystery of life) ..

In light of perceptions of genetics, that the inheritance of the individual left untouched nuclear material of living cells for reproduction and genetic traits due to microscopic precision parts which genes. In light of this decline perception Darwinian view that the changes and the qualities that earned by the animal as a result of experience and practice, or to interact with the ocean or type of food can be transmitted by inheritance to his descendants. Consequently, the researchers moved to the emergence of species is by mutations by observing some aspects of the sudden change.

□ strategy Science and Technology and Education

• require formatting and identify Futures Education and planning the following:

- Education and its relationship to society (social demand).

- To provide the country hand-eligible high (the need for labor).

- The rate of return account (another indicator of the surplus or rarities).

- The role of education in meeting the needs of the labor market.

• Technology and Education

Dealing with this relationship:

- Major technological changes and its effects on education such as information and communication and improve communication (Ansan- family).

- The effects of these technological changes on education through changes that challenged on the economy (the contents of the professions and methods of practice professions).

- The impact of technological change on education through change education tools.

- The impact of technological change on education through curriculum change

- The impact of technological changes on education through the improvement of educational research.

- The impact of technological change on education through computers, like the computer in education and the Internet in education and learning as a computer.

- An integrated policy for the educational systems interlocks Ktdakhl science and technology, education (curriculum, research, and scientists Ttaiwiralbagesan) and the overlap of science and technology and the labor market.

Strategic alternatives for the development of science and technology in the education sector

There are a number of strategic alternatives that pose a wide choice in the use of science and technology which is focused on the potential offered by science activities and technology, including the alternative continuity rationalization and alternative reformist and most importantly, where are some of the current school aspects of renovation and planning education and management, which depends on:

• Increased use of modern educational methods and technologies in education.

• the introduction of new forms of technology vocational training.

• curriculum reform.

□ strategy for the modernization of education and training

When developing a strategy for modernization of education and training requires a lot of reforms:

• A description of how to develop a course for this type of education.

• Clarify how to prototype and evaluation and improvement before adoption.

• diagnose the conditions of education and training in their relationship to work and the economy.

• comprehensive reform of curriculum and technology.

• reform based on the school to develop the educational institution and enable each school to improve her performance.

• human resources and professional development program to provide qualified personnel capable of achieving the ministry.

• institutional rehabilitation of decentralization for the treatment of big government and a review of the organizational structure for workers program.

• administrative system program and a review of the organizational structure for workers.

• technological development and information systems program Astkham increase of technology in education, such as distance education and self-education.

• Calendar and follow-up program and the aim is to establish a complete system for follow-up, calendar and calculator in all aspects of the educational process.

• school buildings program to provide the necessary preparations.

• Develop kindergarten program.

• reform of the educational component for Basic Education program.

**Psychological foundations**

There are a number of ideas that are consistent with the psychological foundations and contribute to the construction of the curriculum and fit with the educational levels of the learners and teachers, including:

• affected by its environment and Iraqi learner generated when it has distinctive characteristics resulting from it.

• the impact of a number of distinguishing factors on the growth of the Iraqi learner including genetics, health status and care resulting from it and the Iraqi environment of the learner in his family and his school and community.

• There are many special features of growth, which is a continuous and integrated process vary from one individual to another with overlapping and distinct phases (of the Nativity and early childhood, and subsequent Alrahqh childhood and maturity stage).

• The growth requirements and the needs of the individual reflected on building curriculum.

**Social foundations**

Iraqi society is characterized by the presence of a number of common elements, including:

• majority of Iraqi society believes in the Islamic faith community.

• Iraqi society part of the Arab country.

• Iraqi society is the history of Old New track achievements.

• Iraqi society suffers from illiteracy and problems related to it.

• Iraqi society is a component of economic and geographic distinct.

• affiliation of teachers in primary schools to the same social environment as much as possible and that this social Privacy reflected on the educational process, including the curriculum.

• Re for the study of history and religious and social education curricula and Iraqi geography teaching in proportion to the situation in Iraq and regional and international contemporary curricula.

It can be addressed through educational foundations:

• Creating public awareness of the concepts of moral and educational and commensurate political and social situation.

• not subject to the educational curricula Atjhat any intellectual or political or narrow partisan.

• Baltnci attention since the early stages (kindergartens and nurseries).

• teacher and teacher social and intellectual stature and benefit.

• schools and institutes, each of which is considered sacred campus of an educational and not the seat of partisan conflicts.

• Focus on educational spammers sophisticated and traditional.

• Strengthening of national belonging concepts in curricula and textbooks during the review to prepare for printing, especially the National Library. The expansion of Arab scientific culture base, and increase cognitive balance of the Arab community and the expansion of scientific knowledge which used in various basic science, and localization of these sciences in the Arab culture, and the balance between them and the Humanities theory, and the vaccination of Arab scientific activity, including in other cultures and civilizations of knowledge and science through translation and localization, transport and supply the Arabic language, and enriching the lexicon of the Arab scientific. introduction to the history of scientific thought in Islam, Ahmed Salim happy, the world of knowledge, the National Council for Culture, Arts and Letters - alkwyt 1988. Ahmed Salim happy (introduction to the history of scientific thought in Islam), 1988, a series of books and cultural monthly published by the National Evacuee culture, arts and literature, Kuwait. Ali Ahmed Shahat (development between science and religion theory), publisher Khanji Foundation in Cairo.

- Arab scholars eating rules curriculum scientific research philosophy reflect the intellectual grasp inductive The curriculum of the Note and the experience and assumptions, Valastqra is not only a stage in the research approach and there, for example, measurement and premise and Sports to address the research value of the intake approach. Ibn al-Haitham measurement methodology has been applied in the phenomenon of light and others in the field of medicine, astronomy.

- This has had a Geber distinctive role in scientific research methodology steps identified four: -

- 1. preview note.
- 2. hypothesis pen.
- 3. scientific experiment.
- 4. determine the scientific concept of the phenomenon.

- Eating contemporary approach, researchers distinguish between the concept and the associated logic, and the concept associated with deprivation, The curriculum is already linked to related logic, experience and the associated determined in accordance with the laws and additional data.

- On the other hand, does not deny what his senior scientists from the likes of Socrates, Plato, and Aristotle in the development of scientific research methods, in various humanities, including natural, even if they did not know this name in those eras, but he was present content and significance.

- Combined with all the peoples and civilizations, each civilization and attributed the characteristics and features, and excludes any civilization in the area, and elevates the status of one over another, he says, for Greece: The Greeks were with ingenious minds ... and they were the owners of wisdom, and they did not coin.

- The scientific research, is the product of the accumulation of civilizations throughout history, even if some of the hallmarks seemed deeper than the other, but mankind has not still learn from each other, each generation who already take. Were it not for this accumulation of knowledge, what king of the human mind not to be reversed to his childhood primitive.

-

ontemporary Islam is not known for its engagement in the modern scientific project. But it is heir to a legendary "Golden Age" of Arabic science frequently invoked by commentators hoping to make Muslims and Westerners more respectful and understanding of each other. President Obama, for instance, in his June 4, 2009 speech in Cairo, praised Muslims for their historical scientific and intellectual contributions to civilization:

- It was Islam that carried the light of learning through so many centuries, paving the way for Europe's Renaissance and Enlightenment. It was innovation in Muslim communities that developed the order of algebra; our magnetic compass and tools of navigation; our mastery of pens and printing; our understanding of how disease spreads and how it can be healed.

- Such tributes to the Arab world's era of scientific achievement are generally made in service of a broader political point, as they usually precede discussion of the region's contemporary problems. They serve as an implicit exhortation: the great age of Arab science demonstrates that there is no categorical or congenital barrier to tolerance, cosmopolitanism, and advancement in the Islamic Middle East.

- To anyone familiar with this Golden Age, roughly spanning the eighth through the thirteenth centuriesA.D., the disparity between the intellectual achievements of the Middle East then and now — particularly relative to the rest of the world — is staggering indeed. In his 2002 book *What Went Wrong?*, historian Bernard Lewis notes that "for many centuries the world of Islam was in the forefront of human civilization and achievement." "Nothing in Europe," notes Jamil Ragep, a professor of the history of science at the University of Oklahoma, "could hold a candle to what was going on in the Islamic world until about 1600." Algebra, algorithm, alchemy, alcohol, alkali, nadir, zenith, coffee, and lemon: these words all derive from Arabic, reflecting Islam's contribution to the West.

- Today, however, the spirit of science in the Muslim world is as dry as the desert. Pakistani physicist Pervez Amirali Hoodbhoy laid out the grim statistics in a 2007 *Physics Today* article: Muslim countries have nine scientists, engineers, and technicians per thousand people, compared with a world average of forty-one. In these nations, there are approximately 1,800 universities, but only 312 of those universities have scholars who have published journal articles. Of the fifty most-published of these universities, twenty-six are in Turkey, nine are in Iran, three each are in Malaysia and Egypt, Pakistan has two, and Uganda, the U.A.E., Saudi Arabia, Lebanon, Kuwait, Jordan, and Azerbaijan each have one.

- There are roughly 1.6 billion Muslims in the world, but only two scientists from Muslim countries have won Nobel Prizes in science (one for physics in 1979, the other for chemistry in 1999). Forty-six Muslim countries combined contribute just 1 percent of the world's scientific literature; Spain and India eachcontribute more of the world's scientific literature than those countries taken together. In fact, although Spain is hardly an intellectual superpower, it translates more books in a single year than the entire Arab world has in the past thousand years. "Though there are talented scientists of Muslim origin working productively in the West," Nobel laureate physicist Steven Weinberg has observed, "for forty years I have not seen a single paper by a physicist or astronomer working in a Muslim country that was worth reading."

- Comparative metrics on the Arab world tell the same story. Arabs comprise 5 percent of the world's population, but publish just 1.1 percent of its books, according to the U.N.'s 2003 Arab Human Development Report. Between 1980 and 2000, Korea granted 16,328 patents, while nine Arab countries, including Egypt, Saudi Arabia, and the U.A.E., granted a combined total of only 370, many of them registered by foreigners. A study in 1989 found that in one year, the United States published 10,481 scientific papers that were frequently cited, while the entire Arab world published only four. This may sound like the punch line of a bad joke, but when *Nature* magazine published a sketch of science in the Arab world in 2002, its reporter identified just three scientific areas in which Islamic countries excel: desalination, falconry, and camel reproduction. The recent push to establish new research and science institutions in the Arab world — described in these pages by Waleed Al-Shobakky (see "Petrodollar Science," Fall 2008) — clearly still has a long way to go.

- Given that Arabic science was the most advanced in the world up until about the thirteenth century, it is tempting to ask what went wrong — why it is that modern science did not arise from Baghdad or Cairo or Córdoba. We will turn to this question later, but it is important to keep in mind that the decline of scientific activity is the rule, not the exception, of civilizations. While it is commonplace to assume that the scientific revolution and the progress of technology were inevitable, in fact the West is the single sustained success story out of many civilizations with periods of scientific flourishing. Like the Muslims, the ancient Chinese and Indian civilizations, both of which were at one time far more advanced than the West, did not produce the scientific revolution.

- Nevertheless, while the decline of Arabic civilization is not exceptional, the reasons for it offer insights into the history and nature of Islam and its relationship with modernity. Islam's decline as an intellectual and political force was gradual but pronounced: while the Golden Age was extraordinarily productive, with the contributions made by Arabic thinkers often original and groundbreaking, the past seven hundred years tell a very different story.

**Education in developing countries**

The recent independence of some countries led to the social and economic backwardness of this state with the back of a number of obstacles that disrupted the education process. What Third World countries continue to suffer illiteracy and backwardness of women's education, poor education opportunities for all, as well as the low spending on education to succeed the economy and the deterioration of productivity and exacerbate unemployment.

Find Iraqi educational philosophy

• Iraqi sources philosophical educational goals

• Iraqi educational philosophy

**Submitting**

The educational system one social system components and derives its philosophy from general philosophy of the community, and can be educational philosophy of the Iraq situation when clarified year social philosophy of the country, and the nature of the composition of society and the multiplicity of intellectual trends when his sons note that the clarity of the social philosophy of the country does not mean dependence on specific social philosophy being adhered to before all other aspects of social activities, including the educational system. The status of educational philosophy in Iraq in accordance with the law legislation can be acknowledged by the legislature in the form of plans and general rules and principles of dealing with aspects of social and educational life.

Iraqi sources educational philosophy • Islamic faith

• Iraqi National Iraqi and personal development

• political and social conditions

• contemporary scientific trends

• global social and economic trends

• contemporary cultural trends

• Iraqi legislation (constitutions and laws)

Iraqi sources educational philosophy

Iraqi educational philosophy in this area depends on the sources can be selected according to the following reflected on finding educational philosophy:

• Islamic faith and other religions.

• The development of the Iraqi National Iraqi figure.

• Arabism, nationalism and heritage.

• political and social conditions.

• global social and economic trends.

• Iraqi legislation (constitutions and laws of educational, etc.).

• contemporary scientific trends and their impact on educational goals.

• contemporary cultural trends.

Islamic faith

It is one of the important sources of Education learned from them when setting its goals, plans and can derive the educational goals of Islam, according to the following:

• Raising the Islamic faith not only in the Islamic era on the mental side, or religious but was integrated breeding bother to mankind in all its aspects.

• honors human reason and science and administration, God says about the virtues of education ((God raises you who believe and those who were given science degrees)) (arguing state 11) and the Almighty said ((Are those who know and those who do not know)).

• • Islam is seen as a human creature to God Almighty to the object and purpose is the worship of God and the reconstruction of the ground and direct life.

• include the meanings of right, justice and goodness and freedom in the administration and the social environment and all the principles and values that transcends humans Semitism.

• adopting the Islamic faith community to help his sons to get rid of the weak Ktaivih, regional and partisan and class loyalties.

• provides references to Islam are considered educational Anmzjh like the Qur'aan and the Sunnah of the Prophet peace be upon him by example in the identification and in the framework of the educational philosophy of landmarks.

Iraqi National

Iraqi National devote rely on the meanings and principles of action and national activity in the ranks of Iraqis through the tread and adherence to the policy of tolerance in society and the rejection of sectarianism, racism and all forms of chronic ethnic division and differentiation between single people.

And therefore it requires reference to some of the main attractions of the ongoing Iraqi national concepts of educational objectives:

• building the Iraqi National doctrine of the students on the basis of the study and insight and understanding, perception and persuasion.

• Develop a national project that aims to unite Iraq's disparate wills to be the will of one community and one begins to adhere to the standardization and gradually and constitutionally.

• Develop a unified cultural, human, social, educational and scientific development projects.

• devote meanings and principles of action and national activity in the ranks of Iraqis by upholding the policy of tolerance in society and the rejection of sectarianism, racism and all forms of the band.

• take advantage of thoroughbreds Iraqi national schools that rely on methodology and thought Progressive social and economic reform and to achieve freedom and independence for the people as a teacher Nadzen Jaafar Abu Taman, for example, but not limited to, communication with the moderation of political methodology school.

• promote the values of citizenship and Equality platform and a guarantor for the present and the future and make it with the Iraqi national foundation of the state.

• acknowledge the rights of all components, especially according to national standards and rehabilitation to full citizenship and respect for privacy.

• the development of the proposed national school and approved based on the economic and national concepts and the rights of citizenship and loyalty to the homeland.

• dimensions of religion from politics and the consolidation of national traditions and styles.

Political and social conditions

Derived from the educational goals of political and social conditions:

• the development of the factors that combine in the framework of the Iraqi society and took linked with in accordance with the Arab and Islamic world and perceptions.

• confirmation on the specifications of Iraqi society and the culture and customs of the value of thoroughbreds with an emphasis on the same Arab and international value.

• adhere to the distribution of population and diversity sectarian, ethnic and national environmental conditions as well as the Iraqi languages of population characteristics.

• Study dilemmas and challenges facing the Iraqi society.

• emphasis on social and political life and the rights of the individual and the adoption of democratic political approach.

• respect for individual and social property of the Iraqi citizen.

**National Heritage**

Arabism and nationalism Mdlolan important supposedly taken into account when formulating educational goals and harmony in accordance with Islam and Iraqi patriotism and according to the following:

• Iraq is a part of the Arab nation, requires consolidation constitutionally and culturally.

• joint Arab action and the need for national and pan to overcome the dilemmas and challenges.

• There are Arab and Islamic heritage accessible to all aspects of scientific, political, economic, educational and developmental life.

• understand the concept of joint Arab relations.

• the discovery of human energies and resources in the Arab countries.

• a qualitative understanding of the challenges facing the Arab nation.

• Arabic language support Arab Kashkagliat need application in different educational levels.

• first, followed by a sense of national belonging Arab identity.

• detect potential in the Arab country.

Contemporary scientific trends and its impact

There are a variety of contemporary scientific trends can be used to:

• the development of means of communication and informatics.

• rapid scientific and technological progress.

• subspecialties that have been developed with the development of science.

• the effects of the development of science and technology on human resources, including:

- The challenges posed by these developments, such as the risk of dependency and to devote the relative underdevelopment.

- Science and technology services for development and human resources development.

• the relationship between the development of science and technological change and human resources that lead to the development of technologies and the change in the needs of qualified human resources.

• lead the development of human resources (Rehabilitation and Health, for example) to a change in the development of science and technology and their concerns and their tracks.

• The use of modern scientific achievements and technological in the dissemination of culture.

• Expanding the use of modern technologies in education to achieve educational goals.

• the introduction of technical and practical studies.

Economic and societal trends

When you put the Iraqi educational philosophy supposed to take into account what is happening in the Arab countries around us as well as the neighbor countries and these countries judged from an economic and societal trends and trying to figure out when this economic and societal trends we note the following:

• speed of social change.

• increasing the provision of social justice claims.

Cultural trends

The advantage of the cultural trends in the development of Iraqi targets Educational, including:

• emphasize the importance of self-education and the acquisition of a private inclinations, skills and attitudes related to the renewed stream of culture.

• continuing education that emphasizes either allow everyone to Mwash Althagafiqi and professional growth.

• diversity of cultural institutions that contribute to the process of education in this day and age, including the school and the means and the capabilities and all the existing institutions both inside and outside Iraq.

• acquisition by individuals and cultural and functional basics.

• build the capacity of individuals and students in the adoption and critical thinking and discussing things.

• acquisition by individuals self-education skills to reach a culture sources.

• openness to the local, Arab, Islamic and world cultures.

• the transfer of cultural heritage from generation to generation because it is the fruit of joint efforts across historical periods without lock to other cultures.

• Adoption of culture as a way for the Iraqi people in the exercise of life in his community belong to the human being alone and represents a cumulative effort and being a case of change and growth constantly.

• that social institutions such as the family, school and other institutions supported by can play a prominent role in children's lives and educate them through social Alncih operations.

• to maintain the cohesion of the people and cultural trends so that detract from the cultural mainstream public for their community.

• Organize exchange for impact, vulnerability process between individuals and the culture to which they belong.

• awareness about the individual development of community culture and seek to accommodate the values that stand behind the cultural vocabulary and elements.

Educational ethnic philosophy

• Features of the Iraqi educational philosophy

• Iraqi educational goals

• Iraqi curriculum and educational philosophy

• educational policy in Iraq, according to the proposed Altrupah Philosophy

• Iraqi science and technology policy

Iraqi educational philosophy

In light of what came in the preceding paragraphs it can determine the features of the Iraqi educational philosophy according to the following:

• Education in Iraq derives its view of man and the universe and life according to the benefit of the Islamic faith.

• develop education targets in Iraq are Iraqi patriotism and brotherhood of Arab and Islamic compatibility of Islamic and national pride and national levels.

• Iraqi education right for all the sons of the Iraqi society, regardless of its origins, gender, religion and membership of a countryside or cities and nationality.

• Education in Iraq contribute smelting different groups and assist in the formation of a national figure one.

• The Iraqi education to create good citizenship without compromising the identity of the individual and uphold all the rights of citizenship.

• Iraqi education an effective contribution to the Iraqi human breeding contribute include his body and his mind and social growth and the absorption of facts and concepts, theories and harnessed to the service of man.

• Iraqi education are the achievement of the ideals in practice in all the individual or collective behavior (the individual and society).

• Iraqi education is to raise the level of general education outcomes.

• Education in Iraq based on the Arab and Islamic heritage of the purified and developed taking into consideration the change through which the societies in the entire world and absorb the heritage elements.

• Education in Iraq keep pace with scientific and technological development and interact with the development of techniques to process and conscious assimilation of technology and skill.

• To achieve harmonization with the labor market contributes to the Iraqi education.

• absorb the facts and concepts of the environment and the investment environment and take advantage of the technology.

**Iraqi educational goals**

Iraqi Education aims to:

• creating and preparing Iraqi citizen to be valid, free and responsible and contribute to education in the care of physically and mentally, emotionally and socially.

• Build Iraqi man as a driving force for the development of his country and its assistance to build multiple values and a hierarchy of criteria it used to his behavior.

• understanding of the natural environment and social aspects.

• Iraqi human educational, cultural, health and sports knowledge development.

• Adoption of the scientific way of thinking and research of the Iraqi people and raising the economic level.

**Iraqi educational curricula and philosophy**

It is known that the curriculum and founded linked to educational goals emanating from a clear educational philosophy and complement each other with the fact that the sources and one with a difference in the nature and mechanism of each and thus always requires the unification of the aims of education and the foundations of the curriculum for the construction of the new curriculum and this requires many studies and research sources to derive goals as well as identifying the general principles for the construction of the curriculum.

Researchers have differed as to the lay the groundwork for building the curriculum and in general are:

• educational foundations

• scientific foundations

• social foundations

• psychological foundations

And Iraq in general adopted for service basis above when setting curriculum as well as its dependence on one characterized by the importance of a signed and Mstrkath, including:

• Religion

• Language

• frame of mind

And when it approaches, especially where the humanitarian situation that adopts the principle of unification of similar elements and useful properties in the planning, and educate students that Iraq is the general framework, social and moral frameworks and controls must be dominant and common law always master of the situation. Moreover, it is not logical methodology cloning experiments and actual other, but must Alastvadhmma is thinking fits in adequate education. According to these perceptions bear the Ministry of Education, the burden of guiding the generations as an educational institution stripped for shapes partisan and intellectual. And bloom for everyone to become a school form of multiple communities trends and stripes and the kind of pluralism oppose the independence of the educational curriculum.

# Chapter Five
# Past , present
# and
# future of Science

**A.Preface**

**B.Arab Science and scientific heritage**

**C.GREEK SCIENCE**

**D. contemporary scientific trends and their impact**

A.Preface

The introduction of past scientific methods in scientific research can accept the hypothesis of further research, for example, that al-Hasan ibn al-Haytham used measurement and explaining vision process, while Abu Bakr al-Razi describes his approach in dealing with, a balance between stability and development and combine them in a creative consistency.

Abbasid Caliphs, had a significant role in the development of what is called Islamic science. Abbasid Caliph Muhammad al-Mahdi was superstitious about astronomy and astrology.
The Arabs not only changed their way of thinking but also their view of the world and their role in it. The origins of sciences can be traced back partly to the scientific heritage of Sumer, Babylon, Egypt, Greece, Persia and India, partly to the inspiration derived from the Qur'an with the spread of Islam in the 7th and 8th centuries, , known as the Islamic Golden Age, lasted until the 13th century. Ancient Egypt made significant advances in astronomy, mathematics and medicine. Their development of geometry was a necessary outgrowth of surveying to preserve the layout and ownership of farmland, which was flooded annually by the Nile river. The 3-4-5 right triangle and other rules of thumb were used to build rectilinear structures, and the post and lintel architecture of Egypt. Egypt was also a center of alchemy research for much of the Mediterranean.

## The Qur'an And Science

The Qur'an does not need unusual characteristics like this to make its supernatural nature felt. In many verses of the Qur'an support this goal including the earth and the alternation of night and day are signs for understanding)) Imran -190-. The 7th century witnessed the intellectual and cultural transformation of the Arab people principally as a result of some unique events that occurred. From the 'Abbasid period onwards, Muslims were avid readers of religion, science and philosophy. Thus scientists such as Jabir ibn Hayyan, al-Kindi, al-Khwarizmi, al-Razi, al-Biruni, al-Farabi and Ibn Sina were combining the religious, sciences such as medicine, philosophy, astronomy or mathematics.

**B.Arab Science and scientific heritage**

The principles and objectives upon the Arab scientific heritage in the field of pure science depend on the following :-

- The link between the scientific movement and activity in the scientific community.
- A belief in the dignity of man and its value and self-respect for his mind.
- Faith estimate of mind which assigned its rights and honor the preference over all other creatures, and appreciation of science and scientists .
- Faith needs openness to beneficial of Science and tolerance campaign and translators of scientific knowledge and books.
- Strengthening the faith of man in God and the Creator, make it more amenable to his orders of the greatness.

**Heritage of science**

Science Heritage is neither fundamental nor experimental is objects, because heritage is not an experiment. In this aspect, the premise of heritage science comes close to social science. Heritage is accessible, in its preserved authentic form or as a (digital) reproduction, .

Heritage science must be based on aninterdisciplinary of knowledge, from fundamental sciences (chemistry, physics, mathematics, biology) to arts and humanities(conservation, archaeology, philosophy, ethics, history, art history etc.), including economics, sociology, computer sciences andengineering.

Since 2010, Master's degree courses in heritage science have become available at University College London] and Queens University Belfast. Many courses include elements of heritage science, e.g. technical art history is often part of art history courses, and natural sciences are often taught in conservation courses.

Stages of science heritage

Uruk, and Babylon, which in 600 BCE was the largest city on earth under King Nebuchadnezzar.especially the monumental architecture of the Ziggurat featured columns, and vaults. . Mesopotamian civilization of Sumer, Assyria and Babylon also gave rise to the Law Code of King Hammurabi). The Sumerians, who were advanced in astronomy, and used a 12-month solar calendar along with a 354 day lunar calendar; but in the 3rd millennium BCE regularly used a 360-day calendar, which had been adopted, in a modified form, by Jews and Muslims. The Babylonians recorded a solar eclipse as early as 763 BCE and devised an instrument to detect when a star or planet was due to appear in the south. The Assyrians used water clocks. King Sargon produced maps in Mesopotamia for the purpose of tax collection. From their beginnings in Sumer (now Iraq) around 3500 BC, the Mesopotamian people began to attempt to record some observations of the world with numerical data. But their observations and measurements were seemingly taken for purposes other than for elucidating scientific laws

.The Egyptian civilization (3000 BCE to 300 CE), has been credited with instituting a 365-day solar calendar (ca 2773 BCE). In 1500 BCE, , Egyptian medicine, practised by the priests in the 2nd millennium BCE, Egyptian became the architect of Memphis. Medical historians believe that ancient Egyptian pharmacology, for example, was largely ineffective. Nevertheless, it applies the following components to the treatment of disease: examination, diagnosis, treatment, and prognosis.

Pure ,

applied and social science

Astronomy

Astronomy can be identified by several terms, such as Science of the Stars . Muslims daily prayer called for a scientific method of fixing he ***qiblah*** according to precise knowledge of mathematical astronomy. ‘Astronomats, including Habash al-Hasib, Jabir ibn Hayyan, Muhammad ibn Musa al-Khwarizmi were credited with writing books on how to construct an astrolabe. " Astronomical observations from China constitute the longest continuous sequence from any civilisation and include records of sunspots , lunar and solar eclipses. By the 12th century, they could reasonably accurately make predictions of eclipses, but the knowledge of this was lost during the Ming dynasty In Babylonian astronomy, records of the motions of the stars, planets, and the moon are left on thousands of clay tablets created byscribes. Even today, astronomical periods identified by Mesopotamian proto-scientists are still widely used in Western calendars such as the solar year and the lunar month. The stars and the planets had astrological significance in that the major heavenly bodies were assumed to “rule” the land when they were in the ascendant (from the succession of these “rules” came the seven-dayweek, after the five planets and the Sun and the Moon), but astronomy was largely limited to the calendrical calculations necessary to predict the annual life-giving flood of the Nile. Mesopotamia was more like China. The life of the land depended upon the two great rivers, the Tigris and the Euphrates, as that of China depended upon the Huang He (Yellow River) and the Yangtze (Chang Jiang).

**Mathematics**

Al-Khwarizmi briefly discussed the mathematics such as the Indian numerals, algebra, trigonometry and alphabetical arithmetic.Initially, Islam inspired the Arabs to apply mathematics in order to resolve the Islamic Law of Inheritance . Al-Khwarizmi's book laid the foundation of modern algebra . Finally, another of al-Khawarizmi's works, dealing with astronomical tables. It was the knowledge of geometry that made a profound impact on Islamic art and architecture, especially in the geometric decoration of windows, and domes and the use of mosaic tiles. Mathematics and astronomy thrived under these conditions. The number system, probably drawn from the system of weights and coinage, was based on 60 (it was in ancient Mesopotamia that the system of degrees, minutes, and seconds developed) and was adapted to a practical arithmetic. The heavens were the abode of the gods, and because heavenly phenomena were thought to presage terrestrial disasters, they were carefully observed and recorded. Out of these practices grew, first, a highly developed mathematics that went far beyond the requirements of daily business, and then, some centuries later, a descriptive astronomy that was the most sophisticated of the ancient world until the Greeks took it over and perfected it.

### Alchemy and Chemistry

Jabir ibn Hayyan (Latin Geber, d. ca 803 CE) was the most famous Arab alchemist, the names of Abu Bakr al-Razi (b.250/864) are also associated with alchemy. Jabir did laboratory work in chemistry and his research has entered the history of science. Al-Razi described his chemical apparatus and laboratory research.

### GEOLOGY

The Qur'an also expresses concepts in the field of geology Nevertheless, we may read the following in the chapter Taa Haa:"(God is the One who) sent down rain from the sky and with it brought forth a variety of plants in pairs." The Qur'an tells all Muslims to face in the direction of Mecca when they pray. However, this would only be possible on a flat Earth because it is not possible to bow down towards the direction of Mecca when you are on the opposite side of the earth. The Qur'an propagates the idea that mountains are crucial in stabilizing the earth when, in fact, the earth would be much more stable and have less earthquakes if mountains did not exist.And He has set up on the earth mountains standing firm, lest it should shake with you; and rivers and roads; that ye may guide yourselves (Qur'an 16:15)

Earth quakes are extremely common along tectonic fault lines and are not a punishment for human behavior but the byproduct of natural forces. The Qur'anic author perpetuates an unscientific understanding of his phenomena in there verses.Do then those who devise evil (plots) feel secure that Allah will not cause the earth to swallow them up, or that the Wrath will not seize them from directions they little perceive(Qur'an 16:45)

## BIOLOGY

Qur'an is dealing with living things, both in the animal and vegetable kingdoms, especially with regard to reproduction and the fact that it is only in modern times that scientific progress has made the hidden meaning of some Qur'anic verses comprehensible to us. There is nothing to indicate that people in the Middle-East and Arabia knew anything more about this subject than people living in Europe or anywhere else.

Physiology The constituents of milk are secreted by the mammary glands which are nourished by the product of food digestion brought to them by the bloodstream. I give you drink from their insides, coming from a conjunction between the digested contents ( of the intestines ) and the blood, milk pure and pleasant for those who drink it." Qur'an, 16:66

## EMBRYOLOGY

It is especially in the field of embryology that a comparison between the beliefs present at the time of the Qur'an's revelation and modern scientific data, leaves us amazed at the degree of agreement between the Qur'an's statements and modern scientific knowledge.

"I fashioned the clinging entity into a chewed lump of flesh and I fashioned the chewed flesh into bones and I clothed the bones with intact flesh.(" Qur'an, 23:14)

The Qur'an contain statements about bodily fluids and the stages of development of the human embryo. Many of these descriptions are vague and unscientific.

The Qur'an describes the formation of a human embryo from fluids emanating from the man (and possibly also of the woman). In fact, semen is the vehicle for the sperm cells, one of which fuses with a woman's ovum in her fallopian tube, and the resulting cell divides and travels back into the womb for implantation. Did We not create you from a liquid disdained? And We placed it in a firm lodging For a known extent

**Fertilization**

. In chapter al-Insaan the Qur'an states:

"Verily, I created humankind from a small quantity of mingled fluids." Qur'an, 76:2. The verse correctly implies that fertilization is performed by only a very small volume of liquid. This is the meaning of the following verse in chapter as Sajdah:"Then He made [ man's ] offspring from the essence of a despised fluid." Qur'an, 32:8

**Medicine** In medicine, Hippocrates (c. 460 BC – c. 370 BC) and his followers were the first to describe many diseases and medical conditions and developed the Hippocratic Oath for physicians, still relevant and in use today. Herophilos (335–280 BC) was the first to base his conclusions on dissection of the human body and to describe the nervous system. Galen (129 – c. 200 AD) performed many audacious operations—including brain and eye surgeries— that were not tried again for almost two millennia.

The Prophet Muhammad's statements regarding cleanliness, diet, sickness and cure were collected together in books, which came to be known as ***Tibb al-Nabawi*** (or the Prophetic medicine). During the Umayyad era, some events of medical significance included the amputation of a leg infected with gangrene.. 'It is clear that Islamic science and medicine developed rapidly in Baghdad under the early 'Abbasid Caliphs, especially al-Mansur, Harun al-Rashid and al-Ma'mun. One early work on ophthalmology was Hunayn ibn Ishaq's which remained a standard for many centuries.

***Political science***

Political science is a late arrival in terms of social sciences]. However, the discipline has a clear set of antecedents such asmoral philosophy, political philosophy, political economy, history, and other fields concerned with normative determinations of what ought to be and with deducing the characteristics and functions of the ideal form of government. The roots of politics are in prehistory. In each historic period and in almost every geographic area, we can find someone studying politics and increasing political understanding.

Later, Plato analyzed political systems, abstracted their analysis from more literary- and history- oriented studies and applied an approach we would understand as closer to philosophy. Similarly, Aristotle built upon Plato's analysis to include historical empirical evidence in his analysis

linguistics emerged as an independent field of study at the end of the 18th century that Sanskrit, Persian, Greek, Latin, Gothic, and Celtic languages all shared a common base. Descriptive linguistics, and the related structuralism movement caused linguistics to focus on how language changes over time, instead of just describing the differences between languages. This effort is based upon a mathematical model of language that allows for the description and prediction of valid syntax. Additional specialties such as sociolinguistics, cognitive linguistics, and computational linguistics have emerged from collaboration between linguistics and other disciplines.

### *Economics*

The basis for classical economics forms Adam Smith's whoSmith criticized mercantilism, advocating a system of free trade withdivision of labour. Karl Marx developed an alternative economic theory, called Marxian economics. Marxian economics is based on the labor theory of value and assumes the value of good to be based on the amount of labor required to produce it., capitalism was based on employers not paying the full value of workers labor to create profit. The Austrian school responded to Marxian economics by viewing entrepreneurship as driving force of economic development.

The above "history of economics" reflects modern economic textbooks and this means that the last stage of a science is represented as the culmination of its history.

### *Psychology*

The end of the 19th century marks the start of psychology as a scientific enterprise. The year 1879 is commonly seen as the start of psychology as an independent field of study.. Freud's influence has been enormous, though more as cultural icon than a force in scientific psychology.

Scientific knowledge of the "mind" was considered too metaphysical, hence impossible to achieve.

The final decades of the 20th century have seen the rise of a new interdisciplinary approach to studying human psychology, known collectively as cognitive science. These new forms of investigation assume that a wide understanding of the human mind is possible, and that such an understanding may be applied to other research domains, such as artificial intelligence.

### *Sociology*

Ibn Khaldun can be regarded as the earliest scientific systematic sociologist. The modern sociology, emerged in the early 19th century as the academic response to the modernization of the world. The aim of sociology was in structuralism, understanding the cohesion of social groups, and developing an "antidote" to social disintegration. Max Weber was concerned with the modernization of society through the concept of rationalization, which he believed would trap individuals in an "iron cage" of rational thought.

### *Anthropology*

Anthropology can best be understood as an outgrowth of the Age of Enlightenment. It was during this period that Europeans attempted systematically to study human behaviour. Traditionally, much of the history of the subject was based on colonial encounters between Western Europe and the rest of the world, and much of 18th- and 19th-century anthropology is now classed as forms of scientific racism.

### C.Greek science

They introduced scientific methods based on reason and observation.. Greek science may have been a continuation of ideas and practices developed by the Egyptians and Babylonians, the Greeks were the first to look for general principles beyond observations. Science before the Greeks, as practised in Babylon and Egypt, consisted mainly of the collection of observations and recipes for practical applications .Aristotle is generally thought to be the father of life sciences. He studied plants and animals. He also wrote on embryology. The greatest Greek contribution to medicine was made by Hippocrates, an author of many books, whose Hippocratic Oath is still used as a code of ethics by the medical profession. He freed medicine from superstition and religion .

There were many differences between ancient Greece and the other civilizations, but perhaps the most significant was religion. What is striking about Greek religion, in contrast to the religions of Mesopotamia and Egypt, is its puerility. Hence, there were no easy answers to inquiring Greek minds. The result was that ample room was left for a more penetrating and ultimately more satisfying mode of inquiry. Thus were philosophy and its oldest offspring, science, born.

**D. contemporary scientific trends and their impact**

The Romantic Movement of the early 19th century reshaped science by opening up new pursuits unexpected in the classical approaches of the Enlightenment. Major breakthroughs came in biology, especially in Darwin's theory of evolution, as well as physics (electromagnetism), mathematics (non-Euclidean geometry, group theory) and chemistry (organic chemistry). The scientific revolution established science as a source for the growth of knowledge. During the 19th century, the practice of science became professionalized and institutionalized in ways that continued through the 20th century. .Contemporary happening

It is the contemporary historical development of modern education in Japan has passed through three periods:

• the first period (1872 - 1939)

The first renaissance in the nineteenth century Japan, is quoted in this period the French system for Educational Administration (a strong central Ministry of Education), and the centrality of the educational philosophy based on the national and the emperor, it turned into a military tendency was able to defeat the Russian army and the Chinese army.

The modernization movement in this period successfully launched rich Japan with a strong army removed expansive. During the second half of the nineteenth century were imported from the West everything was Western technology and the Japanese spirit. Japan has been celebrated in 1896 that there is no illiteracy in Japan and became illiterate in Japan is far from the computer not mastered.

In the early twentieth century, education was in the early stages traditionally while in upper stages became limited and is available for all and set up some universities empire that helped in its creation Germans mismatch universities for females as well as the missionaries opened the educational role of the University of females in middle and an active education levels of education first after the end First World War.

• the second period (1940 - 1945)

This period was marked by the issuance of a new education law in 1941 emphasizes the principle of sacrifice for the benefit of the state and the emperor, it led to a strong imperial army and to defeat it.

• the third period (1946 - until now.)

In this period, which represents the US occupation and restore freedom in 1951 was characterized by different educational models.

Japan has become a demilitarized state under US tutelage and attempted to US occupation forces during the five years of occupation (1946 - 1951) change:

• image of the Holy Emperor

In general this experience in a new period of occupation are without military rule and adopt a peaceful way and not the collision of the building development has maintained her privacy through and maintain a set of traditions and heritage of the dialogue. United States of America as an occupier of Japan and has also mentioned the development of a new philosophy represented by:

• the prohibition of the educational activity that is Balaskarah.

• re-organization of the Japanese education in the light of the new philosophy.

Notes for speed in the local and global variables lead to pressure on education in many countries of the world and led to the emergence of reports on education in many countries of the world refers to the size of the danger resulting from the weakness of the conditions of education and inability to innovate and change. Perhaps the US report A Nation at Risk "A Nation at Risk" in 1983 is one of the most important reports, which revealed failure of the US education and called for the development and excellence, as well as document (the United States in 2000 - a strategy for Education), a Bush document in 1991. As well as a close US President Bill Clinton in February

1997 and which focused on the ten routers for Education in America for twenty-century atheist.

US funding for education

Education funding is different in the United States from state to state and are considered sources of education and its financing, including the Federal different source and the state and local levels and styles of taxes.

Reform experiments in the US educational system

There were several reform experiences of the US educational system and of which we mention the following:

• the practice of racial minorities to education experience.

• Repair Nation at Risk in 1956 as a result of the launch of the Soviet Union's first satellite.

• Education Reform in the eighties (Education for All).

• Education Reform in the nineties (the teacher attention and increase its capacity and interest in the curriculum).

• reform in 2000 (the national project for development).

The history of science is marked by a chain of advances in technology and knowledge that have always complemented each other. Technological innovations bring about new discoveries and are bred by other discoveries, which inspire new possibilities and approaches to longstanding science issues

The Scientific Revolution is traditionally held by most historians to have begun in 1543, were first printed. The period culminated with the publication of the *Philosophiæ Naturalis Principia Mathematica* in 1687 by Isaac Newton, representative of the unprecedented growth of scientific publications throughout Europe.

**Natural sciences**

***Physics***

The scientific revolution is a convenient boundary between ancient thought and classical physics. This was followed by the first known model of planetary motion given byJohannes Kepler in the early 17th century, which proposed that the planets follow ellipticalorbits, with the Sun at one focus of the ellipse. Galileo ("*Father of Modern Physics*") also made use of experiments to validate physical theories, a key element of the scientific method

In 1687, Isaac Newton published, detailing two comprehensive and successful physical theories: Newton's laws of motion, which led to classical mechanics; andNewton's Law of Gravitation, which describes the fundamental force of gravity.

During the early 19th century, the behavior of electricity and magnetism was studied byFaraday, Ohm, and others. These studies led to the unification of the two phenomena into a single theory of electromagnetism, by James Clerk Maxwell (known as Maxwell's equations).

. Beginning in 1900, Max Planck, Albert Einstein, Niels Bohr and others developed quantum theories to explain various anomalous experimental results, by introducing discrete energy levelsThe theory of general relativity, proposed by Einstein in 1915, showed that the fixed background of spacetime, on which both Newtonian mechanics and special relativity depended, could not exist. In 1925, Erwin Schrödinger formulated quantum mechanics, which explained the preceding quantum theories.

The atomic bomb ushered in "Big Science" in physics.

In 1938 Otto Hahn and Fritz Strassmann discovered nuclear fission with radiochemical methods, and in 1939 Lise Meitner and Otto Robert Frisch wrote the first theoretical interpretation of the fission process, which was later improved by Niels Bohr and John A. Wheeler. Further developments took place during World War II, which led to the practical application of radar and the development and use of the atomic bomb.

### *Chemistry*

Mendeleev

Modern chemistry emerged from the sixteenth through the eighteenth centuries through the material practices and theories promoted by alchemy, medicine, manufacturing and mining. Other important steps included the gravimetric experimental practices of medical chemists like William Cullen, Joseph Black, Torbern Bergman and Pierre Macquer and through the work of Antoine Lavoisier (*Father of Modern Chemistry*) on oxygen and the law of conservation of mass, which refuted phlogiston theory.. Dalton also formulated the law of mass relationships. In 1869, Dmitri Mendeleev composed his periodic table of elements on the basis of Dalton's discoveries.

The synthesis of urea by Friedrich Wöhler opened a new research field, organic chemistry, and by the end of the 19th century, scientists were able to synthesize hundreds of organic compounds. The later part of the 19th century saw the exploitation of the Earth's petrochemicals, after the exhaustion of the oil supply from whaling. By the 20th century, systematic production of refined materials provided a ready supply of products which provided not only energy, but also synthetic materials for clothing, medicine, and everyday disposable resources. Pauling's work culminated in the physical modelling ofDNA, *the secret of life* (in the words of Francis Crick, 1953). In the same year, the Miller–Urey experiment demonstrated in a simulation of primordial processes, that basic constituents of proteins, simple amino acids, could themselves be built up from simpler molecules.

French biologist Louis Pasteur. Pasteur was able to link microorganisms with disease, revolutionizing medicine. He also devised one of the most important methods in preventive medicine, when in 1880 he produced a vaccine against rabies. Pasteur invented the process of pasteurization, to help prevent the spread of disease through milk and other foods.

Darwin proposed that the features of all living things, including humans, were shaped by natural processes over long periods of time. The theory of evolution in its current form affects almost all areas of biology. Mendel's laws provided the beginnings of the study of genetics, which became a major field of research for both scientific and industrial research. By 1953, James D. Watson, Francis Crick and Maurice Wilkins clarified the basic structure of DNA, the genetic material for expressing life in all its forms. In the late 20th century, the possibilities of genetic engineeringbecame practical for the first time, and a massive international effort began in 1990 to map out an entire human genome (the Human Genome Project).

***Ecology***

The discipline of ecology typically traces its origin to the synthesis of Darwinian evolution and Humboldtian biogeography, in the late 19th and early 20th centuries. Equally important in the rise of ecology, however, were microbiology and soil science—particularly the cycle of life concept, prominent in the work Louis Pasteur and Ferdinand Cohn. The word *ecology* was coined by Ernst Haeckel, whose particularly holistic view of nature in general (and Darwin's theory in particular) was important in the spread of ecological thinking.

Computer science, built upon a foundation of theoretical linguistics, discrete mathematics, and electrical engineering, studies the nature and limits of computation. Subfields include computability, computational complexity, database design, computer networking, artificial intelligence, and the design of computer hardware. Contemporary computer science typically distinguishes itself by emphasising mathematical 'theory' in contrast to the practical emphasis of software engineering.

Environmental science is an interdisciplinary field. It draws upon the disciplines of biology, chemistry, earth sciences, ecology, geography, mathematics, and physics.

## Chemistry and biology

Amazing developments have taken place in the chemical sciences particularly during the second half of the century, including implicit and other interfaces. Developments on the implicit content and the vocabulary and mechanisms are known in chemistry and provide improved or new interpretation of events and phenomena and chemical reactions, as a result also of new subjects and disciplines within the science of chemistry itself. These developments have led to the opening of new channels in scientific research and technological innovations such as chemical industries to create new chemicals, or chemical industries, and new techniques.

The developments of the second type of chemical sciences interface had addressed the disciplines of science linking chemical sciences and applied various treatments. These developments have led to the developments of science or the new terms of reference were not known before.

## The recent trends in the chemistry of life

Bio-chemistry studies the chemical and physical characteristics of the components of the cell and features of the life systems of the components, as well as the interpretation of what these systems in the cell Biochemistry provided a lot of accomplishments, it has helped to clarify the mechanism of medicine and contributed to the diagnosis and treatment of many diseases and provided techniques which could be used measure the level of many of the compounds in vivo.

Biochemistry lasted over the age of a century in different disciplines, some with a study of the materials that make up plant cell and then called the chemistry of plant life, and then which is related the animal cell which is called chemistry of animal life if the human cell is the target.

Chemistry has expanded to clinical biochemistry that includes chemistry of life, becoming a physical, organic and biochemistry and inorganic chemistry as well as nutrition. Interested in chemistry, life functions of the modern systems of life, have contributed to the means of study in the last century with the observation of these systems directly during the work, either at the present time which has changed the picture and it became possible to obtain the most desirable observations by the development of viable technologies (electron microscope, radioactive isotopes, Immunology, spectrum).

The scientists believed at the end of the nineteenth century that it is possible to obtain some information relating to the systems of life, by studying the chemistry of cells and for decades was followed by chemists adopted the chemical methods available and succeeded in obtaining useful developments. Significant improvements to the technical methods such as the use of chemical isotopes have greatly increased the sensitivity of diagnosis of different types of molecules of life and others, and when it is necessary to separate the components of the chemical reaction through life and is very sensitive, then used deportation electric traditionally.

When the attention has turned physicists, chemists, physicists about the science of life (and perhaps due to the ability of living cells to configure the system, although the laws of physics, emphasizes the universe there is a tendency towards non-attendance) then emerged the technical methods of physical, chemical, physical, such as spectroscopy, diffraction to be applied an the field of biology.

The progress achieved in the chemistry of life has begun to acknowledge that the livelihood systems containing small particles interested m organic chemistry to study and clarify as well as large molecules called macroscopic particles which are not molecular weights less than 100 million times the mass of one atom of hydrogen. The importance of macroscopic particles of the life systems in its ability to privacy in life interactions composition of building blocks, and can say clearly that he had made in the past years considerable effort to characterize the annexation of macroscopic particles as well as the reactors that occur between them and the need for advanced methods of separation and purification and characterization of macroscopic particles in order to obtain information on structural composition of the macroscopic molecule.

The objective of biochemistry for nearly half a century is to collect and organize interactions that occur in living cells. The motivation for this major effort is that a significant number of the attributes of living cells can be understood through these interactions that are typically characterized by the formation or breaking covalent bonds. It is been clarified on the liberalization of energy as a result of break chemical transformation processes as well as molecules of life and mutual assembly operations amino acids, sugars and fats to form macroscopic particles.

During the last thirty years clearly demonstrated that the reactors that occur between molecules due to physical, those that are not or break covalent bonds have the same importance of chemical reactions, for example, that the organization of chemical reactions (i.e., the degree of permitted them to occur) performed by the physical changes that occur in the structure of (construction) of large molecules, as well as the creation of active centers in these molecules and the resulting interdependence of the non-covalent small molecules, in addition to, many of the qualities of a macroscopic aggregates molecules in cells or in the organism (the cell membrane and walls of cells and chromosomes).

Plurality of molecules of life structure consists of installation of the first structural molecular structures of multiple different types of units place (serial), for example, the sequence of amino acids found in proteins and sequence by chemical analysis. The secondary structural composition which involves the formation of a complex three-dimensional structures is called to direct all of the units for multi-particles to other units and is called the secondary structural composition tradition or (body and image) or the status of the foundation structure or backbone of multiple chains. The forms, which consist of surfaces and different types of these mixed forms, and called on the direction of (position) of side chains relative (amino acids, nucleic acids or bases) triangular structural composition. A lot of multiple molecules of life with each other to be as complex as the structures of several structural units viruses, membranes and capillaries bonds and are usually in one level, where you specify the types of bilateral structures of proteins. On the other hand that includes the alpha carbon to allow for many types of structural combinations. The two phosphate ester bonds in nucleic acids are subject to sag as well, because the flexible rule and hate water and one level surrounded by a few of so they are usually located one above the other, thus reducing the adhesion of water, and this increases the structural rigidity of installation.

The multi-life linear molecules, which has no free rotation about the bonds, which do not interact aggregates side is called the file is not a random combination structurally specific dimensions or size of distinct wraps by Brownian movement. Size can be measured by the value equal to the rate of rotation of the radius around a point or an axis.

**Nucleic acids-the mystery of the mysteries of life**

A nucleic acid represent the brain of the cell brain cell with a specific developed program, to be issued through the instructions for that cell fusion and installation of the life and death and plan for the future.

There are two types of nucleic acids (DNA, RNA) both of their differences centered a long chain molecules composed of nucleotides and position of certain forms.

In 1953 Crick and Watson was able, who have previously received the Nobel Prize in 1962 developed a model for the DNA structure, consisting of two strands of units of the four nucleotides arranged in orderly fashion, and every one of them is a multi-helical nucleotides wrapped around a common axis to form the double helix right direction.

The models explained by x-ray, have two sections through the longitudinal axis of the first of 0,34 nm and the second 3,4 nm.

Model of "Watson & Crick Model" Both Watson and Kirk in 1953 the first specialist in genetics and the second physicist to develop a model structure that represents the structural basis of DNA. Jarkav in the light of studies on the percentage of nitrogen bases, X-ray dimensions, as well as the fact that adenine = thymine and Guanine = Cytosine

**Developments in the synthesis and structural forms of DNA**

Features of the model of Watson and Crick, this is called beta-form and include:

- The DNA is composed of two strands of dioxin nucleotides wrapped through multi-spiral system.
- The nucleotide chains are connected by diester bonds within one strand.
- The bases purines and pyrimidines facing each other, so that in particular, adenine faces thymine and guanine faces cytosine through hydrogen bonds.
- The order of nitrogenous bases one in series vary from other.
- The levels of sugar rings parallel to the axis of and phosphate groups abroad.
- DNA is divide in two parts the first is called water hating (nitrogenous bases), located internally and a second which is externally faced of the surrounding water molecules containing phosphate groups.

The second half of the nineteenth century witnessed a series of discoveries of life such as, a serious (cell) theory at the hands by Matthias in the plant and then Theodorishvan in the animal. Since of plant and animal are composed of cells, that evolution (cell theory) and their development is considering critical stage in the progress of life science similar to atomic energy The human Physiology science as one of the branches of the life science that refers to the amazing facts explaining the greatness of the Creator and accuracy of the details and secrets. The digestive system for example (the greatest chemical plant in the world) including, the methods of food analysis of chemical analysis of various surprising and distribution of fairly Safe food distributed to millions of living cells. In view of these living cells the issue of causal efficacy and secret of life that fills justify the astonishment and admiration for self-cell, while adapting to the requirements of their position and circumstances.

If we explore the science of life, we will find another secret of that biggest secrets, the secret of the mysterious life, which fills the moral conscience of mankind, with the concept of divine fear and faith, firmly established in it. The theory of self-regeneration was collapsed at the depth but the unequivocal scientific experiments, demonstrated the invalidity of the theory of self-regeneration. The material basis of life science was examined and then basically spread the idea of elements. The atoms are spread better for the basic materials of the universe and second nature that the elements consist of a central core electrons of the nucleus orbit (negative) and the nucleus contains protons and neutrons. Attempts were made to alter the material to absolute energy, no electric charge. In other words removing character from element in the light of the theory of relativity of Einstein, where the body mass is relative, not fixed, and increase with the speed according to Einstein equation energy = mass of × square of the speed of light and mass = energy ÷ square of the speed of light. As a result, the atom, including of protons and electrons are condensed energy. Appeared in various forms and multiple images, whereas materials has been converted into energy and energy to the material.

It follows from the views put forward that the original materials the world-life the reality show one common in various forms, and the physical properties of compounds are accidental such as the liquidity of water is incidental, not self-evidence, since it is consisted of two toms and possible separation these two elements from each other and the status of water disappear completely.

The characteristics of the simple elements themselves are not self-rule but are incidental to the material. That such material characteristics become the light of the above facts incidental, it is encroached to be among the identified energy and philosophically, the presumption of the material in the world of life on the top reason capable for denial, as well as of effectiveness.

**Genetics**

Mendel discovered the basic principles of heredity and passed him a head later by scientists. As he concluded after mating successive generations of peas plants, that split successor inherits the characteristics of, according to a mathematical formula that could be the secret to life and then called the laws of Mendel. Then genetics born at the beginning of the twentieth century after the principles of, which were designated as Mendelian inheritance. Followed by several changes altered the traditional characters of life science to prolong and settled genetics then rolled up on the basis of discoveries that quoted life science version of the conventional version of the description and classification.

The Darwinian concept, was retreated which relics on the theory of evolution that the changes and the characters that we can get as a result of practice or reaction with the environment can be transferred by inheritance to the descendants. The hypothesis of evolution of species has been the trend due to the mutations, some aspects of the sudden change in the number of cases that called for the assumption that the diversity of animal emerged from mutations, and some of these changes may have inherited.

After that, the transfer life science version of the traditional formula and tradition description, classification, manifestations and modalities of organic evolution and the cell in its entirety to the life science microscopy, which focuses on exploration of the nucleus molecules and chemical structures. It is a mass of spherical material or oval that looks heavier than around it. Then it emerged that in the nucleus, clusters of fine particulate organic form, renamed chromosome or chromosomes that contain the genetic factors mentioned by Mendel each factor was called gene. The cell contain genes and for each type of species there is a special number of chromosomes in every cell of the human body, forty-six (46) chromosomes except in the female reproductive cell (egg) and sperm in the male sperm, each containing two (23) chromosomes. But there is one chromosome in a set of chromosomes of the male sperm determines sex of the fetus generated from the fertilized egg, it may be x or y. Not only the impact of these chromosomes to determine the sex of newborn, but also that genes determine the hereditary characteristics of male and female, then was found that the nucleic acids present in the nucleus of cells issued instructions for their growth and break apart and there are two types, DNA, RNA. Crick and Watson (1962) managed to develop an acceptable model for this structure which is composed of two bands of nucleic acid units (with four bases adenine, thymine, guanine, cytosine) in a corresponding sequential arrangement to RNA and such a model and a specific genetic information is transferred to RNA which controls the composition of proteins.

**Evolution**

There are different opinions about the evolution of such as that living organisms in all its forms and types are fixed and does not change, but some of the scientists are not convinced with validity of this opinion, and expressed the possibility of changing living organisms, or that living organisms are not static but in constant change, depending on the prevailing natural conditions. In the following number of views on development with theories that have been presented each gained continuity and the other stopped until:

- Both humans and animals, with a single installation consists of degrees of nature non-living then plants, animals and low organisms and marine animals and finally humans.
- The organisms, which represent high-degree, can not arise from the low-level organism but, it was created on this picture.

- Hiraclt

Imagine this Greek philosopher, originator of the dialectic image that everyone is marching and everyone is being changed.

- Akhavan Safa (brothers of safa)

  Akhavan Safa pointed to explain the evolution and the doctrine of evolution that the worlds of animals, plants and inanimate are and one separated from each other within a sharp border.

**Genetic Engineering**

Genetic engineering caused major developments in life sciences, including applications in medicine, which includes diagnosis and treatment.

The concept of the genetic engineering and technologies based on multi-splitting of DNA By special enzymes work on specific sites and then linking the pieces formed with DNA from other sources, is then the proliferation of hybrid is able to reproduce on vehicle “Cloning”, including bacteria and viruses which have been used to develop genetic engineering techniques and other such as electrophoresis and auto radiography.

This technology can be used for the preparation of special DNA sensors for the purpose of searching for specific genes or specific parts of the DNA technology to clone parts of the DNA in large quantities by the use of the enzyme "PCR".

Applications of genetic engineering:

- Gene therapy: It is used in the treatment of some diseases, including the treatment of brain tumors and in the reduction of cholesterol in the blood.
- Genetic mapping in humans: one can look at the matter in the future after the location of many human genes.
- Stimulation the immune system to produce antibodies more efficient and accurate (a vaccine against the virus of hepatitis "B").
- Production of proteins of medical importance.
- Determination of the nature and location of genes for some genetic diseases.
- Diagnosis of genetic diseases before birth.

**Biotechnology**

Biotechnology caused enormous developments human, including those which were aimed at human use:

- Industrial electronics.
- Products of biotechnology space.
- Environmental treatment
- Extracting of oil by microorganism
- Medicinal plants.

The transfer of chromosomes from one cell to another is of chemical and living concept for the process of cloning, and its backbone. The genetic chromosomal transfer is not considered new, it has been exercised during the implantation of an egg in the uterus of a female in (1978) (the birth of a tube child - Louise).

Specifications and features of the cloning by the cloning is characterized with specific characteristics of chemical features of some new and some old concepts. These specifications are as follows:

- Converting an adult cell (totally grown) to a cell could reproduce without vaccination.
- Converting the reproducing cell to full a creature breeders (replica of the mother).
- Transfer of mature cell grown to fully grown live immature egg in the uterus of another object (the sheep, for example).
- Development of a new principle called the principle of Wilmurt can be expressed as follows: (very specialized mature cell from an animal that awakens static sophisticated genetic information in the chromosome to become a source of a whole new creature).
- This process of reproduction is not a sexual in the traditional sense (without sexual contact between male and female).
- The genetic cell cannot be used through challenge and breeding.
- The cloning from adult stem cells facilitate the researcher to wait to see the nature of the thing itself before proceeding to reproduction.

Features applied to clones (the experiment of Ian's death)

- Converting a cell of the nipple of the gland of sheep (full growth-extremely) to a cell with a capacity of breeding and without sexual relationship.
- Transfer the mature full cell to fully live the immature egg in the uterus of a sheep.
- The cloned sheep of inherited properties, from the donor mother (the birth of Dolly the sheep).
- The sheep (Dolly) is not the nascent daughter and her mother, can be its twin.
- The sheep of the cell prepared for a clone of the breast, containing all the genes necessary for a complete sheep.

**Bio communications of life**

Communications carry out a number of operations which include successive generations such bioelectrical-communication and chemical, reflects the continuity of generations, which represents the survival of reproduction in different ways. The mainstream way is connection in the process of sexual relationship resulting in the integration of the male cell (sperm) female cell (egg). Reproduction may occur without convergence between male and female, as in the cloning processes, which we have mentioned, or, as occurs in the primitive animals, with single cell in a simple dichotomy, as do animal and ameba, parameseyoum. In the process of sexual convergence the number of chromosomes go back to the original number in all of the cells of the human body and becomes a forty-six chromosomes (23 from the nucleus + 23 from the sperm). Therefore, the characteristics of many genes on the chromosomes are transferred. It is worth mentioning that the vaccination is not described in the same way it may be externally or internally in the fish and some other animals, the sperm and eggs put in the water in which they live. The artificial insemination is carried out by sperm from a male and put in the female's vagina to cause pregnancy for the purpose of the transfer of the qualities required to a large number of females.

The other type of communication named electrical contact is carried out via the nervous system, which includes the central nervous system (brain and nervous system or spinal cord) and peripheral nervous system (nerves of the brain, nerves and spinal cord) nervous system (self-sympathetic and para sympathetic). The brain, which represents the main component of the central nervous system, contains about 12 billion nerve cells that do not divide, and when they die does not change with other cells. The nerve cells for example, connect impulses and the installation of a specific architecture. It is worth noting that electric current flows in a given direction and one inside the neuron, and to understand how we entry into nerve impulses in the nerve cell should be conceived that in all cells there is a difference in voltage between the inner and outer surfaces of the cell membrane surrounding the cell. The chemical communication is carried out by glands that produce hormones by channels "Ductless Glands" then they carry out considerable influence on many of the functions of the body.

There are concepts of life in different formats pave the way to clarify the mysteries of life, including cell and technologies and weaving techniques and giant molecules. The secret of the evolution of life in this area, where many scientists believe that life appeared on earth since about 2000 million years and that the earth has been more than 2000 million a year devoid of life. Some have said that life must have originated from water and others say that the air more than water and is flexible others remember that life is generated from the mud while Alkulliat believed that the origin of the universe comes from the fire, was able to turn to air and then to water and the last change to dry and then back into water, then air to fire. The Ambdoukls has reached to the theory of the four elements; the origin of all beings is earth, water, fire and air. Therefore, the thinking of these theories can be deduced from scientific conceptions and we can continue with it to the path of breeding that have been fabricated intellectually by Alersai first and Pasteur II. Humans have many and great works of intellectual and experimental to underline his perception of this universe, have attempted to link his idea of theoretical and experimental trends. In final his perception was confirmed that life arises only from life, as well as to the truth of God still plays a significant role. The early stage study of the body carried out by the microscope and chemical analysis and other techniques, then developed later, first has allowed the researcher to see fine details, arrangement of atoms can not know by this technology, while the chemical analysis refers to compounds that make up the body and the components of these compounds.

Later several techniques developed, including electron microscope, where it was possible to obtain precise details of the minutes. To reflect the elements of body composition other technique called X-ray have been used to study, the order of atoms in many of the biological compounds.

The isotope techniques have continued as a useful tool in the search for the secret of life is as well as a means of treatment. Researchers have manufactured hundreds of radioactive isotopes, generated from the non-radioactive elements in nature, including sodium, sulfur, calcium, chlorine, copper, cobalt, gold, iron, mercury and silver. The main uses of these element the process of photosynthesis, as well as follow- up to 14C in the development of a new technology called radio- immunoassay which can be used to determine the concentrations of compounds found in very small quantities, especially hormones.

Used several techniques to separate, including electrical and deportation chromatography. The first has been used to isolate many of the vehicles life and purification, and the first uses and which still constitute the mainstay of the first structural study of structure of proteins and amino acids, thanks to the use of positive ion equivalent.

## contemporary scientific trends and their impact

**Quantum chemistry**[

This is the first application of quantum mechanics to the diatomichydrogen molecule, and thus to the phenomenon of the chemical bond. In the following years much progress was accomplished by Edward Teller, Robert S. Mulliken, Max Born,J. Robert Oppenheimer, Linus Pauling, Erich Hückel, Douglas Hartree, Vladimir Aleksandrovich Fock, to cite a few.[ Still, skepticism remained as to the general power of quantum mechanics applied to complex chemical systems].

Hence the quantum mechanical methods developed in the 1930s and 1940s are often referred to as theoretical molecular or atomic physics to underline the fact that they were more the application of quantum mechanics to chemistry and spectroscopy than answers to chemically relevant questions. In 1951, a milestone article in quantum chemistry is the seminal paper of Clemens C. J. Roothaan on Roothaan equations.

. In 1925, Austrian-born physicist Wolfgang Pauli developed the Pauli exclusion principle, which states that no two electrons around a single nucleus in an atom can occupy the same quantum state simultaneously, as described by four quantum numbers. Pauli made major contributions to quantum mechanics and quantum field theory - he was awarded the 1945 Nobel Prize for Physics for his discovery of the Pauli exclusion principle - as well as solid-state physics, and he successfully hypothesized the existence of theneutrino. In addition to his original work, he wrote masterful syntheses of several areas of physical theory that are considered classics of scientific literature.

**The Schrödinger equation**

In 1926 at the age of 39, Austrian theoretical physicist Erwin Schrödinger produced the papers that gave the foundations of quantum wave mechanics. In those papers he described his partial differential equation that is the basic equation of quantum mechanics and bears the same relation to the mechanics of the atom as Newton's equations of motion bear to planetary astronomy Some view the birth of quantum chemistry in the discovery of the Schrödinger equation and its application to the hydrogen atom is often recognised as the first milestone in the history of quantum chemistry.

.

.

**key structural features of DNA**

in 1953 when James Watson and Francis Crick deduced the double helical structure of DNA by constructing models constrained by and informed by the knowledge of the chemistry of the constituent parts and the X-ray diffraction patterns obtained by Rosalind Franklin.

Mendeleev's periodic table

An important breakthrough in making sense of the list of known chemical elements (as well as in understanding the internal structure of atoms) was Dmitri Mendeleev's development of the first modern periodic table, or the periodic classification of the elements. Mendeleev, a Russian chemist, felt that there was some type of order to the elements and he spent more than thirteen years of his life collecting data and assembling the concept, initially with the idea of resolving some of the disorder in the field for his students. Mendeleev found that, when all the known chemical elements were arranged in order of increasing atomic weight, the resulting table displayed a recurring pattern, or periodicity, of properties within groups of elements.

**The cell as secret of life**

The human digestive system which is an arm of the human physiology of man, explain grandeur of the Creator and the accuracy in detail of the multiple and various secrets. The digestive system is sophisticated chemical laboratory having different methods of food analysis then the food will be distributed equitably to millions of cells that make up the human body consideration involving the secret of life and admiration for the cell. These cells are different technologies for tissue engineering and in the digestive system nearly two hundred thousand reaction within 24 hours. Some of which are the heart muscle shrinks and flattens millions of times during the whole year tirelessly, to obtain the necessary energy for thinking and movement and speech, including also the disposal of waste and toxins in the body, looking at the cell that its approach is one of the secrets of life. These secrets are adopted according to serious lay cell; it is one of the discoveries, theory at the hands of Haydn and Schwan then cell theory, considered a critical stage in the progress of life science, similar to atomic theory in chemistry. The cell in the body of an organism is also similar to personnel in the communities or the living cell act as technique specific to perform a particular job. The cell becomes a plant or accurate chemical plant.

The nerve cell act as a system, for example electrochemical transformation of chemical energy to electrical energy and electrical energy to mechanical energy or kinetic energy. There are also some cells manufacture of the of hormones and other life products used as system used defensive attack their products all exotic and cells in the process of purification and filtration and the cells that absorbs. Furthermore arises from a single fertilized cell, tissues are various heterogeneous tissues, the different organs and different functions, the bones, muscles, cartilage and twigs, leather, and the blood vessels. Then a living cell had made specific technologies in particular, for example, nerve cell is electrochemical system that can shift the chemical energy into electrical energy and the last to mechanical or dynamic or may become a cell laboratory or chemical plant carried hundreds of chemical processes complex and there are cells specialized in the manufacture of hormones and the other to produce biological weapons of and cells of the nomination and purification.

It is clear that the tissues that originate from a single fertilized cell then divided into thousands of cells to materialize into the bones and muscles, cartilage, twigs, leather, and the blood vessels, these tissues are formed in the early embryos and mutate into organs and systems in a stand- alone, but integrated in the performance of its functions.

According to the information of originated from the cell that there is strange power lies in living cells. Walker, professor of Plant Physiology say (that components of a cell arranged in a strange way in which life emerged). But researchers still are unable to make blood cell components and accurate knowledge of this so-called the mystery of life.

**Nucleus and secret of life**

There are at the center of the cell mass of material in the spherical or oval clusters of objects in the body and continues to represent the mysteries of life and plans, regulations, and ideas of life.

In the cell of the human body forty-six chromosomes except the egg (cell reproduction female) sperm and egg, each containing 23 chromosomes (half the number in the human cell non-reproductive). The secret of life is due to this process of somatic cell fusion where each chromosome separated into two parts, it becomes in each of the cells of the fission 46 chromosomes. The chromosome (multi-genes), each genetic factor arranged in two strands one received from the mother and the other from the father.

The nucleus synthesize nucleic acids, seen in the central nucleus filamentous structures and spread over its surface of granules of quick- impact dyes include the nuclear network and the network should be clear when they are not in the case of splitting and dividing at smaller and thickening of these lines is called chromosome. The chromosomes in a cell division looks like similar pairs of fixed shape and fixed number for one type of living organisms.

(Russell and Wallace) says that the cell nucleus is not chemical but structure if analyzed and during the processes of analysis of the most secret mysteries of life may be lost.

Molecules of life, which are building the organism

The space of cell is containing the liquid water containing the various ions and compounds with molecular weights of small, medium and macroscopic, and it is possible to measure the ion composition in each cellular organelle, where each one of them has different ionic compositions.

The sodium ion "Na +" ion is the main ion out the cell in which the 140 mM / L is also found a positive ion in the fluid cell in the Interior position that the , is potassium "K +" Cation cell procedure. There is magnesium ion "2 + Mg" in all cellular spaces inside and outside but with lower concentrations of sodium, potassium and chloride is "CL-" ,the main negative ion outside the cell, with hydrogen carbonate ions “and” small amounts of phosphate and sulfate, and the proteins carry a negative charge at pH 7,4 in the tissue fluids.

All living cells contain a different chemical components of water 70-90% and 2-5% of inorganic ions such as sodium, potassium, chloride and sulfate, and magnesium ,carbonate molecules of life, as well as small, medium and macroscopic molecule that constitute 8-25%.

It has been proven that all the elements in the periodic table of Mendeleev's constitute in the composition of the organism divided into small elements and large. The carbon, oxygen, hydrogen and nitrogen constitute 96% of the elements in the cell, while calcium and phosphorus constitute 3% and each of potassium and sulfur, iron, sodium and chlorine 1% There are very small quantities of the elements iodine, magnesium, copper, manganese, cobalt, boron, zinc, fluorine, selenium and molydenom.

The chemical side is concentrated in the molecules of life on the carbon, which constitutes about 50% by weight bio- molecules are characterized by life-covalent bonds, four of which related to carbon stubs and have different angles of particular value from one carbon atom to another in different molecules of life and because of that there are different types of building structures with three-dimensional, these structures contribute to clarify the complexity of the cellular composition with particular reference to its failure, as well as various forms. In addition organic compounds are characterized by free rotation. The tetrahedral bonds emphasizes on individual carbon atom of the very important properties of organic molecules and the presence of four different groups or different atoms connected to carbon and the last become non-symmetrical (a carbon-atom covalently bonded with four different groups) and composed (which every one of which is mirror images of each other) with a symmetric arrangement in space and called isomers mirror of light for the chemical similarity of the interactions but differ in physical properties of the rotation of polarized light.

**Chemistry and biology**

Amazing developments have taken place in the chemical sciences particularly during the second half of the century, including implicit and other interfaces. Developments on the implicit content and the vocabulary and mechanisms are known in chemistry and provide improved or new interpretation of events and phenomena and chemical reactions, as a result also of new subjects and disciplines within the science of chemistry itself. These developments have led to the opening of new channels in scientific research and technological innovations such as chemical industries to create new chemicals, or chemical industries, and new techniques.

The developments of the second type of chemical sciences interface had addressed the disciplines of science linking chemical sciences and applied various treatments. These developments have led to the developments of science or the new terms of reference were not known before.

**The recent trends in the chemistry of life**

Bio-chemistry studies the chemical and physical characteristics of the components of the cell and features of the life systems of the components, as well as the interpretation of what these systems in the cell Biochemistry provided a lot of accomplishments, it has helped to clarify the mechanism of medicine and contributed to the diagnosis and treatment of many diseases and provided techniques which could be used measure the level of many of the compounds in vivo.

Biochemistry lasted over the age of a century in different disciplines, some with a study of the materials that make up plant cell and then called the chemistry of plant life, and then which is related the animal cell which is called chemistry of animal life if the human cell is the target.

Chemistry has expanded to clinical biochemistry that includes chemistry of life, becoming a physical, organic and biochemistry and inorganic chemistry as well as nutrition. Interested in chemistry, life functions of the modern systems of life, have contributed to the means of study in the last century with the observation of these systems directly during the work, either at the present time which has changed the picture and it became possible to obtain the most desirable observations by the development of viable technologies (electron microscope, radioactive isotopes, Immunology, spectrum).

The scientists believed at the end of the nineteenth century that it is possible to obtain some information relating to the systems of life, by studying the chemistry of cells and for decades was followed by chemists adopted the chemical methods available and succeeded in obtaining useful developments. Significant improvements to the technical methods such as the use of chemical isotopes have greatly increased the sensitivity of diagnosis of different types of molecules of life and others, and when it is necessary to separate the components of the chemical reaction through life and is very sensitive, then used deportation electric traditionally.

When the attention has turned physicists, chemists, physicists about the science of life (and perhaps due to the ability of living cells to configure the system, although the laws of physics, emphasizes the universe there is a tendency towards non-attendance) then emerged the technical methods of physical, chemical, physical, such as spectroscopy, diffraction to be applied an the field of biology.

The progress achieved in the chemistry of life has begun to acknowledge that the livelihood systems containing small particles interested m organic chemistry to study and clarify as well as large molecules called macroscopic particles which are not molecular weights less than 100 million times the mass of one atom of hydrogen. The importance of macroscopic particles of the life systems in its ability to privacy in life interactions composition of building blocks, and can say clearly that he had made in the past years considerable effort to characterize the annexation of macroscopic particles as well as the reactors that occur between them and the need for advanced methods of separation and purification and characterization of macroscopic particles in order to obtain information on structural composition of the macroscopic molecule.

The objective of biochemistry for nearly half a century is to collect and organize interactions that occur in living cells. The motivation for this major effort is that a significant number of the attributes of living cells can be understood through these interactions that are typically characterized by the formation or breaking covalent bonds. It is been clarified on the liberalization of energy as a result of break chemical transformation processes as well as molecules of life and mutual assembly operations amino acids, sugars and fats to form macroscopic particles.

During the last thirty years clearly demonstrated that the reactors that occur between molecules due to physical, those that are not or break covalent bonds have the same importance of chemical reactions, for example, that the organization of chemical reactions (i.e., the degree of permitted them to occur) performed by the physical changes that occur in the structure of (construction) of large molecules, as well as the creation of active centers in these molecules and the resulting interdependence of the non-covalent small molecules, in addition to, many of the qualities of a macroscopic aggregates molecules in cells or in the organism (the cell membrane and walls of cells and chromosomes).

Plurality of molecules of life structure consists of installation of the first structural molecular structures of multiple different types of units place (serial), for example, the sequence of amino acids found in proteins and sequence by chemical analysis. The secondary structural composition which involves the formation of a complex three-dimensional structures is called to direct all of the units for multi-particles to other units and is called the secondary structural composition tradition or (body and image) or the status of the foundation structure or backbone of multiple chains. The forms, which consist of surfaces and different types of these mixed forms, and called on the direction of (position) of side chains relative (amino acids, nucleic acids or bases) triangular structural composition. A lot of multiple molecules of life with each other to be as complex as the structures of several structural units viruses, membranes and capillaries bonds and are usually in one level, where you specify the types of bilateral structures of proteins. On the other hand that includes the alpha carbon to allow for many types of structural combinations. The two phosphate ester bonds in nucleic acids are subject to sag as well, because the flexible rule and hate water and one level surrounded by a few of so they are usually located one above the other, thus reducing the adhesion of water, and this increases the structural rigidity of installation.

The multi-life linear molecules, which has no free rotation about the bonds, which do not interact aggregates side is called the file is not a random combination structurally specific dimensions or size of distinct wraps by Brownian movement. Size can be measured by the value equal to the rate of rotation of the radius around a point or an axis.

References

1- Al- Saadi, 2/ 5/ 2006: Iraq's National vision, strategy, and policies: strategic insights, vol. v. _(20

John Livingston, "Western Science and Educational Reform in the Thought of Shaykh Rifas Al-Tahtawi" International Journal of Middle East Studies, Vol.28 (1966), pp.543.

(3) Baker, R. 2004: Iraq and human development: culture, education and the globalization of hope.

4-European Commission – Directorate General for Research, key figures 2003 -2004, towards a European research area: science, technology, & innovation, p 23.

5-Fossum, D., Painter, L.S., Eisman, E., Ettedgui, e. and Adamson, D.M. 2004. Vital Assets: Federal Investment in Research and Development at the Nation's Universities and Colleges. Rand Corporation, Santa Monica, CA.

6-Ghali, H. 2005: The destruction of Iraq's Educational system under us occupation, center for research on Globalisations.

7-International Education standard for professional accounts 7 (2004): continuing development: A program of lifelong learning and contissing development of professional competence.

8-National academy of Science. 1997, preparing for the 21st century: Science and engineering research in a changing world. National Academy of science. 12p.

9- National Institutes of Health. 2004, Summary of the FY 2005 president's budget. 12p.

10-National Science Foundation b. 2005, The budget for fiscal year 2005. www.nsf.gov pp323- 329.

11- National Science Foundation, 2002. Survey of R &D Expenditures at Universities & Colleges, Fiscal Year 2002.

12-Pharmaceutical Industry Profile. 2003, Dramatic growth of research and development, (chapter 2): pp 10-22, www.phrma.org/publicationprofile2002-2003.

13-Qasim Subhi. 1998, Research and development in the Arab States. Development of S&T Indicators, UNESCO, (Cairo) ESCWA.

14-Science Watch. 2003, Middle Eastern Nations Making Their Mark, Vol.14, No.6, www.sciencewatch.com/nov-dec2003.

15-UNESCO Institute for Statistics. 2003, Immediate, Medium and Long – term Strategy in Science and Technology Statistics. UNESCO Institute for Statistics, Montreal.

16-UNESCO Institute for Statistics. 2004, A decade of investment in research and development (R&D): 1990-2000. UIS Bulletin on Science and Technology Statistics. Issue No.1:1-4.

17-Iraq 2004: Iraq, Education overview.

18-Word Education services – Canada (WES), WEP- Iraq, 2004: Iraq Higher Education.

www.ingramcontent.com/pod-product-compliance
Lightning Source LLC
Chambersburg PA
CBHW080647190526
45169CB00006B/2017
*9781535476317*